Practical Statistics for Geographers and Earth Scientists

Practical Statistics for Geographers and Earth Scientists

Nigel Walford

WILEY-BLACKWELL

A John Wiley & Sons, Ltd., Publication

This edition first published 2011, © 2011 John Wiley & Sons Ltd

Wiley-Blackwell is an imprint of John Wiley & Sons, formed by the merger of Wiley's global Scientific, Technical and Medical business with Blackwell Publishing.

Registered office: John Wiley & Sons Ltd, The Atrium, Southern Gate, Chichester, West Sussex, PO19 8SQ, UK

Editorial Offices:
9600 Garsington Road, Oxford, OX4 2DQ, UK

The Atrium, Southern Gate, Chichester, West Sussex, PO19 8SQ, UK

111 River Street, Hoboken, NJ 07030-5774, USA

For details of our global editorial offices, for customer services and for information about how to apply for permission to reuse the copyright material in this book please see our website at www.wiley.com/wiley-blackwell

The right of the author to be identified as the author of this work has been asserted in accordance with the UK Copyright, Designs and Patents Act 1988.

Library of Congress Cataloguing-in-Publication Data

Walford, Nigel.
    Practical statistics for geographers and earth scientists / Nigel Walford.
       p. cm.
    Includes index.
    ISBN 978-0-470-84914-9 (cloth) – ISBN 978-0-470-84915-6 (pbk.)
  1. Geography–Statistical methods.   I. Title.
    G70.3.W36 2011
    519.5–dc22
                                                    2010028020

A catalogue record for this book is available from the British Library.

This book is published in the following electronic formats: ePDF: 978-0-470-67001-9

Set in 10.5 on 12.5 pt Minion by Toppan Best-set Premedia Limited

First Impression   2011

To Ann

Contents

**Plate section: Statistical Analysis Planner and Checklist falls
between pages 172 and 173**

Preface

The quantitative revolution has a lot to answer for, not least in establishing a need for students to learn about statistical techniques and how they can be applied. The majority of undergraduate students in Geography and the Earth Sciences are expected to demonstrate this understanding through carrying out independent projects as part of their degree programmes. Although it is now some 50 years since these disciplines were struck by the quantitative revolution, there remains a continuing need to reinterpret statistical techniques and to demonstrate their application in relation to contemporary issues for successive cohorts of students. Information technology in the form of powerful computers and sophisticated software has, to a large extent removed the drudgery of hand calculation that dogged the early years of quantification. In those days datasets for students were necessarily limited to small samples of observations because the time required to calculate a mean, standard deviation, correlation coefficient, regression equation or some other statistical quantity would otherwise seem interminable.

Information technology has seemingly made the application of statistics to help with answering research questions a relatively straightforward task. The entry of data into computer software is often still a time-consuming task requiring care and attention to detail in order to minimize or preferably remove any error. However, once the data have been captured in this way it is now comparatively straightforward to select from the range of statistical techniques available and to produce some results: so the need for students to undertake calculations in order to apply statistical techniques has now been removed. However, looking at how they are derived helps to understand their purpose and use. This text does not cover the details of how to use different software, how to select from the alternative ways of presenting results or how to discuss results in a piece of written work. These topics are addressed in other places. Instead, it focuses on the practicalities of choosing statistical techniques for different situations when researching a topic in Geography and the Earth Sciences. There is a wide range of techniques available and even the comprehensive introduction given here cannot hope to include everything.

This book has been written in recognition that three features are common to many undergraduate degrees in Geography and the Earth Sciences. Students are commonly expected:

1. to develop a level of understanding and competence in statistical analysis;

2. to undertake a well-defined, limited-scale independent research investigation involving the qualitative and/or quantitative analysis;

3. to be able to interpret the results of statistical analyses presented in journal papers and other published material.

This book helps students to meet these expectations by explaining statistical techniques using examples drawn from teaching first-degree level students for over 20 years and datasets that in some cases relate directly to student projects on field-work. The chapters in the book progress from consideration of issues related to formulating research questions, collecting data and summarizing information through the application of statistical tests of hypotheses and the analysis of relationships between variables. Geography and Earth Science students are faced with not only coming to terms with classical, nonspatial statistics, but also techniques specially intended for dealing with spatial data. A range of such spatial statistics are included from simple measures describing spatial patterns to ways of investigation spatial autcorrelation and fitting surfaces. The format of the chapters includes boxed sections where the spatial or nonspatial technique is explained and applied to one of a series of project datasets. These data sets are available on the book's website. The chapters also include some questions for students to consider either on their own or in conjunction with their tutors. The Glossary defines various statistical terms used throughout the book and acts as a constant reference point for clarifying topics. The Selected Reading assembles a range of sources into one location to enable readers to delve more deeply in the topics and techniques in the preceding chapters.

Nigel Walford
April 2010
Kingston upon Thames

Acknowledgements

I would like to thank Fiona Woods at Wiley for her patience and gentle reminders while the chapters slowly drifted in over the last few years and to others in the production and editorial offices for steering the manuscript through to publication.

Most of the figures and diagrams have been produced by Claire Ivison in the School of Geography, Geology and the Environment at Kingston University. I have been constantly mystified and astounded at how she has taken my rough drafts on scrappy bits of paper and turned them into illustrations of consistent and complete clarity.

I also acknowledge the support and assistance from colleagues at Kingston University and thank the students whose questions over the years prompted me to attempt to write this book as they attempt to understand the complexities of statistical analysis.

Living with someone trying to write a book about statistical analysis must be very difficult most of the time and I would like to offer a very sincere and heartfelt thank you to Ann Hockey, who has now had to endure this three times over the years.

Glossary

A Posteriori **Probability** A probability that takes into account additional 'after the event' information.

A Priori **Probability** Probability determined from past evidence or theory.

Acceptance Region The section of the range of probability associated with acceptance of the null hypothesis in hypothesis testing (e.g. 1.00 to >0.05).

Alternative Hypothesis The opposite of the Null Hypothesis.

Analysis of Variance (ANOVA) A statistical procedure used to examine whether the difference between three or more sample means indicates whether they have come from the same or different populations.

Attribute A unit of categorical data relating to an observation. Also refers to information about features or elements of a spatial database.

Autocorrelation Correlation between observations that are separated by a fixed time interval or unit of distance.

Bias Systematic deviation from a true value.

Binomial Distribution A probability distribution relating to a series of random events or trials each of which has two outcomes with known probabilities.

Bivariate Analysis where there are two variables of interest.

Bivariate Normal Distribution A joint probability distribution between two variables that follow the normal distribution and are completely uncorrelated.

Canonical Correlation Analysis A method of correlation analysis that summarizes the relationship between two groups of variables.

Categorical Data Data relating to the classification of observations into categories or classes.

Categorical Data Analysis A collection of statistical techniques used to analyse categorical data.

Census The collection of data about all members of a (statistical) population, which should not contain random or systematic errors.

Central Limit Theorem Theorem underpinning certain statistical tests (e.g. Z and t tests) that states the sampling distribution of the mean approaches normality as the sample size increases, irrespective of the probability distribution of the sampled population.

Central Tendency (Measures of) A summary measure describing the typical value of a variable or attribute.

Centroid (with spatial data) The centre of an area, region or polygon and is effectively the 'centre of gravity' in the case of irregular polygons.

Chi-square Distribution Probability distribution relating to continuous variable that is used in statistical testing with frequency count data.

Chi-square Statistic Statistic measuring the difference between observed and expected frequency counts.

Chi-square Test Statistical test used with discrete frequency count data used to test goodness-of-fit.

Cluster Sampling The entire population is divided into groups (clusters) before randomly selecting observations from the groups or a random set of the groups.

Coefficient of Variation Defined as the standard deviation divided by the mean of a variable.

Confidence Intervals (Limits) A pair of values or two sets of values that define a zone around an estimated statistic obtained from a sample within which the corresponding population parameter can be expected to lie with a specified level of probability.

Coordinate System A set of coordinate axes with a known metric.

Coordinates One, two or three ordinates defined as numeric values that are mutually independent and equal the number of dimensions in the coordinate space.

Correlation Coefficient A measure of the strength and direction of the relationship between two attributes or variables that lies within the range -1.0 to $+1.0$.

Correlation Matrix Correlation coefficients between a group of variables that is symmetrical either side of the diagonal where the correlation of the variables with themselves equal 1.0.

Correlogram A plot of spatially or temporally linked correlation coefficients.

Covariance The expected value of the result of multiplying the deviations (differences) of two variables from their respective means that indicates the amount of their joint variance.

Cross-sectional Data Data relating to observations at a given point in time.

Crosstabulation Joint frequency distribution of two or more discrete variables.

Data (datum) Recorded counts, measurements or quantities made on observations (entities).

Data Mining Discovery of hidden patterns in large datasets.

Datum The origin, orientation and scale of a coordinate system tying it to the Earth.

Degrees of Freedom The smallest number of data values in a particular situation needed to determine all the data values (e.g. $n - 1$ for a mean).

Dependent Variable A set of one or more variables that are functionally depend on another set of one or more independent variables.

Descriptive Statistics A group of statistical techniques used to summarize or describe a data set.

Dichotomous An outcome or attribute with only two possible values.

Discrete Distribution A probability distribution for data values that are discrete or integer as opposed to continuous.

Discrete Random Variable A random variable with a finite (fixed) number of possible values.

Dispersion (measures of) A measure that quantitatively describes the spread of a set of data values.

Distance Matrix A matrix of measurements that quantifies the dissimilarity or physical distance between all pairs of observations in a population or sample.

Error Difference between the estimated and true value.

Estimation Use of information from a sample to guess the value of a population parameter.

Expected Value The value of a statistical measure (e.g. mean or count) expected according the appropriate random probability distribution.

Explanatory Variable An alternative name for an independent variable.

F Distribution A group of probability distributions with two parameters relating to degrees of freedom of the numerator and the denominator that is used to test for the significance of differences in means and variances between samples.

Feature An abstract representation of a real-world phenomenon or entity.

Geocoding Allocation of coordinates or alphanumeric codes to reference data to geographical locations.

Geographically Weighted Regression A form of regression analysis that fits different regressions at different points across a study area thus weighting by spatial location.

Georeferencing Assignment of coordinates to spatial features tying them to an Earth-based coordinate system.

Geospatial Data Data relating to any real world feature or phenomenon concerning its location or in relation to other features.

Hypothesis An assertion about one or more attributes or variables.

Hypothesis Testing Statistical procedure for deciding between null and alternative hypotheses in significance tests.

Independent Events (Values) The situation in which the occurrence or nonoccurrence of one event, or the measurement of a specific data value for one observation is totally unaffected (independent of) the outcome for any other event or observation.

Independent Variable A set of one or more variables that functionally control another set of one or more dependent variables.

Inferential Statistics A group of statistical techniques including confidence intervals and hypothesis tests that seek to discover the reliability of an estimated value or conclusion produced from sample data.

Interquartile Range A measure of dispersion that is the difference between 1st and 3rd quartile values.

Interval Scale A measurement scale where 0 does not indicate the variable being measured is absent (e.g. temperature in degrees Celsius), which contrasts with the ratio scale.

Inverse Distance Weighting A form spatial data smoothing that adjusts the value of each point in an inverse relationship to its distance from the point being estimated.

Join Count Statistics Counts of the number of joins or shared boundaries between area features that have been assigned nominal categorical values.

Judgemental Sampling A nonrandom method for selecting observations to be included in a sample based on the investigators' judgement that generates data less amenable than random methods to statistical analysis and hypothesis testing.

Kolmogorov–Smirnov Test (one sample) A statistical test that examines whether the cumulative frequency distribution obtained from sample data is significantly different from that expected according to the theoretical probability distribution.

Kolmogorov–Smirnov Test (two samples) A statistical test used to determine whether two independent samples are probably drawn from the same or different populations by reference to the maximal difference between their cumulative frequency distributions.

Kriging Kriging uses inverse distance weighting and the local spatial structure to predict the values and points and to map short-range variations.

Kruskal–Wallis Test The nonparametric equivalent of ANOVA that tests if three or more independent samples could have come from the same population.

Kurtosis A measure of the height of the tails of a distribution: positive and negative kurtosis values, respectively, indicate relatively more and less observations in comparison with the normal distribution.

Lag (Spatial or Temporal) A unit of space or time between observations or objects that is used in the analysis of autocorrelation.

Level of Significance Threshold probability that is used to help decide whether to accept or reject the null hypothesis (e.g. 0.05 or 5%).

Linear Regression Form of statistical analysis that seeks to find the best fit linear relationship between the dependent variable and independent variable(s).

Local Indicator of Spatial Association A quantity or indicator, such as Local Moran's I, measuring local pockets of positive and negative spatial autocorrelation.

Mann–Whitney U Test A nonparametric statistical test that examines the difference between medians of an ordinal variable to determine whether two samples or two groups of observations come from one population.

Mean The arithmetic average of a variable for a population or sample is a measure of central tendency.

Mean Centre The central location in a set of point features that is at the intersection of the means of their X and Y coordinates.

Mean Deviation Average of the absolute deviations from a mean or median.

Measurement Error Difference between observed and true value in the measurement of data usually divided into random and systematic errors.

Median Middle value that divides an order set of data values into two equal halves and is the 50th percentile.

Median Centre A central location in a set of point features that occurs at the intersection of the median values of their X and Y coordinates or that divides the points into four groups of equal size.

Mode Most frequently occurring value in a population or sample.

Moran's I A quantitative measure of global or local spatial autocorrelation.

Multiple Regression Form of regression analysis where there are two or more independent variables used to explain the dependent variable.

Multivariate Analysis where there are more than two variables of interest.

Nearest-Neighbour Index An index used as a descriptive measure to compare the patterns of different categories of phenomena within the same study area.

Nominal Scale Categories used to distinguish between observations where there is no difference in magnitude between one category and another.

Nonparametric Tests Form of inferential statistics that makes only limited assumptions about the population parameters.

Normal Distribution Bell-shaped, symmetrical and single-peaked probability density curve with tails extending to plus and minus infinity.

Normality Property of a random variable that conforms to the normal distribution.

Null Hypothesis Deliberately cautious hypothesis that in general terms asserts that a difference between a sample and population in respect of a particular statistic (e.g. mean) has arisen through chance.

Observed Value The value of a statistical measure (e.g. mean or count) obtained from sample data.

One-Tailed Test Statistical test in which there is good reason to believe that the difference between the sample and population will be in a specific direction (i.e. more or less) in which case the probability is halved.

Ordinal Scale Data values that occur in an ordered sequence where the difference in magnitude between one observation and another relates to its rank position in the sequence.

Ordinary Least Squares Regression Most common form of ordinary linear regression that uses least squares to determine the regression line.

Outlier An observation with an unusually high or low data value.

P Value Probability that random variation could have produced a difference from the population parameter as large or larger than the one obtained from the observed sample.

Paired-Sample Data Measurement of observations on two occasions in respect of the same variable or single observations that can be split into two parts.

Parameter A numerical value that describes a characteristic of a probability distribution or population.

Parametric Tests Type of inferential statistics that make stringent assumptions about the population parameters.

Pearson Correlation Coefficient A form of correlation analysis where parametric assumptions apply.

Percentile The percentage of data values that are less than or equal to a particular point in the percentage scale from 1 to 100 (e.g. the data value at which 40% of all values are less than this one is 40th percentile).

Poisson Distribution A probability distribution where there is a discrete number of outcomes for each event or measurement.

Polygon A representation of area features.

Polynomial A function, for example a regression equation containing N terms for the independent variable raised to the power of N.

Population A complete set of objects or entities of the same nature (e.g. rivers, human beings, etc.).

Precision The degree of exactness in the measurement of data.

Quadrat A framework of usually regular, contiguous, square units superimposed on a study area and used as a means of counting and testing the randomness of spatial patterns.

Quartile The 1st, 2nd and 3rd quartiles correspond with the 25th, 50th and 75th percentiles.

Random Error The part of the overall error that varies randomly from the measurement of one observation to another.

Random Sampling Sometimes called simple random sampling, this involves selecting objects or entities from a population such that each has an equal chance of being chosen.

Range A measure of dispersion that is the difference between the maximum and minimum data values.

Rank Correlation Coefficient A method of correlation analysis that is less demanding than Pearson's and involves ranking the two variables of interested and then calculating the coefficient from the rank scores.

Raster Representation of spatial data as values in a matrix of numbered rows and columns that can be related to a coordinate system in the case of geospatial data.

Ratio Scale A scale of measurement with an absolute zero where any two pairs of values that are a certain distance apart are separated by the same degree of magnitude.

Regression Analysis A type of statistical analysis that seeks the best fit a mathematical equation between dependent and independent variable(s).

Regression Line Line that best fits a scatter of data points drawn using the intercept and slope parameters in the regression equation.

Relative Frequency Distribution Summary frequency distribution showing relative percentage or proportion of observations in discrete categories.

Residuals The differences between observed and predicted values commonly used in regression analysis.

Root Mean Square (RMS) The square root of the mean of the squares of a set of data values.

R-squared The coefficient of determination acts as a measure of the goodness of fit in regression analysis and is thought of as the proportion or percentage of the total variance accounted for by the independent variable(s).

Sample A subset of objects or entities selected by some means from a population and intended to encapsulate the latter's characteristics.

Sample Space The set of all possible outcomes from an experiment.

Sample Survey A survey of a sample of objects or entities from a population.

Sampling The process of selecting a sample of objects or entities from a population.

Sampling Distribution Frequency distribution of a series of summary statistics (e.g. means) calculated from samples selected from one population.

Sampling Frame The set of objects or entities in a population that are sampled.

Scatter Plots (Graphs) A visual representation of the joint distribution of observed data values of two (occasionally three) variables for a population or sample that uses symbols to show the location of the data points on X, Y and possibly Z planes.

Simple Linear Regression Simplest form of least squares regression with two parameters (intercept and slope).

Skewness A measure of the symmetry (or lack of it) of a probability distribution.

Smoothing A way of reducing noise in spatial or temporal datasets.

Spline The polynomial regression lines obtained for discrete groups of points that are tied together to produce a smooth curve following the overall surface in an alternative method of fitting a surface.

Standard Deviation A measure of dispersion that is the square root of the variance and used with interval and ratio scale measurements.

Standard Distance A measure of the dispersion of data points around their mean centre in two-dimensional space.

Standard Error A measure of the variability of a sample statistic as it varies from one sample to another.

Standard Normal Distribution A version of the normal distribution where the mean is zero and the standard deviation is 1.0.

Standard (Z) Score The number of units of the standard deviation that an observation is below or above the mean.

Statistic Either a number used as a measurement or a quantity calculated from sample data.

Statistical Test A procedure used in hypothesis testing to evaluate the null and alternative hypotheses.

Stratified Random Sampling A method of selecting objects or entities for inclusion in a sample that involves separating the population into homogenous strata on the basis of some criterion (e.g. mode of travel to work) before randomly sampling from each stratum either proportionately or disproportionately in relation to the proportion of the total entities in each stratum.

Systematic Error A regular repeated error that occurs when measuring data about observations.

Systematic Sampling A method of selecting objects or entities from a population for inclusion in some regular sequence (e.g. every 5th person).

t-Distribution A bell-shaped, symmetrical and single-peaked continuous probability distribution that approaches the normal distribution as the degrees of freedom increase.

Time Series Data that includes measurements of the same variable at regular intervals over time.

Time Series Analysis Group of statistical techniques used to analyse time series data.

Trend Surface Analysis The 'best fitting' of an equation to a set of data points in order to produce a surface, for example by means of a polynomial regression equation.

t-statistic Test statistic whose probabilities are given by the t-distribution.

t-test A group of statistical tests that use the t-statistic and t-distribution that are in practice rather less demanding in their assumptions than those using the normal distribution.

Two-Tailed Test Most practical applications of statistical testing are two-tailed, since there is no good reason to argue that the difference being tested should be in one direction or the other (i.e. positive or negative) and the probability is divided equally between both tails of the distribution.

Type-I Error A Type-I error is made when rejecting a null hypothesis that is true and so concluding that an outcome is statistically significant when it is not.

Type-II Error A Type-II error is committed when a null hypothesis is accepted that is false and so inadvertently failing to conclude that a difference is statistically significant.

Univariate Analysis where there is one variable of interest.

Variance A measure of dispersion that is the average squared difference between the mean and the value of each observation in a population or sample with respect to a particular variable measured on the interval or ratio scale.

Variance/Mean Ratio A statistic used to describe the distribution of events in time or space that can be examined using the Poisson distribution.

Variate An alternative term for variable or attribute, although sometimes reserved for things that are numerical measurements (i.e. not attribute categories).

Variogram (Semivariogram) Quantification of spatial correlation by means of a function.

Vector Representation of the extent of geographic features in a coordinate system by means of geometric primitives (e.g. point, curve and surface).

Weighted Mean Centre The mean centre of a set of spatial points that is weighted according to the value or amount of a characteristic at each location.

Weights Matrix A matrix of numbers between all spatial features in a set where the numbers, for example 0 and 1, indicate the contiguity, adjacency or neighbourliness of each pair of features.

Wilcoxon Signed Ranks Test A nonparametric equivalent of the paired sample t-test that tests a null hypothesis that difference between the signed ranks is due to chance.

Wilk's Lambda A multivariate test of difference in mean for three or more samples (groups).

Z distribution Probability distribution that is equivalent to the standardized normal distribution providing the probabilities used in the Z test.

Z Statistic Test statistic whose probabilities are given by the Z distribution.

Z Test A parametric statistical test that uses the Z statistic and Z distribution that makes relatively demanding assumptions about the normality of the variable of interest.

Z-score A Z-score standardizes the value of a variable for an observation in terms of the number of standard deviations that it is either above or below the sample or population mean.

Section 1
First principles

1

What's in a number?

Chapter 1 provides a brief review of the development of quantitative analysis in Geography, Earth and Environmental Science and related disciplines. It also discusses the relative merits of using numerical data and how numbers can be used to represent qualitative characteristics. A brief introduction to mathematical notation and calculation is provided to a level that will help readers to understand subsequent chapters. Overall this introductory chapter is intended to define terms and to provide a structure for the remainder of the book.

Learning outcomes

This chapter will enable readers to:

- outline the difference between quantitative and qualitative approaches, and their relationship to statistical techniques;

- describe the characteristics of numerical data and scales of measurement;

- recognize forms of mathematical notation and calculation that underlie analytical procedures covered in subsequent chapters;

- plan their reading of this text in relation to undertaking an independent research investigation in Geography and related disciplines.

Practical Statistics for Geographers and Earth Scientists   Nigel Walford
© 2011 John Wiley & Sons, Ltd

1.1   Introduction to quantitative analysis

Quantitative analysis comprises one of two main approaches to researching and understanding the world around us. In simple terms quantitative analysis can be viewed as the processing and interpretation of data about things, sometimes called phenomena, which are held in a numerical form. In other words, from the perspective of Geography and other Earth Sciences, it is about investigating the differences and similarities between people and places that can be expressed in terms of numerical quantities rather than words. In contrast, qualitative analysis recognizes the unique-ness of all phenomena and the important contribution towards understanding that is provided by unusual, idiosyncratic cases as much as by those conforming to some numerical pattern. Using the two approaches together should enable researchers to develop a more thorough understanding of how processes work that lead to variations in the distribution of phenomena over the Earth's surface than by employing either methodology on its own.

If you are reading this book as a student on a university or college course, there will be differences and similarities between you and the other students taking the same course in terms of such things as your age, height, school level qualifications, home town, genetic make-up, parental annual income and so on. You will also be different because each human being, and for that matter each place on the Earth, is unique. There is no one else exactly like you, even if you have an identical twin, nor is there any place exactly the same as where you are reading this book. You are different from other people because your own attitudes, values and feelings have been moulded by your upbringing, cultural background and physical characteristics. In some ways, it is the old argument of nature versus nurture, but in essence we are unique combina-tions of both sets of factors. You may be reading this book in your room in a university hall of residence, and there are many such places in the various countries of the world and those in the same institution often seem identical, but the one where you are now is unique. Just as the uniqueness of individuals does not prevent analysis of people as members of various different groups, so the individuality of places does not inhibit investigation of their distinctive and shared characteristics.

What quantitative analysis attempts to do is concentrate on those factors that seem to be important in producing differences and similarities between individual phe-nomena and to disregard those producing aberrant outcomes. Confusion between quantitative and qualitative analysis may arise because the qualitative characteristics of people and places are sometimes expressed in numerical terms. For example, areas in city may be assigned to a series of categories, such as downtown, suburbs, shopping mall, commercial centre, housing estate and so on, to which numerical codes can be attached, or a person's gender may be labelled as male or female with equivalent numerical values such as 1 or 2. Just as qualitative characteristics can be turned into numerical codes, so too can numerical measurements be converted, either singly or in combination, into descriptive labels. Many geodemographic descriptions of neigh-bourhoods, such as (e.g. *Old people, detached houses; Larger families, prosperous suburbs; Older rented terraces;* and *Council flats, single elderly*) are based on having taken a selection of different socioeconomic and demographic numerical counts for different locations and combined them in an analytical melting pot so as to produce a label describing what the places are like.

The major focus of this book is on quantitative analysis as applied to Geography and other Earth Sciences. However, this should not be taken to imply that quantitative analysis is in some definitive sense regarded as 'better', or even more scientific, than qualitative analysis. Nor are these approaches mutually exclusive, since researchers from many disciplines have come to appreciate the advantages of combining both forms of analysis in recent years. This book concentrates on quantitative analysis since for many students, and perhaps researchers, dealing with numbers and statistics is difficult, and trying to understand what these can usefully tell us about the 'real' Geography or Earth Science topics that interest us is perplexing. Why should we bother with numbers and statistics, when all we want to do is to understand the process of globalization in political economy, or to explain why we are experiencing a period of global temperature increase, or to identify areas vulnerable to natural hazards?

We can justify bothering with numbers and statistics to answer such research questions in a number of different ways. As with most research, when we actually try to explain things such as globalization, global warming and the occurrence of natural hazards, the answers often seem like commonsense and perhaps even rather obvious. Maybe this is a sign of 'good' research because it suggests the process of explaining such things is about building on what has become common knowledge and understanding. If this is true, then ongoing academic study both rests upon and questions the endeavours of previous generations of researchers, and puts across complex issues in ways that can be generally understood. Yet despite the apparent certainty with which the answers to research questions might be conveyed, these are often underlain by an analysis of numerical information that is anything but certain. It is likely that the results of the research are correct, but they may not be. Using statistical techniques gives us a way of expressing this uncertainty and of hedging our bets against the possibility that our particular set of results has only arisen by chance. At some time in the future another researcher might come along and contradict our findings.

But what do we really mean by the phrase 'the results of the research'. For the 'consumers' of research, whether the public at large or particular professional groups, the findings, results or outcomes of research are often some piece of crucial factual information. The role of such information is to either confirm facts that are already known or believed, or to fulfil the unquenchable need for new 'facts'. For the academic, these detailed factual results may be of less direct interest than the implications of the research findings, perhaps with regard to some overarching theory. The student undertaking a research investigation as part of the programme of study sits somewhere, perhaps uncomfortably between these two positions. Realistically, many undergraduate students recognize that their research endeavours are unlikely to contribute significantly to theoretical advance, although obviously there are exceptions.

Yet they also recognize that their tutors are unlikely to be impressed simply by the presentation of new factual information. Further, such research investigations are typically included in undergraduate degree programmes in order to provide students with a training that prepares them for a career where such skills as collecting and assimilating information will prove beneficial, whether this be in academia or more typically in other professional fields. Students face a dilemma to which there is no simple answer. They need to demonstrate that they have carried out their research in a rigorous scientific manner using appropriate quantitative and qualitative techniques, but they do not want to overburden the assessors with an excess of detail that obscures the implications of their results.

In the 1950s and 1960s a number of academic disciplines 'discovered' quantitative analysis and few geography students of the last five decades can fail to have heard of the so-called 'quantitative revolution' in their subject area, and some may not have forgiven the early pioneers of this approach. There was, and to some extent still is, a belief that the principles of rigour, replication and respectability enshrined in scientific endeavour sets it apart from, and possibly above, other forms of more discursive academic enquiry. The adoption of the quantitative approach was seen implicitly, and in some cases explicitly, as providing the passport to recognition as a scientific discipline. Geography and other Earth Sciences were not alone, although perhaps were more sluggish than some disciplines, in seeking to establish their scientific credentials. The classical approach to geographical enquiry followed on from the colonial and exploratory legacies of the 18th and 19th centuries. This permeated regional geography in the early 20th century and concentrated on the collection of factual information about places. Using this information to classify and categorize places seemed to correspond with the inductive scientific method that served the purpose of recognizing pattern and regularity in the occurrence of phenomena. However, the difficulty of establishing inductive laws about intrinsically unique places and regions led other geographers to search for ways of applying the deductive scientific method, which was also regarded as more rigorous. The deductive method involves devising hypotheses with reference to existing conditions and testing them using empirical evidence obtained through the measurement and observation of phenomena.

Geography and to a lesser extent the other Earth Sciences have emerged from a period of self-reflection on the scientific nature of their endeavour with an acceptance that various philosophies can coexist and further their collective enterprise. Thus, many university departments include physical geographers and Earth scientists, adhering to generally positivist scientific principles, working alongside human geographers following a range of traditions including humanism, Marxism and structuralism as well as more positivist social science. The philosophical basis of geographical and Earth scientific enquiry has received a further twist in recent decades on account of the growing importance of information and communications technology (ICT). Hence, students in academic departments need to be equipped with the skills not only to undertake research investigations in these areas, but also to handle geographical and spatial data in a digital environment.

Johnston (1979) commented that statistical techniques provide a way of testing hypotheses and the validity of empirical measurements and observations. However, the term statistics is used in a number of different ways. In general usage, it typically refers to the results of data collection by means of censuses and surveys that are published in books, over the Internet or on other media. The associated term 'official statistics' is usually reserved for information that has been collected, analysed and published by national, regional or local government and are therefore deemed to have a certain authority and a connotation of conveying the 'truth'. This belief may be founded upon a presumption of impartiality and rigour, although such neutrality of method or intent cannot realistically be justified or assumed in all instances. Statistics also refers to a branch of mathematics that may be used in scientific investigations to substantiate or refute the results of scientific research. In this sense statistics also has a double meaning, either comprising a series of **techniques** ranging from simple summarizing measures to complex models involving many variables, or the term may refer to the numerical **quantities** produced by these techniques. All these senses of the term statistics are relevant to this text, since published statistics may well contribute to research investigations, and statistical techniques and the measures associated with them are an essential part of quantitative analysis. Such techniques are applied to numerical data and serve two general purposes: to confirm or otherwise the significance of research results towards the accumulation of knowledge with respect to a particular area of study; and to establish whether empirical connections between different characteristics for a given set of phenomena are likely to be genuine or spurious.

Different areas of scientific study and research have over the years carved out their own particular niches. For example, in simplistic terms the Life Sciences are concerned with living organisms, the Chemical Sciences with organic and inorganic materials, Political Science with national and international government, Sociology with social groups and Psychology with individuals' mental condition. When these broad categories have become too general then often subdivision occurred with the emergence of fields in the Life Sciences such as cell biology, biochemistry, physiology, etc. Geography and the other Earth Sciences do not seem to fit easily into this seemingly straightforward partitioning of scientific subject matter, since their broad perspective leads to an interest in all the things covered by other academic disciplines, even to the extent of Earth Scientists transferring their allegiance to examine **terrestrial** processes on other planetary and celestial bodies. No doubt if, or when, ambient intelligent life is found on other planets, 'human' geographers will be there investigating its spatial arrangement and distribution. It is commonly argued that the unifying focus of Geography is its concern for the spatial and earthly context in which those phenomena of interest to other disciplines make their home. Thus, for example human geographers are concerned with the same social groups as the sociologist, but emphasize their spatial juxtaposition and interaction rather than the social ties that bind them, although geographers cannot disregard the latter. Similarly, geochemists focus on the chemical properties of minerals not only for their individual characteristics but also for how assemblages combine to form different rocks in distinctive locations. Other disciplines may wonder what Geography and Earth Science add to their own academic endeavour, but geographers and Earth scientists are equally certain that if their area of study did not exist, it would soon need to be invented.

The problem that all this raises for geographers and Earth scientists is how to define and specify the units of observation, the things that are of interest to them, and them alone. One possible solution that emerged during the quantitative revolution was that geography was pre-eminently the science of spatial analysis and therefore it should be concerned with discovering the laws that governed spatial processes. A classical example of this approach in human geography was the search for regions or countries where the collections of settlements conformed to the spatial hierarchy anticipated by central place theory. Physical geographers and Earth scientists also became interested in spatial patterns. Arguably they had more success in associating the occurrence of environmental phenomena with underlying explanatory processes, as evidenced by the development plate tectonics theory in connection with the spatial pattern of earthquake and volcanic zones around the Earth. According to this spatial analytic approach, once the geographer and Earth scientist have identified some spatially distributed phenomena, such as settlements, hospitals, earthquakes or volcanoes, then their investigation can proceed by measuring distance and determining pattern.

This foray into spatial analysis was soon undermined, when it became apparent that exception rather than conformity to proposed spatial laws was the norm. Other approaches, or possibly paradigms, emerged, particularly in human geography, that sought to escape from the straightjacket of positivist science. Advocates of Marxist, behavioural, politicoeconomic and cultural geography have held sway at various times during the last 40 years. However, it is probably fair to say that none of these have entirely rejected using numerical quantities as a way of expressing geographical difference. Certainly in physical geography and Earth Science where many would argue that positivism inevitably still forms the underlying methodology, quantitative analysis has never receded into the background. Despite the vagaries of all these different approaches most geographers and Earth scientists still hold on to the notion that what interests them and what they feel other people should be reminded of is that the Earth, its physical phenomena, its environment and its inhabitants are differentiated and are unevenly distributed over space.

From the practical perspective of this text, what needs to be decided is what constitutes legitimate data for the study of Geography and Earth Science. Let us simply state that a geographical or Earth scientific dataset needs to comprise a collection of data items, facts if you prefer, that relate to a series of spatially distributed phenomena. Such a collection of data needs to relate to at least one discrete and defined section of the Earth and/or its immediate atmosphere. This definition deliberately provides wide scope for various types and sources of data with which to investigate geographical and Earth scientific questions.

The spatial focus of Geography and the Earth Sciences has two significant implications. First, the location of the phenomena or observations units, in other words where they are on or near the Earth's surface is regarded as important. For instance investigations into landforms associated with calcareous rocks need to recognize whether these are in temperate or tropical environments. The nature of such landforms including their features and structures is in part dependent upon prevailing climatic conditions either now or in the past. Second, the spatial arrangement of phenomena may in itself be important, which implies that a means of numerically quantifying the position of different occurrences of the same category of observation may be required. In this case, spatial variables such as area, proximity, slope angle, aspect and volume may form part of the data items captured for the phenomena under investigation.

1.2 Nature of numerical data

We have already seen that quantitative analysis uses numbers in two ways, as shorthand labels for qualitative characteristics to save dealing with long textual descriptions or as actual measurements denoting differences in magnitude. Underlying this distinction is a division of data items into **attributes** and **variables**. Williams (1984, p. 4) defines an attribute as 'a quality ... whereby items, individuals, objects, locations, events, etc. differ from one another.' He contrasted these with variables that 'are *measured* ... assigned numerical values relative to some standard – the *unit of measurement*' (Williams, 1984, p. 5). Examples of attributes and variables from Geography and other Earth Sciences are seemingly unbounded in number and diversity, since these fields of investigation cover such a wide range of subject areas. Relevant attributes include such things as rock hardness, soil type, land use, ethnic origin and housing tenure, whereas stream discharge, air temperature, population size, journey time and number of employees are examples of geographical variables. The terms attribute and variable are sometimes confused and applied interchangeably, although we will endeavour to keep to the correct terminology here.

The subdivision of numerical data into different types can also be taken further, into what are known as the **scales of measurement**. The four scales are usually known as **nominal**, **ordinal**, **interval** and **ratio** and can be thought of as a sequence implying a greater degree of detail as you progress from the nominal to the ratio. However, strictly speaking the nominal is not a scale of measurement, since it only applies to attributes and therefore is an assessment of qualitative difference rather than magnitude between observations. The other three scales provide a way of measuring difference between observations to determine whether one is smaller, larger or the same as any other in the set. Box 1.1 summarizes the features of each of these measurement scales and shows how it is possible to discard information by moving from the interval/ratio to the nominal scale. Such collapsing or recoding of the values for data

Box 1.1a: Scales of measurement: location of a selection of McDonalds restaurants in Pittsburgh, USA.

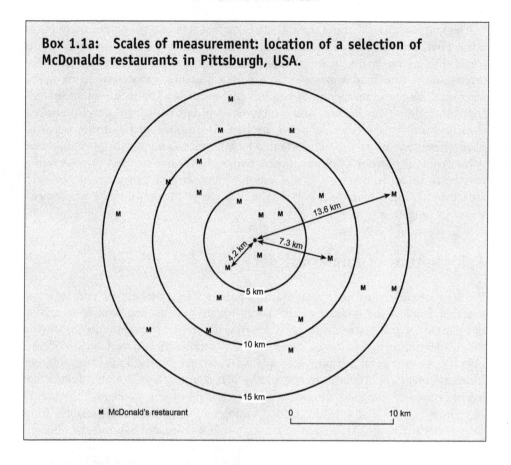

Box 1.1b: Scales of measurement: definitions and characteristics.

Nominal scale attributes record a qualitative difference between observations.

Ordinal scale variables rank one observation against others in the set with respect to the ascending or descending data values of a particular characteristic. These are particularly useful when analysing information obtained from secondary sources where the raw data are unavailable.

Interval scale variables record differences of magnitude between observations in a set where the measurement units have an arbitrary zero.

Ratio scale variables record differences of magnitude between observations in a set where the ratio between the values for any two observations remains constant, irrespective of the measurement units which have an absolute zero.

For most practical purposes the interval and ratio scales, sometimes referred to as scalar measurements, are treated as equivalent with respect to various analytic procedures. The four scales of measurement are listed in order of increasing sophistication and it is possible collapse the detail for a characteristic measured on the interval/ratio scales to the ordinal and to the nominal. This process results in a loss of detail as information is discarded at each stage, but this simplification may be necessary when analysing data from across different measurement scales. Box 1.1c illustrates that progressive loss of detail occurs with respect to the distance measurements between the centre of Pittsburgh and a selection of the McDonalds restaurants in the city. Distance is measured on the ratio scale, with an absolute zero in this case indicating that a restaurant is located exactly at the centre. Sorting these raw data measurements in ascending order and assigning rank scores converts the information into an ordinal scale variable. There are two restaurants closest to the centre at 2.2 km from this point and the furthest is 18.4 km. Further loss of detail results from allocating a qualitative code to denote whether the restaurant is located in the inner, middle or outer concentric zone around the city centre. Although it is feasible to collapse the detailed raw data of ratio/interval variables in this way, it is not possible to go in the other direction from qualitative attributes, through ordinal to interval/ratio scale variables.

Box 1.1c: Collapsing between scales of measurement: distance of a selection of McDonalds restaurants from centre of Pittsburgh, USA.

Restaurant No.	Distance to centre Ratio variable	Sorted distance	Rank score Ordinal variable	Concentric zone Nominal attribute
1	17.5	2.2	1	1 (Inner)
2	15.1	2.2	1	1 (Inner)
3	14.2	2.8	3	1 (Inner)
4	10.3	3.6	4	1 (Inner)
5	6.0	4.2	5	1 (Inner)
6	8.2	6.0	6	2 (Middle)
7	9.2	6.4	7	2 (Middle)
8	10.8	7.0	8	2 (Middle)
9	4.2	7.2	9	2 (Middle)
10	3.6	7.3	10	2 (Middle)
11	2.2	8.1	11	2 (Middle)
12	7.0	8.2	12	2 (Middle)
13	2.2	9.2	13	2 (Middle)
14	2.8	10.3	14	3 (Outer)
15	11.2	10.8	15	3 (Outer)
16	8.1	11.2	16	3 (Outer)
17	6.4	11.2	16	3 (Outer)
18	7.2	13.6	18	3 (Outer)
19	7.3	13.9	19	3 (Outer)
20	11.2	14.2	20	3 (Outer)
21	13.9	15.1	21	3 (Outer)
22	13.6	17.5	22	3 (Outer)
23	18.4	18.4	23	3 (Outer)

items may be carried out for various reasons, but may be necessary when combining a qualitative data item, such as housing tenure, with a quantitative one, such as household income in your analysis.

In most investigations the data items will come from more than one of these scales of measurement, and may possibly include all four. This makes it even more important to be able to recognize the different categories, because some types of statistical quantity and technique are appropriate for data items on one measurement scale and not others. In other words, statistical analysis is not just a question of 'pick and mix', more 'horses for courses'. An important distinction is between discrete and continuous measurements, respectively separating the nominal and ordinal from the interval and ratio scales. Discrete data values occur in situations where observations can only take certain values, usually integers (whole numbers), whereas theoretically any value is allowed with continuous measurements except where there is an absolute zero. Information technology has helped investigators to carry out research that applies statistical analysis to numerical data, but generally computer software is not intelligent enough to know whether a given set of integers are category labels for qualitative attributes, an ordinal sequence of integers or real data values on the interval or ratio scales. Investigators need to intervene to select appropriate types of analysis and not just tell the software to produce the same set of analyses for all their attributes and variables.

Investigations using attributes and variables usually involve holding the data in a numerical format. In the case of nominal attributes this usually involves simple integer numbers such as 1, 2, 3 up to J, where J equals the number of categories and there is a one-to-one correspondence between the number of categories and integers. However, in some cases negative integers may be used. An example would be attitudinal scales, where respondents might be asked to say how they felt about a series of statements in terms of Strongly Disagree, Disagree, Neutral, Agree or Strongly Agree. These qualitative labels may be linked to the numbers −2, −1, 0, 1 and 2, rather than 1, 2, 3, 4 and 5. The ordinal scale is based on ranking observations in order from smallest to largest, or vice versa. These are often converted into a rank score (first, second, third, etc.) corresponding to either the largest or smallest value depending upon whether they are sorted in ascending or descending numerical order. These rank scores are usually stored as integers, although values are repeated if two or more observations are tied (see Box 1.1c). Thus, there may be three observations tied at 15 and no 16 or 17, and the next observation is 18, although arguably the three observations could all be ranked 16. Interval and ratio scale variables are usually recorded to a certain number of decimal places, 2 or 3 for example, even if particular values happen to be integers (e.g. 5.0 km). Computers used to analyse your data may store the values with a far higher degree of precision, for example to 25 decimal places. However, such precision may be spurious, since you may only have captured the data values to one decimal place or perhaps only as whole numbers. An example will illustrate the problem. This book follows the convention of using the International System of metric units (kilometres, degrees Celsius, hectares, etc.) employed in scientific research. However, many people in the USA, UK and some other countries will be unfamiliar with this system and will give distances in miles, temperatures in Fahrenheit and areas in acres. It is possible to convert between these two sets of scales, for instance 1 hectare equals 2.4691 acres (4 decimal places). If we apply this conversion to a specific acreage we may end up with what appears to be a highly accurate number of hectares, but in reality is an entirely spurious level of precision, for example 45 acres divided by 2.4691 equals 18.2252642661138618119962739 ha to 25 decimal places, which could for most practical purposes be shortened to 18.2 ha without any significant loss of information.

Before leaving our consideration of the characteristics of numerical data, it is as well to recognize that there are other types of data item that we may wish to use in our analysis. Dates, for example relating to a specific day, month or year, are another useful category of information that provides investigations with a temporal dimension. It is worth distinguishing between two different types of date depending upon how this temporal reference is obtained. One is recorded by the investigator without directly consulting the observation units and refers to the date when the specific data were collected. An example of this would be the day, month and year, and the start and end times when an individual survey questionnaire was completed or when a field measurement was recorded. The other type of date is obtained from the data subjects themselves and is more common in investigations involving people. Again in a survey of company directors respondents may be asked the date when it was founded, farmers may be asked when they acquired or disposed of land, or car drivers may be asked the time when they started their journey. Although the subjects of an Earth scientific investigation are unlikely to be capable of responding to such questions, this kind of date variable may still be determined. An example would be the dating of lake sediment cores by means of microfossils for the purpose of environmental reconstruction or perhaps more simply by counting the annual growth rings from cores bored in tree trunks.

Another type of measurement requiring special consideration is those recording financial values in a currency (e.g. US $, £ Sterling, € Euro, etc.), which are often needed in investigations concerned with people and institutions. In some respects a currency is just another unit of measurement like metres or litres, which are expressed to two decimal places. Many currencies are made up of units with 100 subunits, for example £8.52 (i.e. 8 whole pounds and 52 pence), which is notionally similar to 8.52 m (i.e. 8 whole metres and 52 centimetres). However, it is notoriously difficult to get accurate monetary figures of this type despite their obvious importance. Individuals and households may not only be reluctant to give details of their income and expenditure, but also may not know the precise figures. How many households in the UK could say **exactly** how much their total gross income was in the last six months? Or equally how many US corporations could report precisely how much they had spent on staff expenses in the last 3 months? Calculations based on rather general figures, such as to the nearest £1000 may produce results that are seemingly more accurate than they really are, for instance an average household income of £15 475.67.

Finally, before leaving our consideration of the characteristics of numerical data, the various spatial variables that are used in geographical and Earth scientific investigations such as distance, length, surface area, volume and so on may not only be analysed on their own, but also be combined with 'standard' nonspatial variables to create composite indicators, such as population per km^2, cubic metres of water discharged per second or point pollution sources per km of a river. Such indicators are similar to rates expressed as per 100 (percentages), 1000 or 1 000 000 insofar as they standardize data values according to some fixed unit, which makes comparison easier. It is simpler to compare population density in different countries or the concentration of chemicals dissolved in water, if the variables representing these concepts are expressed as relative values rather than absolute figures, even if the latter are collected as raw data in the first place. In other words, investigators should be prepared to analyse not only the raw data that they collect, but also additional variables that can be calculated from these, possibly in conjunction with some information obtained from another published source.

1.3   Simplifying mathematical notation

It is difficult to avoid the fact that quantitative analytical techniques depend on the calculation of mathematical values and on underlying statistical theory. It would no doubt please many students of Geography, Earth and Environmental Science if this were not the case, since it is likely to be such substantive topics and themes as those mentioned previously that drew them to these subjects rather than the prospect of encountering mathematical equations. Unfortunately, it is difficult to understand what these statistical techniques are doing, when they should be used without some understanding of their underlying mathematical basis and how they contribute to our understanding of various substantive topics. This inevitably raises the question of how to explain these principles without providing an excess of equations and formulae that might deter some readers from progressing beyond the first chapter and that others might choose to ignore. This text employs two strategies to try and overcome this problem: first, a simple basic principles introduction to the mathematical notation and calculation processes that will be encountered in later chapters; and secondly, wherever possible, restricting equations and formulae to Boxes to illustrate how the calculations work.

The notation used in mathematical equations can be thought of as a language, just as it is possible to translate from English to French, and French into German, it is also feasible to convert mathematical notation into English or any other language. The problem is that a typical equation expressed in prose will use many more words and characters than standard mathematical notation. For example:

$Y = \sum X^2 - \sum X$	translates as	the value of Y equals the result of adding together each value of X multiplied by itself and then subtracting the total produced by adding up all the values of X.
5 'words'		31 words

There is clearly a saving in terms of both words and characters (excluding spaces) between these two 'languages', with the equation requiring 5 'words' (8 characters) and the English translation 31 words (133 characters). The purpose of a language is to enable communication so that concepts and information can be conveyed from one person to another, although with the development of software capable of recognizing speech, this should perhaps be amended to say the transfer may also occur between people and computers. When taking notes in a lecture or from published literature, students will often develop their own abbreviated language to represent common words or phrases. For example, hydrology may be shortened to hyd. or government to gov., which are relatively easy to guess what they mean, but incr. and ↑ might be used interchangeably to denote increase. So a language is most useful when both parties understand what the symbols and abbreviations mean. Some mathematical and statistical notation has made the transition into everyday language. So books, newspapers and other media will often use %, for instance in respect to the national unemployment rate, to indicate a percentage. Many people will understand what the % symbol means, if not necessarily how the figure was calculated. However, much mathematical and statistical notation is like foreign language to many people.

The basic principle of a mathematical equation is that the value on the left side is produced once the calculations denoted on the right have been carried out. In one sense it does not really matter what numbers are substituted for the symbols on the right-hand side, since you can always perform the calculation and obtain a matching value on the left-hand side. In simple mathematics the numbers do not have any intrinsic meaning, you could just as easily derive 11 by adding 3 to 8, or subtracting 3 from 14. However, when using numerical data and applying statistical techniques in particular subject areas, such as Geography and the Earth Sciences, the numbers quantifying our attributes and variables have a meaning with respect to each of the different observation units in the set. If one of the variables in a survey of farms is land area, then each observation could have a different value for this variable. Variations in these values between the different farms actually have meaning in the sense that they convey something about farm size. In other words they are not just a series of numbers, but they signify something seemingly important, in this case that there is variability in the size of farms and some farms have more land than others.

The four sections in Box 1.2 provide a simple guide to explain how to read the different mathematical symbols used in later chapters of this text. First, some general symbols are presented to indicate labelling conventions for variables, attributes and statistical quantities. Secondly, the standard elementary mathematical operators indicate how to compute one number from another. Thirdly, mathematical functions specify particular computations to be carried out on a set of numbers. Finally, comparison or relational operators dictate how the one value on one side of an equation compares with that on the other.

The second strategy for introducing and explaining the mathematical background to the statistical techniques presented in this text is the use of Boxes. These are included to provide a succinct definition and explanation of the various techniques.

Box 1.2: Guide of mathematical notation and conventions and their demonstration in boxes.

Symbol	Meaning	Examples
General		
Decimal places	Number of digits to the right of the right of the decimal point	3 decimal places = 0.123, 5 decimal places 0.000 45
X, Y, Z	Attributes or variables relating to a statistical population	X represents the values of the variable distance to city centre measured in kilometres for each member of the statistical population
x, y, z	Attributes or variables relating to a statistical sample	x represents the values of the variable customers per day for each member of a sample
Greek letters	Each letter represents a different parameter calculated with respect to a statistical population	α (alpha), β (beta), μ (mu) and λ (lambda)
Arabic letters	Each letter represents a different statistic calculated with respect to a statistical sample	R^2, d and s
()	Brackets used, possibly in a hierarchical sequence to subdivide the calculation – work outwards from innermost calculations	$10 = 4 + (3 \times 2)$ $10 = 4 + ((1.5 \times 2) \times 2)$
Mathematical Operators		
+	Add	$10 = 4 + 6$
−	Subtract	$-2 = 4 - 6$
×	Multiply	$24 = 4 \times 6$
/or ——	Divide	$0.667 = 4/6$ $0.667 = \dfrac{4}{6}$
Mathematical Functions		
X^2	Multiply the values specified variable by themselves once – the square of X	$5^2 = 25.00$, $9.3^2 = 86.49$

Symbol	Meaning	Examples
X^3	Multiply the values of the specified variable by themselves thrice – the cube of X	$5^3 = 125.00$, $9.3^3 = 804.507$
X^n	Multiply the values of the specified variable by themselves n times – exponentiation of X to n	$5^5 = 3125.00$, $9.3^5 = 69\,568.837$
$\sqrt{X}$	Determine the number which, when multiplied by itself, will produce X – the square root of X	$\sqrt{25.00} = 5$, $\sqrt{86.49} = 9.3$
$\sum X$	Add together all the values of the specified variable	3 data values for X (4.6, 11.4, 6.1) $\sum X = 4.6 + 11.4 + 6.1 = 22.1$
$X!$	Factorial of the integer value X	$5! = 5 \times 4 \times 3 \times 2 \times 1 = 120$
$\lvert (X - Y) \rvert$	Take the absolute difference between the value of X and the value of Y (i.e. ignore any negative sign)	$\lvert (12 - 27) \rvert = 15$ (not -15)
Comparison or Relational Operators		
$=$	Equal to	$10 = 4 + 6$
$\neq$	Not equal to	$11 \neq 4 + 6$
$\approx$	Approximately equal to	$7 \approx 40/6$
$<$	Less than	$8 < 12$
$\leq$	Less than or equal to	$8 \leq 8$
$>$	Greater than	$5 > 3$
$\geq$	Greater than or equal to	$5 \geq 5$

They illustrate the calculations involved in these techniques by using a series of case-study datasets relating to various countries and to topics from Geography and the Earth Sciences. These datasets include attributes and variables referring to the spatial and nonspatial characteristics of the particular observation units. Box 1.3 provides an example of the general design features of the Boxes used in subsequent chapters, although the figures (data) are purely illustrative. Most Boxes are divided into different sections labelled (a), (b) and (c). If appropriate, the first of these comprises a diagram, map or photographic image to provide a geographical context for the case-study data. The second section is explanatory text describing the key features and outline interpretation of the particular statistical concept. The third section defines the statistical quantity in question in terms of its equation, tabulates the data values for the observation units and details the calculations required to produce a particular statistic (see example in Box 1.3c). Inevitably some of the Boxes vary from this standard format, and some of the datasets introduced in the early chapters are reused and possibly joined by 'companion' data in later chapters to enable comparisons to be made and to allow for techniques involving two or more attributes/variables (bivariate and multivariate techniques) to be covered. In addition to the Boxes and other text,

Box 1.3: Guide to conventions in demonstration in boxes.

Population symbol; sample symbol.

The symbols used to represent the statistical measure covered in the box will be given in the title section, if appropriate.

(a) Graphic image – photograph, map or other diagram

Boxes introducing some of the case study data will often include a graphic image to provide a context for the data and to help with visualizing the significance of the results.

(b) Explanatory text

This section provides a relatively short two or three paragraph explanation of how the particular statistical technique 'works', and a summary of how it should be calculated and interpreted.

(c) Calculation table

This section explains how to calculate particular statistical measure(s), in some instances comparing different methods. The first part provides the equation for the statistical quantity the elements of which are then used as column/row headings.

The following example, although based on the calculation of a particular statistic (the variance), is provided simply to illustrate the general format of these computation tables.

Identification of individual observation units	$s^2 = \sqrt{\dfrac{n(\Sigma x^2) - (\Sigma x)^2}{n(n-1)}}$		Definitional equation
	x	x^2	Column headings: x = the values of variable X; x^2 = the values of X squared
1	3	9	Arrows here indicate that mathematical operations applied to values in one column produce new values in
2	5	25	another column
3	6	36	Shading to lead the eye down a column of numbers to the calculations
$N = 3$	$\Sigma x = 14$	$\Sigma x^2 = 70$	Column totals
	$s^2 = \dfrac{3(70) - (14^2)}{3(3-1)}$		Numbers substituted into equation
	$s^2 = 2.33$		Result
Sample mean	$\dfrac{\Sigma x}{n} = 4.67$		Other related statistics

there are a series of self assessment or reflective questions scattered through the chapters. The aim of these is to encourage readers to reflect on the points discussed. These questions also provide the opportunity for tutors and students to discuss these issues in workshops or tutorials.

1.4 Introduction to case studies and structure of the book

The majority of this text is intended as an introduction to statistical techniques and their application in contemporary investigations of the type typically undertaken by undergraduate students during the final year of study in Geography and other Earth Sciences. Since the days of hand calculation, albeit using calculators has long since passed, the book assumes students will undertake statistical analysis in a digital environment (i.e. using computer software). However, it is not intended as a substitute for consulting online help or other texts that provide detailed instruction on how to carry out particular procedures in specific software (e.g. Bryman and Cramer, 1999).

The book has three main sections. The present chapter is in the *First Principles* section and has provided an introduction to quantitative analysis and outlined the features of numerical data and elementary mathematical notation. Also in this section Chapters 2 and 3 focus on collecting and accessing data for your analysis and considers the differences between statistical populations and samples, and strategies for acquiring data. Chapters 4 and 5 examine ways of reducing numerical data into manageable chunks either as summary measures or as frequency distributions. Chapter 5 also considers probability and how the outcome of events when expressed in numerical terms can be related to probability distributions. The *First Principles* section concludes with defining research questions and devising hypotheses accompanied by an introduction to inferential statistics, in other words statistical procedures from which you want to be able to reach some conclusion about your research questions. Separating the first and second sections is a pair of diagrams and a short series of questions that guide you through the decisions that need to be taken when carrying out your research investigation.

These diagrams refer to the statistical techniques covered by the chapters in the second and third sections. The second section, *Testing Times*, includes two chapters with a common structure. These, respectively, concentrate on parametric and nonparametric statistical tests and progress through situations in which the analysis focuses on one, two or three or more different attributes or variables. The final section, as its title suggests, *Forming Relationships*, goes beyond testing whether samples are significantly different from their parent populations or each other, and examines ways of expressing and measuring relationships between variables in terms of the nature, strength and direction of their mathematical connections. Examination of the explanatory and predictive roles of statistical analyses covers both bivariate and multivariate situations. Spatial and nonspatial analytical techniques are incorporated into the chapters in both of these sections and make use of a series of case-study datasets. The attributes and variables in these datasets are, by and large entirely genuine 'real' data values, although details that would possibly have permitted the identification of individual observations have been removed and in some cases the number of cases has been restricted in order to limit the Boxed displays. Nevertheless, the datasets can reasonably be regarded as illustrating different types of investigation that might be carried out by undergraduate students either during the course of their independent studies or during group-based field investigations.

2
Geographical data: quantity and content

Chapter 2 starts to focus on geographical data and looks at the issue of how much to collect by examining the relationships between populations and samples and introduces the implications of sampling for statistical analysis. It also examines different types of geographical phenomena that might be of interest and how to specify attributes and variables. Examples of sampling strategies that may be successfully employed by students undertaking independent research investigations are explained in different contexts.

Learning outcomes

This chapter will enable readers to:

- outline the relationship between populations and samples, and discuss their implications for statistical analysis;

- decide how much data to collect and how to specify the attributes and variables that are of interest in different types of geographical study;

- initiate data-collection plans for an independent research investigation in Geography and related disciplines using an appropriate sampling strategy.

2.1 Geographical data

What are geographical data? Answering this question is not as simple as it might at first appear. The geographical part implies that we are interested in things – entities, observations, phenomena, etc. – immediately above, below and on, or indeed

Practical Statistics for Geographers and Earth Scientists   Nigel Walford
© 2011 John Wiley & Sons, Ltd

comprising, the Earth's surface. It also suggests that we are interested in where these things are, either as a collection of similar entities in their own right or as one type of entity having a spatial relationship with another type or types. For example, we may wish to investigate people in relation to settlements; rivers in respect of land elevation; farms in association with food-industry corporations; or volcanoes in connection with the boundaries of tectonic plates. The data part of the term 'geographical data' also raises some definitional issues. At one level the data can be thought of as a set of 'facts and figures' about 'geographical things'. However, some would argue that the reduction of geographical entities to a list of 'facts and figures' in this fashion fails to capture the essential complexity of the phenomena. Thus, while it is possible to draw up a set of **data items** (attributes and variables) relating to, for example, a river, a farm, a settlement, a volcano, a corporation, etc., we can only fully understand the nature of such phenomena by looking at the complete entity. It is really a question of the cliché that the whole is greater than the sum of the parts.

The collection and acquisition of geographical data raises two sets of issues: how much data should you collect to answer your research question(s); and what means do you use to obtain the desired amount of data? This chapter examines the issues surrounding the first of these questions, whereas Chapter 3 explores the means of obtaining geographical data from primary and secondary sources and the ways of locating where geographical 'things' are in relation to the Earth's surface. From a theoretical perspective all geographical entities are unique, there is only one occurrence of a particular type of entity in a specific location, nevertheless, unless our interest is with such individuality, we usually focus on comparing and contrasting at least two and often many more phenomena of the same or different types. Consequently, one of the first 'data quantity' issues needing to be addressed is how many instances of a particular category of phenomena should be included in our study. The second main issue relates to specifying a set of data items, often called attributes or variables that are chosen because it is believed that they will allow us to answer out research questions. Together, selected entities and data items comprise what is commonly referred to as a **dataset**. In other words, this is a coherent collection of information (facts, opinions, measurements, etc.) for each of a chosen set of geographical entities. The remainder of this chapter will examine these issues and show how we can assemble datasets that are capable of answering research questions of the type asked by undergraduate students of Geography and other Earth Sciences.

2.2 Populations and samples

One of the first issues to address when planning an investigation involving the collection of original data and/or the assembly of information from existing sources is to decide how many observation units to include in your dataset. There are obviously two initial, extreme responses to this question – gather data relating to as many or to as few observations as possible. Obtaining data for all observations avoids the potential problems of leaving out some individual items that might be idiosyncratically important, whereas the minimalist approach saves time and money, and cuts out seemingly unnecessary duplication and waste of effort. It would be easy to decide how many observation units to include if we knew all the information there was to know about them in advance: we could select the minimum number required to exactly reproduce these overall characteristics. However, if we were in this fortunate position of perfect knowledge in advance then we might with some justification decide to abandon the investigation altogether. It is similar to the situation in which the winning numbers of the UK's National Lottery™ were known in advance. There would be two logical responses either for everyone to buy a ticket and mark it with the known set of winning numbers, with the probable result that over 50 million people aged 16 and over in the UK population would win an equal share of prize fund. Alternatively, everyone would decide to abstain from taking part in the draw on the basis that they are only likely to win a few pence when the prize fund is divided between some 50 million people, since for every £1 ticket purchased 28 pence goes into the prize fund. The winners would be the people who knew what everyone else was going to do and then did the opposite themselves.

Unfortunately, just as when playing The National Lottery™, when undertaking research we do not normally have such perfect knowledge in advance of carrying out an investigation and therefore it is necessary to devise some strategy for selecting sufficient observations to provide reasonable insight into the whole set but not so many such that we are wasting too much time and effort. When playing The National Lottery™ many people use numbers related to memorable dates or events, such as birth dates, weddings, number of grandchildren, etc. Such numbers serve as no more of a reliable guide to selecting entities for inclusion in a research study as they do for choosing the winning numbers in a game of chance. The observations or individuals selected for inclusion in a research study where quantitative techniques are to be employed should, as a group, be representative of the complete set.

There are some occasions when it is possible or desirable to collect data for all observations in a set. The term **census** is usually reserved for situations where a count of all members of a human or animal population is carried out, but the origin of the word census simply means 100 per cent of observations are enumerated. Often, such censuses are carried out on a regular basis by government or other national organizations, for example most countries under United Nations guidance carry out a Population Census every 10 years or in some instances every five years. Remote sensing carried out by satellites and, for smaller areas of the Earth by aircraft or airborne platforms, in principle can also be thought of as censuses, since they collect data in digital or analogue form for all parts of the surface in their overview. However, just as censuses of human and animal populations may accidentally omit some individuals, so intervening clouds and shadows may mean that such remotely sensed data are incomplete in their coverage.

Another problem with enumerating and measuring all observation units is that their data only remains current for a fixed, possibly relatively short period of time, since the various phenomena that are of interest to us are subject to fluctuation and modification. Air and water temperature, people's age and location, and corporations' market valuation and workforce are theoretically and practically in an almost constant state of flux. The problem originates from the perpetual interaction between space and time. Only one observed geographical entity can occupy a particular portion of space at a given moment in time. Since for all practical purposes time is unidirectional moving forwards and not backwards, we can only imagine or visualize how that unit of space appeared in the past or rely on the accuracy of historical records, we cannot return or advance to a different time. Despite the possible omission and the certainty of change in observation units, censuses perform the important function of providing a benchmark against which to compare the results from a selected subset of observations on a later occasion. Thus, while censuses do not provide comprehensive data about all observations in advance, they may contain information that will help us to decide how many to select for more in-depth enquiry.

In the context of statistical analysis a complete set of observations is referred to as a **population** and the subset selected from this is known as a **sample**. Some investigations in Geography and the Earth Sciences will use data for a population of observation units, such as in the case of analysing Agricultural or Population Census statistics covering all farm holdings and all individuals/households, respectively aggregated to certain spatial units. Similarly, investigators of ice sheets or vegetation at a continental scale may analyse data captured remotely by satellite-mounted sensors for every square metre of their area of interest. However, it is unusual for academic researchers to have the resources to initiate an investigation that involves collecting original data on this scale for a statistical population and, as these examples illustrate, analysis of all observation units usually involves use of data from secondary sources. More often than not, and almost invariably in the case of undergraduate project work, it is necessary to select a sample of observations for analysis and on which to base our conclusions.

There are five main reasons why collecting original, primary data about an entire population is an unrealistic proposition for most researchers – cost, time, indeterminacy, unavailability and contamination of the entities (see Box 2.1). These difficulties almost invariably mean that, if the collection of original data is considered important for answering the research questions, a strategy for selecting one or more samples of observation units has to be devised. A sample, provided it is selected with careful planning, can give just as clear a picture of what the population is like, but at a fraction of the cost, time and effort. There are two main sampling strategies: one based on using probability to achieve a random selection of units; and the other covering nonprobabilistic procedures to choose observations in a purposive or relatively subjective fashion. These are examined in the following sections, although the problems involved with achieving a sampling strategy based on purely random selection should be recognized at the outset.

Box 2.1:   Five reasons for sampling.

Time	Primary data collection is time consuming and the amount of time that can be assigned to this task has to be scheduled alongside all the other activities required for completing an investigation. An important part of project planning is to specify the amount of time available – days, weeks or months – to collect data from either primary or secondary sources.
Cost	Every means of data collection has direct and indirect costs. Administration of questionnaires by post, email or face-to-face interview, the gathering of field specimens and samples, the surveying of land use, landforms, slope and altitude and the counting of vehicles, cyclists or public-transport passengers, and species of fauna or flora all cost money in terms of travel and equipment expenses. Indirect or opportunity costs are those associated with not being able to carry out some other activity. For example, the loss of earnings by a student from not being in paid employment or the reduced number of publications by an academic from not publishing past research.
Indeterminacy of entities	Some populations contain an infinite or fluid (variable) number of observations and thus, unless the required data can be collected instantaneously for those observations existing at a particular time, sampling is necessary. This problem is common when the units of observation are part of a flow, such as water in an ocean, lake or river, gases in the atmosphere, vehicles on a road or pedestrians in a shopping centre, and are constantly changing.
Unavailability of entities	Some observations in a population may not be available, even if they can theoretically be defined as existing now or having existed in the past. If investigators can only find those observations that have been left behind they cannot be certain whether those remaining were differentially privileged allowing them to survive processes of decay and transformation. For example, historical documents, oral history interviewees and small-scale landscape features (e.g. mediaeval field systems) are those that have survived deterioration, destruction and obliteration and not necessarily all of the entities that once existed. The researcher cannot feasibly decide whether those observations that remain are in some indeterminate way different from the population that existed formerly.
Contamination of entities observations	In some circumstances the method and/or quantity of data collected can produce negative feedback with respect to the observation units causing their characteristics to change or be inaccurately measured. Some landforms or physical features may be disturbed by the approach of human investigators, for example walking over a beach or dunes can alter the morphology and mineralogy of small scale and microscale features. Similarly, interviewees can give mixed responses if they become confused with excessive or seemingly pointless questioning.

2.2.1 Probability sampling techniques

Most statisticians would regard sampling based entirely on probability as providing the ideal means of representing a statistical population in all its typicality as well as its diversity and idiosyncrasy. The basis of probability sampling is that chance, not some biased whim of the investigator or historical accident, determines whether or not a particular observation is included. Initially, it might seem rather perverse to base your sampling strategy on chance dictating which observations are included or excluded from the investigation. However, the value of this approach is that it recognizes the innate variability of 'real-world' events. Many such events when considered in aggregate terms tend towards displaying a particular pattern. For example, people living on the same housing estate or in the same part of a city tend to have similar socioeconomic characteristics in respect of attributes such as income, marital status, education level, occupation, and so on. However, this does not mean that these individuals are identical, simply that they are more similar to each other than they are to people living on a different housing estate or in another part of the city. Similarly, pebbles lying in the same area of a beach in respect of high and low mean water levels are likely to be similar to each other in terms of size, angularity, lithology and shape. Again they will not be identical, but simply more like each other in these respects than they are to pebbles from a different area of the beach. A sample of people in the first case and a sample of pebbles in the second based on probability is likely to represent the innate variability of their respective populations more closely than one based solely on subjective judgement.

The 'gold standard' sampling strategy based on probability, known as **simple random sampling**, gives each and every observation in the population an equal chance of inclusion and exclusion at all stages in the sample selection process. The strategy depends on being able to identify all observations in the population and then selecting at random from this **sampling framework** the required number to make up the sample. If the population is finite in size, in other words has a fixed and unchanging number of observations within the time period covered by the investigation, their identification may be entirely possible. But, if the population is infinite in size or fluid in composition, as illustrated above, then it is not possible to compile a complete list of observations. In these circumstances, the investigator can only realistically assume that those observations capable of being identified are not substantively different from those that are unknown. Once the 'population' of observations has been identified and listed, sample members can be selected at random. Apart from using rather imprecise and unsatisfactory manual methods to achieve this, such as sticking a pin in the list of observations, the standard approach involves assigning a unique sequence number to each observation and using random numbers to select sufficient observations to make up the desired sample size. Traditionally a table of random digits from 0 to 9 has been used to determine which observations should be included in the sample. With each observation in the population having been assigned a unique identification number, groups of 2, 3, 4 or however many digits as necessary can be read from the table within the required range by moving diagonally, horizontally or vertically from a randomly determined starting point. Any sequences of digits forming a number larger than the total observations in the population would be ignored. A simpler approach is to use the random number function in software such as PASW (SPSS) (originally Statistical Package for the Social Sciences), Minitab or MS Excel to generate the requisite random numbers. Table 2.1 illustrates the procedure for generating random numbers in a selection of software programs used for statistical analysis.

Whatever method is chosen to select a set of random numbers that can be mapped onto the corresponding numerical observation identifier, there remains one further question to consider. Each observation in the population is unique and has a different identifier. However, there is nothing to prevent a particular random number occurring more than once. Unfortunately, this produces the seemingly illogical situation of an item being included in the sample several times. Simple random sampling in its purest form aims to give each and every observation in the population an equal chance of selection at all stages in the process, otherwise some bias is present because the probability of an observation being selected changes. An observation left in the population towards the end of the sampling process has a greater chance of being selected

Table 2.1 Procedures for generating random numbers using a selection of software programs (population total observations = 150).

Software	Procedure
PASW/SPSS	In PASW/SPSS select *Compute Variable* from the *Transform* menu item. Specify a Target Variable name (RANDNUMS) into which the list of random numbers will be inserted and apply the UNIFORM function with the total count of observations in the population in brackets. Note that it is necessary to have open a Data Editor window with a column of nonblank numbers equal to the required sample size.

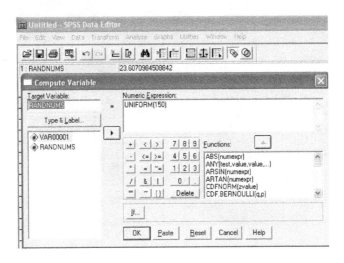

Table 2.1   *Continued*

Software	Procedure
Minitab	In Minitab Select *Random data* … followed by Integer from the *Calc* menu item. Specify a name (e.g. RANDNUMS) for the column into which the list of random numbers will be inserted and enter how many random numbers are to be generated, and the minimum and maximum values of the observation identifiers.

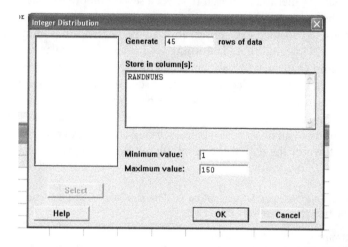

MS Excel	Duplicate Excel *RANDBETWEEN* command in successive cells as necessary to equal desired sample size. A new random number is returned every time the worksheet is calculated. Note that the function requires the Data Analysis Pack Add-in to be available.

than they all did at the start. It comes down to whether the selection process views sampled observations as being 'physically' transferred from a list of the total known observations onto a list of sampled items, or whether it copies the randomly selected identifiers from one list to another. In essence, there are two problems to deal with: what to do about duplicate random numbers that refer to the same observation more than once; and, if the answer is to allow items to be included only once, how do you assess the bias arising from the changing likelihood of inclusion at the second as each subsequent stage or iteration in the selection process? From a statistical analysis perspective these problems relate to whether you should sample with or without replacement and are illustrated in Box 2.2.

Box 2.2a: Simple random sampling with and without replacement.

Sampling with replacement means that once a randomly selected observation has been included in the sample, it is returned to the population and has a chance of being re-selected at any subsequent stage in the sampling process, if its identification number so happens to be chosen again. This means that the size of the population from which the sample is taken remains constant each time an observation is selected, but individual observations might feature more than once. Sampling without replacement effectively removes sampled observations from the population as if they are physically transferred into the sample. In practice, substitute numbers are used to replace any duplicates. Thus, each time an observation is chosen the size of the population reduces by one and the chance of selection for each of the remainder increases slightly.

Boxes 2.2b and 2.2c illustrate the implications of sampling with or without replacement and are illustrated below with respect to selecting a sample of 20 observations from a population of 100 items. Only the stages selecting the first five observations, and then the 19th and 20th are shown in order to limit the size of the tables. The 20 random numbers, constrained to integers between 1 and 100, were generated using statistical software (PASW/SPSS), which resulted in three duplicates (12, 64 and 70) when sampling with replacement. Sampling without replacement avoided this problem by requiring that, at any stage, if a random number selected had already been chosen then another was used as a substitute. This prevented any duplicates and resulted in the numbers 2, 71 and 84 being added to the initial list of 20 random numbers in place of the second occurrences of 12, 64 and 70. If a different set of random numbers was to be generated then such duplication might not occur, but three duplicates occurred without any attempt to 'fix the result' and hence illustrates the problem.

Comparing the rows showing the percentage of the population sampled at each stage in Boxes 2.2b and 2.2c indicates that this remained constant at 1% when sampling with replacement, but reduced by a steady 0.01% as every new observation was transferred into the sample when sampling without replacement. Thus, the first observation chosen had a 1 in 100 (1.00%) chance of selection, but by the time of the 20th this had reduced to 1 in 81 (1.23%). Provided that the size of the sample is small in relation to the number of observations in the population, which may of course be infinite, this change does not cause excessive difficulties in normal academic research. However, suppose the sample size in this example was raised to 50, then the 50th observation would have a 1 in 51 chance of selection or would be 1.96% of the sampled population when sampling without replacement. This is nearly double what it was at the start of the process.

Box 2.2b: Sampling with replacement (includes duplicates in final set).

Stage (iteration)	1	2	3	4	5	...	19	20
Population (remaining)	100	100	100	100	100	...	100	100
Accumulated sample size	1	2	3	4	5	...	19	20
% population sampled	1.00	1.00	1.00	1.00	1.00	...	1.00	1.00
Probability of selection	0.01	0.01	0.01	0.01	0.01	...	0.01	0.01
1st number selected	64	64	64	64	64	...	64	64
2nd number selected		39	39	39	39	...	39	39
3rd number selected			90	90	90	...	90	90
4th number selected				64	~~64~~	...	~~64~~	~~64~~
5th number selected					05	...	05	05
6th number selected						...	51	51
7th number selected						...	92	92
8th number selected						...	18	18
9th number selected						...	73	73
10th number selected						...	70	70
11th number selected						...	88	88
12th number selected						...	52	52
13th number selected						...	73	73
14th number selected						...	48	48
15th number selected						...	70	70
16th number selected						...	92	92
17th number selected						...	34	34
18th number selected						...	12	12
19th number selected						...	23	23
20th number selected						...	76	76

Box 2.2c: Sampling without replacement (excludes from final set).

Stage (iteration)	1	2	3	4	5	...	19	20
Population (remaining)	100	99	98	97	96	...	82	81
Accumulated sample size	1	2	3	4	5	...	19	20
% population sampled	1.00	1.01	1.02	1.03	1.04	...	1.22	1.23
Probability of selection	0.01	0.01	0.01	0.01	0.01	...	0.01	0.02
1st number selected	64	64	64	64	64	...	64	64
2nd number selected		39	39	39	39	...	39	39
3rd number selected			90	90	90	...	90	90
4th number selected				~~64~~		...		~~64~~
5th number selected					05	...	05	05

6th number selected	...	53	51
7th number selected	...	92	92
8th number selected	...	18	18
9th number selected	...	73	73
10th number selected	...	70	70
11th number selected	...	88	88
12th number selected	...	52	52
13th number selected	...	73	73
14th number selected	...	48	48
15th number selected	...		~~76~~
16th number selected	...	92	92
17th number selected	...	34	34
18th number selected	...		~~12~~
19th number selected	...	23	23
20th number selected	...	76	76
1st replacement number	...		71
2nd replacement number	...		84
3rd replacement number	...		02

Most investigations in Geography and the Earth Sciences use sampling without replacement to avoid the seemingly ludicrous situation of potentially including an observation more than once. This is obviously particularly acute when carrying out surveys of people, who might be puzzled, let alone reluctant, if they were asked to answer the same set of questions twice, thrice or even more times. Provided that the **sampling fraction**, the proportion or percentage of the population included in the sample is relatively small, but nevertheless large enough to provide sufficient observations for the intended analysis, then the issues raised by the reducing population size and consequently increasing the chance of selection with each stage (iteration) of the sampling process is unimportant. Figure 2.1 illustrates the relationship between the sampling fraction and population size. The lines depict the sampling fraction of two sample sizes (40 and 100 observations) selected from a population total that increments by 1 observation for each data point graphed along the horizontal (X) axis. Thus, the first point plotted for each line represents the situation where the population and sample sizes are equal; thereafter the sampling fraction decreases, as the sample total becomes an ever-smaller proportion of the population, which continues to diminish beyond the 2500 shown in Figure 2.1. The shape of the graph is similar for both sample sizes with an exponential decrease in sampling fraction as the population size increases. Once the population total reaches 2500 the sampling fractions are virtually indistinguishable, varying between 1.54% and 3.85%, although up to this point there is some variation. The two sample sizes, which would not be untypical of those used in undergraduate investigations, pass through the 10.0% sampling fraction

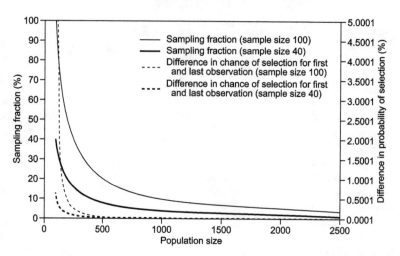

Figure 2.1 Relationship between sampling fraction and difference in probability for first and last sampled observation with population size.

level at population totals of 400 and 1000. The dashed lines in Figure 2.1 plot the difference in the probability for the first and last observation selected for inclusion in the sample, which extends the more limited case study in Box 2.2c. The general form of the curves is similar to those for the decrease in sampling fraction with considerable difference between the 40 and 100 size samples up to a population total of approximately 750. In general terms, sampling without replacement is entirely feasible when the population has at least 1000 observations and the sample fraction does not exceed 10.0%.

It is often not possible to develop a sampling strategy based purely on simple random sampling, irrespective of the issue of sampling with or without replacement. There are frequently legitimate reasons for regarding the population of observations as being differentiated into a series of classes (i.e. the population is subdivided). The justification for defining these strata is usually a theoretical presumption that the observations of interest are fundamentally not all the same. Households living in a city may be subdivided accorded to the tenure of their accommodation for example as owner occupiers, renting from private landlord or renting from local authority or housing association; soil and vegetation specimens may be taken from different types of location, for instance according to aspect and altitude; small and medium-size enterprises may be selected on the basis of where they operate, such as rural, suburban or city-centre areas. When such strata can reasonably be assumed to exist with respect to the entire population **stratified random sampling** may be used to ensure that observations from each stratum are randomly selected in such a way that each is fully represented. In essence, this involves associating unique identification numbers to each observation in the different strata and then generating sets of random numbers corresponding to the ranges of the identifiers in the strata. In other words, a simple random sample is selected from each stratum.

However, there are a several ways of achieving this: an equal number of observations can be selected at random from each stratum; observations can be chosen in the same proportion as each stratum's percentage of the total population; or a simple random sample of observations from the whole population can be subdivided according to the stratum to which they belong. Each approach has its strengths and weaknesses. Equalizing the 'subsamples' comes up against the problem that strata with few observations overall will be over-represented and those with large numbers will be under-represented. Sampling proportionately avoids this problem, but requires knowledge about the number of observations in each of the strata. The most risky of the three options is to take a single simple random sample and then subdivide, since this could easily result in no observations being selected from some strata, especially those with relatively few members. Stratified random sampling is widely used, for example by organizations carrying out public opinion polls, and the decision on whether to use proportionate or disproportionate selection depends on the anticipated extent of variation between the strata. However, given that there is unlikely to be sufficient information available in advance of the survey to assess the extent of variation and that it will almost certainly vary between the different attributes and variables then a proportionate approach is normally advised.

Which strata might be suitable for a sample survey of farmers in your country? How easy would it be to obtain the information needed to identify the populations of farms in these strata?

The basic assumption of stratified random sampling is that there are sound theoretical reasons for subdividing the overall population into a series of strata. Equally there are occasions when an investigator has no prior justification for stratifying the population in this way. Nevertheless, if the population comprises a very large number of observations, has a wide geographical distribution or is fluid in the sense that individuals flow in and out, subdivision into groups for the purpose of **cluster sampling** may be useful. The observations are allocated to mutually exclusive groups or clusters according to their geographical location or temporal sequence. A subset of clusters is selected at random and then either all observations or a simple random sample is chosen from each. This procedure assumes that the geographical location and/or temporal sequence of the observations do not produce any substantive differences between them in respect of the topic under investigation. This may be risky assumption since many geographical phenomena exhibit Tobler's (1970, p. 236) first law of Geography: 'Everything is related to everything else, but near things are more related than distant things.' In other words, observations of the same type (i.e. from one population) that are located relatively close together are more likely to be similar to each other than they are to those of the same type that are further away. Since cluster sampling works by randomly excluding a substantial number of observations from consideration when selecting the sample, it should be seen as a way of helping to reduce the amount of time and money required to collect original data.

Cluster sampling has been used widely in Geography and the Earth Sciences, for instance to select regular or irregular portions of the Earth's surface or 'slices' of time in which to obtain data relating to vegetation, soil, water, households, individuals and other phenomena. There are three problems with cluster sampling that are especially important in respect of Geography and the Earth Sciences. First, the focus of interest is often on discovering whether there is geographical (spatial) and/or temporal differentiation in the topic under investigation, which makes nonsense of assuming that where observations are located and when they occur is irrelevant to selecting a sample. Secondly, there is no way of being certain that the sampled observations are truly representative of the overall population, since the final sample is based on selection from a set of arbitrary clusters. These could be defined in many different ways. Time may be divided into various combinations of temporal units (e.g. seconds, minutes, hours, months, years, decades, centuries or millennia), or into convenient epochs, for example, when a political party held power, when an individual led an organization or when the effects of a natural disaster persisted. Geographical space may be similarly dissected into portions on the basis of units of physical distance or area (e.g. metres, hectares, kilometres, etc.), of administrative and political convenience (e.g. census tracts, counties, regions, states, continents, etc.), or with cultural connotation (e.g. buildings, communities, gang territories and tribal areas). Thirdly, the number of clusters and their 'subpopulation' totals has a critical influence on the size of the sample. If clusters are defined to equalize the number of observations then it is reasonably easy to determine sample size in advance, but random selection of clusters with uneven numbers of observations will produce an unpredictable sample size.

The principle of cluster sampling may be taken further by means of **nested sampling**, which involves successive subdivision and random selection at each stage down the tiers of a hierarchy. This system can be illustrated with the example of selecting areas of deciduous woodland to form a sample in which a count of flora species will be made. A regular grid of 100 km squares placed over the country under investigation can be subdivided into 10 km and 1 km units. First, a random group of the largest units is chosen, then another of the intermediate units within the first subset and finally some of the 1 km squares. The areas of woodland lying in these squares constitute the sample. In addition to selecting a sample of observations with respect to a nested hierarchy of spatial units, it is also possible to select from a hierarchy of time slices. For example, two months in a year might be chosen at random, then five days in each month and then four hours on those days. The problems associated with cluster sampling also apply to nested sampling, in particular the number of clusters chosen randomly at each stage strongly influences the eventual sample size and the extent to which it represents the population.

One further strategy for selecting a sample that introduces an element of chance is to use **systematic sampling**. The procedure involves selecting observations in a regular sequence from an initial randomly determined starting point, for example every 50th person encountered in a shopping mall, every house number ending in a 3 on a city's residential estates or every 15th individual met on a tourist beach. Systematic sampling is particularly useful when drawing a simple random sample is beyond the resources of the investigator and the total number of observations is difficult to identify and list. The method assumes that the observations themselves are distributed randomly in time and space, and, since regular systematic selection starts from a random point in this distribution, the resulting sample should be reasonably representative. The problem is that it may be difficult to determine whether the observations are randomly distributed and that no obscure, underlying pattern is present that might introduce undetected bias.

This difficulty is exacerbated on many geographical and Earth scientific investigations, since they are often underpinned by a belief that spatiotemporal variation is of crucial importance. In other words they are seeking to explore the reasons why observations are distributed according to certain spatial and temporal patterns. In these circumstances it is hardly appropriate to sample on the basis that the observations are in fact randomly distributed, when they clearly are not. However, provided that the presence and potential impact of an underlying pattern in the population of observations is recognized and that the systematic sample does not coincide with this regularity, so that only one subset of observations are included and not the full range, then systematic sampling is a feasible option. The presence of such patterns only really becomes problematic when their potential impact on the sample is ignored.

The random component that gives strength to simple random sampling clearly diminishes on moving through stratified, cluster, nested and systematic sampling. Boxes 2.3 and 2.4, respectively, illustrate the application of these strategies to human and physical geographic research topics. The illustration in Box 2.3 depicts the concourse of major railway terminus in London with people passing through on their daily commute to and from work, *en route* to and from airports and seaports, when visiting shops in the station or for other reasons. There are many entrances and exits to the concourse resulting in a constant flow of people arriving and departing, connecting with other modes of transport or using the station for various other reasons. The statistical population of observations (individuals) formed of the people on the station concourse in a 24-hour period, even if not strictly infinite, is fluid. This makes simple random sampling, where all observations can be identified, impractical for selecting a sample of people for interview in a survey. The examples of applying stratified, cluster, nested and systematic sampling to this situation are intended to be illustrative rather than prescriptive.

Box 2.3:   Sampling strategies in transport geography: Liverpool Street Station, London.

Each of the alternatives to simple random sampling seeks to obtain a sample of observations (people) from the concourse at Liverpool Street Station in London. Passengers continually enter and leave the station concourse, which has a number of entrances and exits, and hence it is impossible to identify a finite population with unique identifier numbers, which rules out the option of simple random sampling.

Stratified random sampling:

The concourse has been divided into three zones based on what appear to be the main throughways for passenger movement and areas where people congregate (part of each stratum is shown in the image). An interviewer has been positioned in each of these zones. A set of 25 three-digit random numbers has previously been generated. Interviewers have been instructed to approach people and ask for their mobile or landline telephone number. If this includes any of the sequences of three-digit random number sequences, the person is interviewed for the survey.

Cluster sampling:

Seven zones in the concourse have been defined arbitrarily based around the entrances and exits. Two of these have been selected at random. Interviewers are positioned within each of these areas and select people by reference to a previously generated set of five random numbers between 1 and 20 that tells the surveyor the next person to approach once an interview has been completed (e.g. 4 denotes selecting the 4th person, 11 the 11th, and so on.

Nested sampling:
The same two randomly selected zones used
  in the cluster sampling procedure are
  used for nested sampling. But now the
  surveyors attempt to interview ALL
  passengers and visitors within these areas
  during a randomly selected 15-minute
  period in each hour.

Systematic sampling:
The team of four interviewers have been
  positioned across the concourse
  (circled locations on image) and have
  been instructed to approach every 8th
  person for interview. This number
  having been chosen at random. The
  interviewers have also been told to
  turn to face the opposite direction
  after each completed interview so that
  interviewees come from both sides of
  the concourse.

Box 2.4 illustrates an area of lateral moraine of Les Bossons Glacier in France in the mid-1990s that was built up during the glacial advances during the so-called Little Ice Age in the 18th century. The glacier's retreat in recent times has exposed the lateral moraines enabling an investigation to be carried out into the extent to which the constituent rock debris has been sorted. The material ranges in size from small boulders and stones to fine particulate matter providing a partial clay matrix. There is a break of slope on the moraine, above which the angle of rest of the partially consolidated material is 55° and below the surface slopes gently at 12°. The upper section is reasonably stable in the short term, although periodically there are falls of small stones. Thus, the observations (stones and pebbles) forming the statistical population of surface material may be regarded as infinite and to some extent changeable as various processes, for example related to temperature change and runoff from rainfall, moves debris across the slopes of the moraine and into the outwash stream. The illustrated applications of stratified, cluster, nested and systematic sampling strategies to this problem represent four of many possible approaches to sampling material from the surface of the moraine.

Box 2.4: Sampling strategies in geomorphology: Les Bossons Glacier, French Alps.

The four alternatives to simple random sampling have been applied by overlaying the chosen study area with a 1 m grid (each small square represents 1 m²). Each sampling strategy seeks to obtain a sample of observations (stones, pebbles and similar rock debris) from the surface of the moraine that can be measured with respect to such morphological variables as size, shape and angularity. The difficulties of identifying a finite population with unique identifier numbers makes simple random sampling impractical.

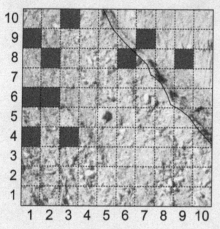 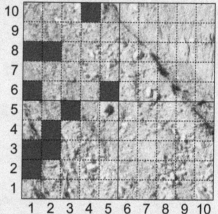

Stratified random sampling:
The study area has two strata according to slope angle: 20% of the surface area (20 m²) is steeper (upper right). A sample of grid cells selected using two sets of paired random numbers covers 10% of the 100 m² with 20% (2 m²) and 80% (8 m²) of these in steeper and gentler areas.
7, 9; 9,8.
2,6; 2,8; 1,6; 3,10; 1, 9; 1,4; 6,8; 3,4.

Cluster sampling:
The study area has been divided into four equal-sized quadrants of 25 m² (solid lines). Two of these have been selected at random and then 2 sets of 5 random number pairs within the range covered by each quadrant provide a 10% sample of grid cells across the study area.
2,8; 5,6; 1,6; 4, 10; 1,8.
1,3; 1,2; 2,3; 2,4; 3,5.

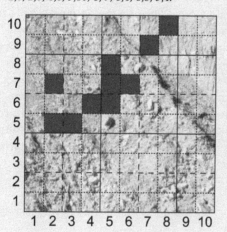 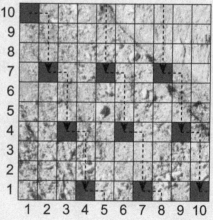

Nested sampling:
The study area has been covered by another grid with 4 whole squares of $16\,m^2$ and parts of 5 other similar units. These are further subdivided into $4\,m^2$ quadrants and then $1\,m^2$ squares. In two randomly selected $16\,m^2$ squares four pairs of random numbers and in one of the part 16-m^2 squares two pairs have been selected. These provide the grid references for the sample of ten 1 m squares representing a 10% sample of the study area.

2, 5; 3, 5; 2, 7; 4, 6.

5, 7; 5, 6; 5, 8; 6, 7.

8,10; 7,9.

Systematic sampling:
A random starting point (1,10) within the grid of $1\,m^2$ squares has been selected. Then moving to the next column and down by a randomly determined number of squares (4) and so on across the columns provides a 10% sample of the overall study area.

What are the strengths and weaknesses of the sampling strategies shown in Boxes 2.3 and 2.4? List some other situations where they could be applied.

2.2.2 Subjective sampling techniques

There is an element of subjectivity in most of sampling techniques, even in those based on probability examined in the previous section, although the element of chance is generally regarded as making them more representative and relatively free from investigator bias. For these reasons they are considered preferable for obtaining original data that is to be analysed using quantitative statistical techniques. However, another perspective is represented by the view that, since we can rarely, if ever, realistically achieve the ideal of simple random sampling and that the inherent failings of the alternatives render them incapable of providing a truly representative sample, we might as well make use of overtly subjective approaches. Subjective sampling techniques place less emphasis on obtaining a sample that is representative of a population, rather preferring to rely on the investigator's perception of what constitutes a selection of sufficiently diverse and typical observations. The difficulty with this approach is that what might seem like an entirely acceptable sample of observations to one person may be regarded as inappropriate by another.

One of the main differences between a sampling strategy based, at least in part, on chance, and one relying on subjectivity is the question of replication. If chance has had a part to play in Investigator A selecting a set of observations, then Investigator B could apply the same strategy, obtain a different sample yet generate similar results so that both reached the same conclusions. Applying a subjective sampling strategy that relies on an individual investigator's or a research team's collective perception and interpretation of the population will almost inevitably result in inconsistency when trying to replicate and verify results. Qualitative analysis does not demand the opportunity for replication, but recognizes the uniqueness of events and the investigator's role in influencing the research process. Proponents of qualitative research often maintain that the objectivity supposedly afforded when employing chance in a sampling strategy is more illusory than real, since ultimately the definition of a statistical population entails subjective judgement.

The sampling strategies illustrated in Boxes 2.3 and 2.4 for selecting samples of observations using an element of chance are indicative of the many situations in which it is difficult to define or identify a statistical population. An alternative option would be to employ **convenience sampling**, in other words to select those observations that are accessible without excessive cost, danger or inconvenience to the investigator. Using convenience sampling to select stones and pebbles for measurement on the lateral moraine of Les Bossons Glacier might result in a daring researcher venturing onto the steeper, dangerous slopes to collect specimens, whereas a less intrepid investigator might only select material while traversing the gentler area near the outwash stream. Similar differences could arise when sampling passengers conveniently in the concourse of Liverpool Street station: one self-assured surveyor might accost anyone as a potential respondent to the survey, whereas a more reticent interviewer might only approach people who appeared 'friendly' and willingly co-operated. Passengers who agree to take part in the survey may be regarded as self-selected volunteers, and the sample could end up comprising docile, amiable people with time on their hands, rather than a representative cross section of visitors to the concourse. These cases illustrate the two-way interaction between the investigator and the individual observations that comprise the population, which makes convenience sampling a risky option when the reactions of surveyors and observations cannot be standardized or at least deemed equivalent.

An alternative method, relying on the opinion of the investigator rather than the convenience of the observations, is **judgmental sampling**. This strategy requires that the investigator can obtain some initial overview of the population, for example by visual inspection of a terrain, by perusal of published statistics or by examination of a cartographic or remotely sensed image, in order to decide what the population is like. This assessment provides the basis for selection of a set of individual observations that the investigator judges will constitute a 'representative' sample. The main problem with this approach is that we are not very good at making unbiased judgments, especially from limited information, people are inclined to overestimate the number of observations of some types and to underestimate others. An investigator casting his/her eyes over the scenes presented in Boxes 2.3 and 2.4 would have difficulty in deciding how many stones of different sizes or how many passengers making various types of journey to select for the respective samples. Figure 2.2 shows another image of people in the concourse of the same London station. Examine the image carefully and see how difficult it is to make a judgement about basic characteristics of the population, such as the proportions of people who are male and female.

Figure 2.2 Passengers on the concourse of Liverpool Street station, London.

Judgemental sampling is often used when only limited information can be found about the nature of the population, but even so it is important its validity is checked and that the investigator's judgement is as free from bias as possible.

Quota sampling can be viewed as the subjective equivalent of stratified random sampling insofar as it is based on the presumption that population legitimately subdivides into two or more separately identifiable groups. However, the similarity ends here, since in other respects quota sampling, by combining features of the convenience and judgmental approaches, introduces the negative aspects of both. The procedure for quota sampling starts from two decisions: how many observations to select and whether to split these equally or disproportionately between the groups. The latter requires the researcher to assess the relative proportions of observations in the different sets, which may be a matter of imprecise judgement given a lack of detailed knowledge about the population. This may inflate the number of observations chosen from some categories and deflate those in others, or it may be part of the research design that certain groups are over represented. Whichever method is required the investigator proceeds by filling the subsamples up to the predetermined limit with conveniently accessible observations. An investigator using quota sampling in the earlier transport geography and geomorphology projects might employ the same groups as outlined previously, but select observations with reference to accessibility and judgement rather than chance. One of the main problems with quota sampling is to decide whether sufficient observations in each group have been included, which essentially comes back to the problem of not having sufficient knowledge about the structure and composition of the population in advance.

One of the most common reasons for an investigator moving away from a strategy based on random selection towards one where subjectivity plays a significant part is the difficulty of adequately identifying and delimiting the population. A carefully designed strategy based on simple random sampling will still fail if the population is incompletely defined. Some members of a single population may be more difficult to track down than others. For example, farmers managing large-scale enterprises may be listed in trade directories, whereas those operating smaller or part-time concerns may be less inclined to pay for inclusion in such listings. Thus, the population of farmers is imperfectly defined. One way of partially overcoming this problem is to use an approach known as **snowball sampling**, which involves using information about the known members of the population to select an initial, possibly random, sample and then augmenting this with others that are discovered by asking participants during the data collection phase. The obvious disadvantage of this approach is that the ultimate composition of the sample is determined by a combination of the subjective judgement of the investigator and the views and suggestions of those people who readily agreed to participate in the survey. For example, an investigator may be told about other potential respondents who are likely to co-operate or to have particularly 'interesting' characteristics, whereas others who may be reluctant to particular or considered 'ordinary' may be not be mentioned.

2.2.3 Concluding comments on sampling

The observations in geographical and Earth scientific investigations are often spatial units or features. In theoretical terms such populations tend to be difficult to identify, because there are an infinite number of points, lines and areas that could be drawn across a given study area and hence a similarly large number of locations where measurements can be taken. The grid squares, used as basic spatial building blocks in Box 2.4, were entirely arbitrary and reflected a decision to use regular square units rather than any other fixed or irregular areas, to arrange these on the 'normal' perpendicular and horizontal axes rather than at any other combination of lines, at right angles to each other, and to define their area as $1\,m^2$. Selecting observations for inclusion in a sample as part of a geographical or Earth scientific investigation needs to distinguish whether it is the spatial units or the phenomena that they contain that are of interest (i.e. are the relevant observations). For example, an investigation into the extent of sorting in lateral moraine material needs to focus on the characteristics of the stones, pebbles and other measurable items rather than the grid squares; and a project on the reasons for passengers and other visitors to a major rail station is concerned with individual people rather than their location in the concourse. However, location, in the sense of where something is happening, frequently forms a context for geographical and Earth scientific enquiries. Thus, the location of the lateral moraine as adjacent to the western flank of Les Bossons glacier in the European Alps and the situation of Liverpool Street station close to the City of London and as a node connecting with North Sea ports and Stansted airport provide geographical contexts in these examples.

No matter how perfect the design of a particular a sample, unforeseen events can occur that lead to the data collected being less than perfect. Consider the example of a company running the bus service in a large city that decides to carry out a simple on-board survey of passengers to discover how many people use a certain route, their origins and destinations and their reason for making the journey. The survey is carefully planned to take place on a randomly selected date with all passengers boarding buses on the chosen route being given a card to complete during the course of their journey, which will be taken from them when they alight. Unfortunately, on the day of the survey one of the underground lines connecting two rail termini is closed temporarily during the morning commuter period due to trains being unexpectedly out of service. An unknown proportion of commuters who usually take the underground between the rail stations decide to travel by buses on the survey route. Unless the survey team become alert to this unanticipated inflationary factor, the estimated number of passengers normally travelling the bus route is likely to be exaggerated. This unfortunate change of conditions is not as fanciful as it might at first seem, since this situation arose when the author participated in such a survey on his normal commute through London when this text was being written.

2.3 Specifying attributes and variables

This section could in theory be endless! The list of attributes and variables that are of interest to Geographers and Earth Scientists in their investigations is potentially as long as the proverbial piece of string. Any attempt to compile such a list would be futile and certainly beyond the competence of a single person or probably even outside the scope of an international 'committee' of Geographers and Earth Scientists. There have been attempts to produce lists of attributes and variables relating to particular theme, such as the 'Dewdney' set of demographic and socioeconomic indicators from the 1981 British Population Census (Dewdney, 1981) or Dorling and Rees' (2003) comparable counts from British Population Censuses spanning the 1971 to 2001. However, in both these cases the researchers' efforts were constrained to those counts that were contained in the published statistics and, while they might have had their personal wish list, this could not be realized because they were referring to data from secondary sources. Similarly, there are sets of indicators of environmental quality. Such lists may not be universally accepted and the exclusion or inclusion of particular indicators is likely to be a contentious issue.

A comprehensive, all embracing list of geographical and Earth scientific attributes and variables that we can call upon when undertaking an investigation may not exist, nevertheless, there are some general principles that can be established to promote more successful research endeavour. Section 2.2 suggested two alternative responses when faced with the question of how many observations to include in a sample – include the maximum or the minimum. Deciding what attributes and/or variables to collect could easily present a similar dichotomy. Collect as many as possible just on the off chance that they may be useful, or seek the minimum necessary to answer the specified research question. The first approach is likely to lead to excessive levels of data redundancy as things are collected that are never used. Whereas the minimalist approach may answer the question, but disallow any opportunity for serendipitous discovery and going back to collect more data is normally out of the question since your cannot recreate the situation exactly as it occurred on the first occasion.

It is obvious that when setting out to undertake some form of research investigation you not only need to have research question(s) that you want to answer, but also some idea of what kind of information would provide answers to these questions. In a different context, if you wanted to discover how many British cricketers had hit six sixes in an over of world class cricket, then you would turn to Wisden (www.wisden.com), The Cricketers Almanack, and not to the records of the Lawn Tennis Association. This chapter started by defining geographical data as attributes and variables, generally called **data items**, about geographical or spatial 'things' – those phenomena that occupy a location under, on or near the Earth's surface. We need to distinguish between geographical entities or phenomena as a whole, such as cities, lakes, sand dunes, factories, glaciers, road networks, national parks and rivers, and the information collected about them. In other words a researcher needs to be clear about the units of observation and the attributes or variables used to express certain characteristics or quantities of these units. Some of these attributes or variables may apply to different categories of entity. For example, we may be interested in the area of a farm or of urban open space, or similarly the volume of water in a lake or frozen in an ice sheet. The examples depicted in Boxes 2.3 and 2.4 have indicated that it would be feasible to collect data relating to some of the people on the station concourse or some of the material on the surface of the moraine and still reach valid conclusions about the phenomena as a whole (i.e. passengers in the train station and debris on the Bossons Glacier moraine). The next stage in planning our research involves deciding exactly what attributes and variables we will seek out in respect of the geographical phenomena of interest.

Geographical phenomena (cities, lakes, sand dunes, etc.) are commonly separated into three types based on their spatial properties, namely points, lines and areas. Additionally, surfaces are sometimes regarded as a further category of geographical feature, although in other respects these may be treated as an extension of area features into three-dimensional entities. A lake can simply be considered as a two-dimensional surface of water, although we may also be interested in its subsurface characteristics. For example, micro-organisms (flora and fauna) are often dispersed at different levels of concentration and systematic sampling throughout the water body would enable this variability to be captured in a (sub-) surface model. In a similar way, people are dispersed at varying levels of density within urban settlements and these patterns may change over time. Quantifying the numbers of people per unit area (e.g. per hectare) allows surface models to be created. However, some features change from being points to areas and from lines to areas depending upon the scale at which they are viewed cartographically and the nature of the research questions being asked. For example, the city of Norwich may simply be viewed and located as a point on a map of all cities in the UK and in a study of urban population change be represented by a graduated symbol whose size depicts a certain level of population growth or decline between two census dates. Figure 2.3a illustrates a range of estimated population growth for selected English cities between 2001 and 2006 and indicates that Norwich's population increased between 5.0 and 9.9%. However, such a study may choose to focus on population change within and between areas of the city in which case the detail of Norwich's urban morphology and the arrangement of its socioeconomic and physical features are likely to be relevant. Figure 2.3b records estimated population

a

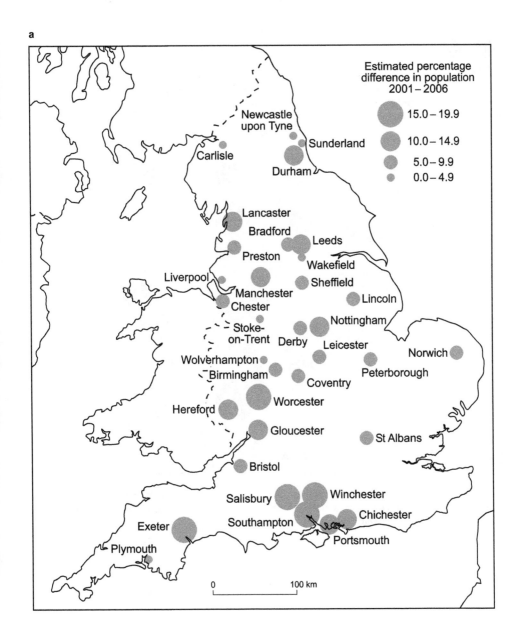

Estimated percentage
difference in population
2001 – 2006

15.0 – 19.9

10.0 – 14.9

5.0 – 9.9

0.0 – 4.9

Newcastle
upon Tyne

Sunderland

Carlisle

Durham

Lancaster

Bradford

Leeds

Preston

Wakefield

Liverpool

Sheffield

Manchester

Lincoln

Chester

Nottingham

Stoke-
on-Trent   Derby

Leicester

Norwich

Wolverhampton

Birmingham

Peterborough

Coventry

Hereford

Worcester

Gloucester

St Albans

Bristol

Salisbury

Winchester

Southampton

Chichester

Exeter

Portsmouth

Plymouth

0   100 km

b

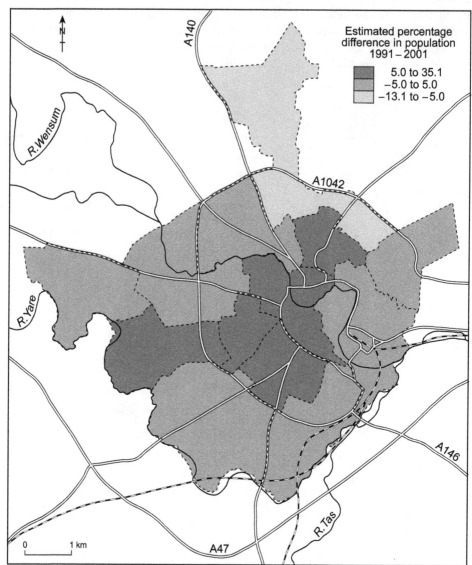

Figure 2.3 Differences of scale when investigating population increase and decrease. (a) Estimated population change in a selection of English cities, 2001–06. (b) Estimated population change in Norwich, England, 1991–2001.

growth and decline in Norwich's wards between 1991 and 2001 adapted to compensate for boundary changes together with selected features of the urban landscape (major roads and buildings). In the present context the relevance of Figure 2.3 lies not in a discussion of the population changes themselves, but in reflecting on how geographical entities and research questions need to be related to the scale of the investigation.

Boxes 2.5, 2.6 and 2.7 illustrate some of the attributes and variables that might be associated with a selection of geographical phenomena of these different types. Chapter 3 will examine how we can locate geographical phenomena and their data items by means of coordinates in a grid referencing system. However, for the present it is relevant to note that point features require one pair of coordinates (X and Y), lines a minimum of two pairs (X_{start}, Y_{start}; and X_{end}, Y_{end}) and areas (polygons) at least three, and usually many more, coordinate pairs to specify their location. Before discussing examples of point, line and area features, it should be noted that the attributes and variables associated with them could relate to their geometrical properties, their spatial dimensions, or to the features themselves. We have already referred to population density where the concentration of people is expressed as a ratio in respect of the area of land they occupy (e.g. persons per km^2). There are many occasions when we want to relate an attribute or variable associated with an entity to one or more of its spatial characteristics. If such measures as population density or discharge from a drainage basin are expressed in a standardized fashion, respectively, as persons per km^2 and cubic metres per second (CUMECs), they become comparable in different contexts and universally applicable.

2.3.1 Geographical phenomena as points

Points represent the simplest form of geographical feature and are, from a theoretical perspective, dimensionless, in the sense that they do not have a linear or areal extent: hence, the importance of being aware of scale in an investigation. Nevertheless, we will often be interested not in a single isolated point, for example just the city of Norwich, but in all or some occurrences of other points in the same category of entities. Rarely are such collections of points located on top of each other, but more commonly occur within a defined area or perhaps along a line and can therefore be related to these rather more complex types of feature. For instance, we may be interested in the density or distribution of points in an area, their distance apart or their regularity along a line. The majority of point features are land-based, and there are relatively few located at sea, and some of those that might be treated as points, such as small rock outcrops may appear and disappear from view with the rise and fall of the tide. Although such features do not vanish when the sea covers them, they do illustrate the temporality or transitory nature of some types of geographical entity. While there may be some features at sea that could be regarded as points, for example oil platforms, although even these have some areal extent, there are none at a fixed nongrounded location in the Earth's atmosphere, although this is not to say that some point features connected to Earth's surface do not extend some considerable distance into the atmosphere, such as aerial masts.

One type of point feature illustrating this characteristic is a skyscraper block, such as those shown in downtown Honolulu in Hawaii in Box 2.5a. Although they are viewed obliquely in the photograph from the 17th floor of one such building and clearly each has a footprint that occupies a physical area, the buildings can be treated as point features with a discrete location. Such areal extent is a less obvious characteristic of the example of point features shown in Box 2.5b. These are some of the trees across the moraines of the Miage Glacier in the Italian Alps. Using the X, Y coordinate location of the skyscrapers and the trees, the spatial variables relate to the density of blocks or trees per unit area (e.g. per km^2 or per m^2) and to the distance between buildings or trees. In the case of the skyscrapers, simple density measurements may be weighted according to number of floors on the basis that higher buildings with more stories will have more impact on the urban landscape. Similarly the density of trees might be weighted according to their height or the width of the canopy.

Box 2.5 also includes some substantive attributes and variables relating to both types of point feature. These are illustrative because the details of which attributes and variables are required will obviously vary according to the aims and objectives of the investigation. The skyscrapers provide an example of how something that might be regarded as a single-point feature may in fact be subdivided in terms of some of its characteristics. As a single geographical feature each tower block has a number of floors, lifts or elevators and a date of completion (age), although in respect of other characteristics, such as the use or ownership of the building, may have multiple values. For example, a single block may be used entirely for offices and occupied by one corporation operating in the financial services sector or a hotel run by one company. In contrast, another building might be in multiple use, with a gym, shop and restaurant on the ground floor, and residential apartments on the remaining floors. Some characteristics, such as the address or post (zip) code of the building, which might initially seem to have one value, could in fact have many different values. If such a tower block in the UK was occupied by several companies, each is likely to have a different 'large-user postcode'. Similarly, if it were an apartment block then each dwelling unit would have its individual identifying number so that the mail for households was delivered correctly. These examples illustrate what is often referred to as hierarchical data, where one characteristic, such as occupier, may have one or many different genuine values. In some cases we may have to deal with multiple hierarchies, such as where a skyscraper not only has multiple uses present (e.g. retailing on the ground floor, financial services on floors 1 to 10 and residential from 11–20), but each of these uses may also be subdivided, with different retail firms on the ground floor, several financial corporations and many households in the residential apartments. Some examples of attribute variables for the trees on the Miage Glacier moraines are included on Box 2.5b. These have been selected from the collection of forest mensuration variables that are of interest to those involved with timber management and production.

Box 2.5:   Geographical features represented as points.

(a) Honolulu skyscrapers

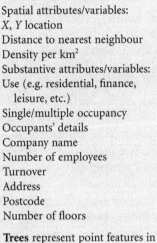

Skyscrapers represent point features in an urban environment.
Spatial attributes/variables:
X, Y location
Distance to nearest neighbour
Density per km^2
Substantive attributes/variables:
Use (e.g. residential, finance, leisure, etc.)
Single/multiple occupancy
Occupants' details
Company name
Number of employees
Turnover
Address
Postcode
Number of floors

Trees represent point features in different types of environment.
Spatial attributes/variables:
X, Y location
Distance to nearest neighbour
Density per km^2
Height
Substantive attributes/variables:
Species
Age
Canopy cover

(b) Trees on the Miage Morraines, Italian Alps

The list of attributes and variables shown in Box 2.5b does not include any that could be regarded as hierarchical (i.e. potentially having multiple values for the same individual specimen). Can you specify any additional attributes or variables associated with the trees that might be hierarchical?

2.3.2   Geographical phenomena as lines

Simple linear features follow a path through or across space and represent a connection between a minimum of one point and another. Often, these are referred to, especially in the context of a Geographical Information Systems (GIS), as 'start' and 'end' points, which implies a direction of flow along the line. However, this may be connected with how the digital coordinate data representing the line were captured and held in the system (i.e. with which terminal point on a line was digitized first), than with the flow of something along the line. Just as the occurrence of a single point is rarely of interest, so it is also unusual to concentrate on one instance of a simple linear feature. Groups of lines can be compared to see if they follow the same bearing or direction, whether some are more sinuous, or some shorter or longer than others. It is also possible to determine the density of lines within an area and whether the lines display a clustered, regular or random pattern. Lines are also inherently land-based features, although again some linear features may describe paths over the surface of the Earth's seas and oceans, for example shipping lanes that are used to help navigation. But such features are not visible on the water's surface, instead they are shown in an abstract form on navigation charts and even the wake from passing vessels is transitory and soon dispersed by the waves, which are also ephemeral linear features moving across the water's surface.

Box 2.6a illustrates the multiple occurrence of a particular type of linear feature. The linear scratches formed by glacial ice dragging subsurface moraine over solid rock on a valley floor, referred to as striations. These often occur in groups where the individual lines are not connected with each other. These striations are microfeatures in the landscape and also help to illustrate the importance of scale in an investigation. In detail, such features as striations may be considered not only as having a linear extent but also depth and width as well as length. Box 2.6a includes some the spatial variables associated with striations, such as orientation and length, which may be used to indicate the direction of flow of the glacier and the pressure exerted on the rock where they occur. Such features are therefore indicative of the erosive force of the glacier that created them as well as the quantity and hardness of the moraine being dragged along by the ice. Box 2.6a also includes some of the substantive attributes or variables that might be relevant in respect of the striations and the situation in which they occur, such as rock type and hardness.

A group of simple lines, whether straight or sinuous, may also form a network, which implies not only a direction of flow along the lines, but also that what flows along one line can pass to another. A flow along a line may be two-way (bidirectional) or constrained to one direction either physically (e.g. by a downward slope allowing the flow of water in a river) or artificially (e.g. by a 'No entry' sign in a transport network). The junctions where lines meet are referred to as nodes, which denote locations in the network where the material in transit can proceed in two or more directions. The regulation of the flow at such nodes may be achieved in different ways depending upon the type of network. For example, in a road-transport network the flow of vehicles may be controlled by signals that prevent drivers from taking a right turn, whereas in drainage networks slope and the velocity of flow may prevent the flow of water from the main channel into a tributary. It is sometimes suggested that there are two main types of linear network known as closed and open. However, totally closed networks are extremely rare and most networks can be regarded as having entry and exit points. In a hydrological network, for example, water flows through from source to destination and there is rarely any significant backward flow, although this is not necessarily the case in other instances of open networks such as a metro system where there will be a two-way flow of passengers and trains along rail tracks going in opposite directions.

Box 2.6b shows some of the watercourses formed by glacial melt water streams from the Lex Blanche Glacier in the Italian Alps. These flow across a relatively flat area composed of a mixture of rocky debris brought down by the streams and boggy wet areas where silt has accumulated and vegetation becomes established. After flowing across this expanse the streams combine and cascade through a narrow gap, towards lower left in the photograph, and thus unified carry on down the valley. Some

---

Box 2.6: Geographical phenomena represented as lines.

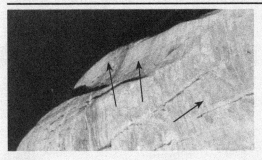

Striations are microfeatures on
  rocks in glaciated environments
Spatial attributes/variables:
Length
Depth
Angle
Density per m²
Substantive attributes/variables:
Rock type
Rock hardness

(a) Striations on an erratic in the European Alps

A **stream or river** represents an
  environmental unit in a
  network
Spatial attributes/variables:
Length
Width
Connectivity
Cross-sectional profile
Substantive attributes/variables:
Water velocity
Concentration of suspended
  matter
Water pH
Water temperature
Air temperature

(b) Streams flowing from the Lex Blanche Glacier,
    Italian Alps

of the streams in the network flowing across this area have been highlighted in the photograph, although there is clearly a fairly high degree of bifurcation and braiding as the main channels divide and reunite as shown lower right in the photograph and is apparent in the distance. This example demonstrates how the notion of a hydrological network as being formed from a series of simple lines may be misleading. Even assuming that the main channels can be identified, as has been attempted with the annotation lines on the photograph, these will vary in width and depth, and in some places the water is relatively stagnant, spreading out into areas of surface water. Thus, the conceptualization of the streams as lines is to some extent a matter of convention and also one of scale. For the purposes of investigating the hydrology of a drainage basin, the linear model applies reasonably well, since it is possible to quantify the capacity of the water to transport different types of material (chemicals, particulate matter, etc.). However, from a geomorphological perspective the dynamics of the streams as geographical features need to be taken into account. Since these streams are fed by melting water from glacial ice, there will be seasonal and diurnal variation not only in flow velocity, but also in the number of streams and the location of their main channel. Thus, the network of streams is a not a static but a dynamic system.

The list of substantive attributes and variables shown in Box 2.6b (water velocity, concentration of suspended matter, etc.) relate to the water flowing through the streams. Assuming suitable field equipment was available to you, where would you set about measuring these variables and how often should such measurements be made?

2.3.3 Geographical phenomena as areas

The discussion and examples of point and linear features has noted that whether these are treated as areas is a matter of the scale and the purpose of the investigation. However, there are many geographical features that are genuine areas with substantial physical extent, which cannot realistically be conceived as points or lines. The fundamental idea behind what makes up an area is that it is a portion of the Earth's surface that is undifferentiated in respect of at least one characteristic. Conceptualizing areas in this way leads to a recognition that there are two ways in which such portions of the Earth's surface might be defined: by means of the boundary drawn around the area, or by identifying where concentrations of particular values of the defining characteristic occur. In the first instance, the presence of a boundary line around a feature implies there is an abrupt change in the characteristic, whereas in the latter there is a gradation across the surface between relatively high and low values. The farm boundaries annotated onto the photograph in Box 2.7a illustrate the former situation, since they denote where the land occupier alters from one person or company to another. Similarly, the boundaries of the London Boroughs in Box 2.7b signify where the responsibility for local government passes from one council to another. However, the upland landscape in Box 2.7c shows part of a slope where there is a change of

Box 2.7: Geographical phenomena represented as areas.

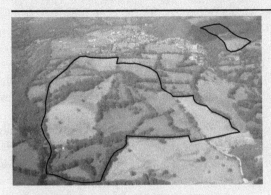

(a) A farm in the Auvergne, France

A **farm** represents an environmental and socioeconomic unit.
Spatial attributes/variables:
$X_1, Y_1...X_n, Y_n$
Area
Number of land parcels
Substantive attributes/variables:
Soil types
Loam (%)
Sand (%)
Clay (%)
Etc.
Crops grown (area)
Cereals (ha)
Potatoes (ha)
Vegetables (ha)
Etc.
Diversification present/absent

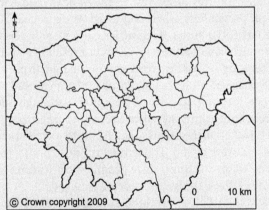

© Crown copyright 2009

0   10 km

(b) London Boroughs

Local authorities represent administrative or political units.
Spatial attributes/variables:
$X_1, Y_1...X_n, Y_n$
Area
Substantive attributes/variables:
Total population
Ruling political party
Male unemployment (%)
Wastepaper recycling (tonnes/household)

(c) Upland landscape in northern England.

Areas of flora represent **vegetation zones.**
Spatial attributes/variables:
$X_1, Y_1... X_n, Y_n$
Area
Altitude
Substantive attributes/variables:
Number of species
Rainfall (mm)
Mean annual temperature (°C)
Soil pH

vegetation type from improved grassland in the foreground to rather more mixed vegetation in the middle distance, where the grass grows in tussocks and is interspersed with other taller species. Drawing a boundary line between these two combinations of vegetation types in order to delimit zones is more problematic than in either of the previous types of area feature, since improved grass dominates the floral assemblage in some parts of the fuzzy border zone between the dashed lines, whereas in others the tussocky grass and taller species are dominant. The dashed line annotations indicate two of the many possible boundaries that might be drawn between these vegetation zones.

Geographical features in the form of areas are sometimes examined individually, for example in studies of a particular mountain range or country. However, such relatively large features are more often subdivided into smaller units, such as the individual mountains and valleys in a mountainous region or the local authorities in a country. An investigation may focus on a sample of such units and further divide these into their constituent components. In general, there has been a trend in recent years for researchers to concentrate on smaller units as the basis of their investigations and in some fields of human geography this has resulted in relatively small numbers of individuals coming under scrutiny. Where the investigation involves a number of areal units, their spatial characteristics, shape, physical area, distribution and juxtaposition may be of interest. Most area features of interest to geographers and Earth scientists are inherently land-based, particularly when conceived as delimited by a boundary line. Thus, although the territorial limits of countries may extend a certain distance from the mean tide line around their coastline, this does not result in a visible boundary out to sea. However, there are occasions when the marine environment seemingly contains area features. Spillage of oil from tankers or wells can spread out across the surface of the water thus forming an area of pollution. Although such features may be ephemeral, only remaining until the salvage teams have secured the flow of oil leaking from the source and applied dispersal chemicals to aid the break up of the area by micro-organisms and wave action. The temporary existence of such an oil slick does not invalidate it as an area feature, since geographical features are rarely permanent, immutable fixture of the Earth's landscape or seascape.

The three geographical area features in Box 2.7 include examples of spatial and substantive attributes or variables. The farm represented in Box 2.7a illustrates how one occurrence of an area feature might comprise a number of physically separate but substantively connected land parcels. This example also shows how a geographical feature of one type (i.e. a farm) may itself be subdivided into another type of entity, in this case fields, which are visible within its boundary. Thus, our geographical data may not only be hierarchical in relation to the substantive attributes and variables, as illustrated with respect to the occupancy of the skyscraper blocks in Box 2.5a, but also in respect of the features themselves having constituent spatial components, potentially with different characteristics. In this case, some of the fields on the farm may grow the same crop, but have different types of soil. The example attributes and variables for the London Boroughs (Box 2.7b) indicate some of the different types of information that may be relevant in a particular investigation, spanning not only demographic and political attributes but also environmental indicators. Finally, the attributes and variables included in Box 2.7c represent some of those that might be used to help with defining the boundary between one vegetation zone and the other. In this we could envisage covering the hill slope with a regular $1\,m^2$ grid and counting the numbers of different floral species or measuring the soil pH in each square and using these characteristics to allocate the squares between each of the vegetation zones.

2.3.4 Closing comments on attributes and variables

The chapter has focused on two important issues with respect to the collection of geographical data – how much and what data to collect. These are not easy questions to answer and part way through an investigation it may well become clear that either not enough or too much data was obtained, or that the appropriate attributes and variables were not obtained or defined carefully enough. Unfortunately, it is rarely possible to go back and remedy these mistakes after the event and, as we shall see in Chapter 3, researchers are often constrained not by their ability to define exactly what data they need, but by being unable to obtain this information in the time and with the resources available. Students will invariably need to undertake some form of sampling when carrying out their independent investigations, since the time and resource constraints are likely to be even more pressing than in funded academic research. The discussion of sampling strategies in this chapter has highlighted that there are many different approaches and exercising careful consideration as to how the resultant data will be analysed is a crucial factor in determining whether the chosen procedure will be satisfactory and achieve the desired outcomes. There has been a trend towards reducing the number of observation units or geographical features as we have referred to them, however, such a reduction in the amount of data obtained is not advisable simply on the basis of saving time and money. Whilst investigations following a qualitative as opposed to a quantitative approach may not seek to ensure that the sample of observations is statistically representative of the parent population, it remains the case that those included in the study should not overemphasize the idiosyncratic at the expense of omitting the average. This is not to dismiss the importance of the unusual, but to remind us of the need to include the typical.

The specification of what information is going to be relevant is another important question facing students charged with the task of undertaking some form of independent investigation. It is all very well to say that the attributes and variables that you need should follow on from the research questions that you are seeking to answer. However, if the questions are imperfectly or ambiguously defined then it is likely that unsuitable attributes and variables will be devised and the study will conclude by saying that if other data had been available then different answers to the research questions would have resulted. This is, of course, not very helpful or informative. It is in the nature of research that one project will lead to and possibly improve on another, but this is not an excuse for failing to specify attributes and variables carefully. The examples of attributes and variables given in Boxes 2.5, 2.6 and 2.7 are purely illustrative and not prescriptive. There are many alternatives that could have been included, but they do indicate that in geographical studies we might need to take account of both the spatial and the substantive characteristics of what is being investigated.

3

Geographical data: collection and acquisition

Chapter 3 continues looking at the issues associated with geographical data but focusing on collection and acquisition. Having discovered in the previous chapter that it is usually not necessary and often impossible to collect data about all observations in a population, we examine some of the different tools available for collecting geographical and spatial data. It also explores the option of obtaining data from secondary sources as a way of avoiding primary data collection altogether or at least in part. Examples of different ways of collecting data from global remote sensing to local-scale questionnaire surveys are examined together with the ways of coding or georeferencing the location of geographical features. In other words how to tie our data to locations on the Earth's surface.

Learning outcomes

This chapter will enable readers to:

- explain the differences between primary (original) and secondary (borrowed) sources of data;

- select appropriate data collection tools and procedures for different types of geographical study;

- recognize different ways of locating observation units in geographical space;

- continue data-collection planning for an independent research investigation in Geography and related disciplines.

Practical Statistics for Geographers and Earth Scientists Nigel Walford
© 2011 John Wiley & Sons, Ltd

3.1   Originating data

Where do geographical data come from? There are many different ways of obtaining geographical data depending upon what research questions are being asked, and there are often a number of different types of data that can be used to address similar topics (Walford, 2002). For example, information on land cover or land-use patterns could be collected by means of a field survey, by the processing of digital data captured by orbiting satellites or by examining planning documents. Each method has its strengths and weaknesses; each is potentially prone to some form of error. However, as illustrated by this example, one common division of data sources is between whether they are primary or secondary, original or borrowed. This distinction is based on the relationship between the person or organization first responsible for collecting the data and the purpose for which they are about to be used. The investigator going into the field in the study area and identifying individual land parcels is able to take decisions on the spot about whether to record a particular parcel's use as agriculture, woodland, recreation, housing, industry, etc., as well as the various subcategories into which these may be divided, but may reach different decisions from another investigator conducting a similar survey in the same location. The researcher using digital data sensed remotely by a satellite and obtained from an intermediate national agency, commercial organization or free Internet resource has no control over whether the sensor is functioning correctly or exactly what wavelengths of electromagnetic radiation are captured (Mather, 2004; Walford, 2002), but is dependent upon someone else having taken these decisions.

In some ways, the primary versus secondary distinction is more transitory than real, since a primary source of data for one person or organization may be a secondary source for another. It is perhaps worth recalling that the word 'data' and its singular 'datum' come from Latin meaning 'having been given', in other words something (facts and figures perhaps) that is just waiting to be collected or gathered up. It seems to suggest that the data used in our investigations would invariably be interpreted and understood in exactly the same way by different people. The problem is that often the very opposite is true. The difference between primary and secondary sources of data is not about the **methods** of data collection (e.g. questionnaire survey, field measurement using equipment and instrumentation, remote sensing, laboratory experimentation or observation and recording), but about the **relationship** between the person(s) or organization undertaking the analysis and the extent of their control over the selection and specification of the data collected. Our investigator of land-cover/use patterns could theoretically, if sufficient funds were available, organize the launch of a satellite with a sensor that would capture reflected radiation in the desired sections of the electromagnetic spectrum. Data obtained from a secondary source, such as a data archive or repository, a commercial corporation, a government department/ agency or over the Internet, will have been collected according to a specification over which the present analyst has no direct control. Sometimes published journal articles, text books and official reports are referred to as secondary sources of data, but this fails to recognize that much of the material in these is argument, debate and discussion based on interpretation of analytical results and other information. This clearly forms part of a literature review but not normally part of your quantitative or qualitative analysis that you carry out. In some cases such publications will include numerical data and results that may be integrated into your own investigation, but this is not quite the same as extracting data from an Internet website where datasets can be downloaded and used for secondary analysis.

The choice between using data from primary or secondary sources in an investigation is often difficult. Researchers frequently opt to combine information from both sources, not through indecision, but because they can potentially complement each other and provide more robust answers to research questions. It is important to recognize that the key difference between primary and secondary sources of data is essentially a question of whether or not you, as a researcher, have control over key aspects of the data-collection process. Although data obtained from secondary sources may be less costly and time consuming to acquire, they will not necessarily be exactly what you need and it may be difficult to determine their quality. In the 1960s a book entitled *Use and Abuse of Statistics* (Reichman, 1961) was published, which attempted to show that unscrupulous or thoughtless selection of data, analytical techniques and results had the potential to allow investigators to reach contradictory conclusions; in effect to argue that black is white and white is black. It is even easier to obtain data and statistics in the era of the Internet than it was when the *Use and Abuse of Statistics* was published, but it also makes it more important that those seeking to answer research questions check the reliability and authenticity of the data sources they employ. Although a distinction is often made between primary and secondary sources of data, the following section will focus on the different means of collecting geographical data irrespective of this division, since most of the methods are not exclusively associated with either source.

3.2 Collection methods

Many geographic and Earth scientific investigations use a combination of different means for collecting and obtaining data about the observations of interest. Therefore, a second question to be addressed when planning an investigation relates to the methods used for this purpose: how the data, the attributes and variables, are obtained from the observations making up the population or sample. This issue transcends the division between primary (original) and secondary (borrowed) sources of data, which refers to whether the data reach the investigator at first, second or even third hand, since most, perhaps even all, data at some time were **collected** from measurable or observable phenomena. The focus in this section is not on the primary/secondary division, but on the methods or means whereby the quantitative and qualitative attributes and variables are obtained from the observations. How do you get observations to give up the data they contain?

The different methods of collecting data can be classified in several ways and Box 3.1 illustrates something of the range of options. A useful starting point is to consider where the data are going to come from. There are essentially three options: to go into the field and collect data by direct contact with human or physical phenomena; to enter the computing or science laboratory and develop mathematical or analogue models of a process or system to generate synthetic data; or to gather data indirectly through remote-sensing devices that measure some characteristics of the phenomena that act as a surrogate for the attributes and variables of interest. A second classification relates to the way in which data are stored, which divides broadly between analogue and digital formats. Data held in an analogue format are represented in a similar way to how our minds perceive it. For example, an analogue 12-hour clock or watch with hour and minute hands portrays the passage of time throughout a day and we refer to time using phrases such as '20 to 11' or '10 minutes past 12'. In contrast, a digital clock or watch shows time as a numerical sequence and would represent these times as 10:40 and 12:10. The difference between analogue and digital is partly a question of how the recording or measuring device operates (e.g. mechanically or electronically) and partly how it stores the data that are obtained. In general terms, a digital format provides greater flexibility for analytical treatment, since it is possible to move such data between different media, for example from a digital camera or scanner to a computer, whereas the equivalent transfer from photographic paper or handwritten questionnaires introduces another stage where errors can enter the process.

3.2.1 Field observation, measurement and survey

One of the most basic methods of collecting data for investigations in Geography, the Earth Sciences and cognate subject areas involves observation, measurement and survey of phenomena in the field. Despite increased competition for research funding and pressure on budgets in teaching departments, the majority of funded projects and undergraduate/postgraduate degree courses in these areas include a period in the 'field'. In the case of taught courses, these visits provide the opportunity for students to develop their skills of observation, measurement and survey and in the case of research they enable the collection of data and for indepth investigation of the area in a way that is simply not possible at a distance. There is a long tradition of using the field sketch as a device for getting students at various levels of geographical education to look at the world around them, to record the locations of and to think about the interrelationships between phenomena. In part, this approach goes back to the philosophy that emphasized the connection between the human and physical domains in the study of Geography. Visualizing the juxtaposition of phenomena and the closeness or distance of their proximity to one another helped to indicate how they might be connected.

Box 3.2 attempts to reproduce such an experience using a photograph to simulate being in the field with a clear view of a glaciated landscape from an elevated vantage

Box 3.1:   Methods and means of collecting geographical data.

Tape or digital recorder (oral interview)

Questionnaire

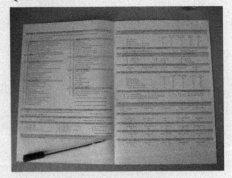

Global Positioning System (GPS) receiver

The Internet

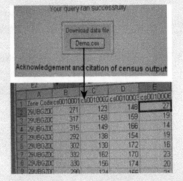

Aerial Photograph

Field Survey

Box 3.2: The use of sketch maps to record key features of the landscape.

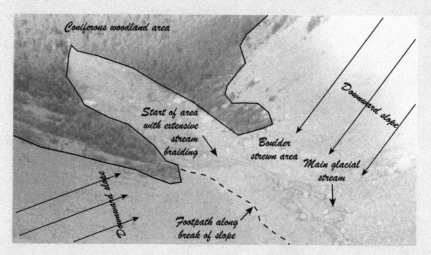

Box 3.2a: Sample sketch map superimposed on photographic image.

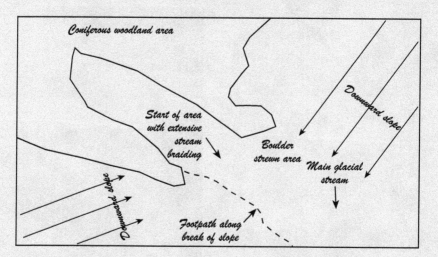

Box 3.2b: Sample sketch map of the area shown in Box 3.2a

point looking down onto the outwash plain of the Bossons Glacier near Chamonix in the French Alps. The use of annotation indicates areas with a particular type of surface feature or vegetation cover, such as the extensive coniferous woodland in the upper part of the photograph. Lines of different types can show the approximate boundaries of such areas and paths can be denoted with different line styles. Arrows pointing down a slope together with labelling aligned with the contours can show variations in elevation. Returning to the notion that field sketches can indicate the juxtaposition of features, it is worth observing that the footpath created by visitors (tourists) to the area has been defined along the valley side away from the incised main channel and area where stream braiding occurs, and also roughly along the line where there is a break of slope. Sketches of this type do not of course provide data capable of statistical analysis by the methods covered in later chapters. However, as an aid to understanding processes of human and physical geography, they provide a useful reminder of the principle, visible relationships evident in the field.

Selection of sites or locations for field observation, survey and measurement is a key issue (Reid, 2003), which in some respects relates to the question of sampling strategy discussed in Chapter 2. One of the primary drivers that should guide investigators in their selection of suitable sites for field survey is to eliminate or 'hold constant' the effect of influences capable of distorting the outcome. However, selection of the ideal, perfect site where the cause and effect relationship between the variables will be manifest is rarely if ever achievable and the investigator should have recourse to the statistical techniques covered in later chapters in order to make statements about the probability that they have discovered how the variables are connected. Making measurements of variables in the field often involves the use of instruments, and just as with any other piece of technology there are often more or less sophisticated versions of any particular type of equipment. Reid (2003) offers a basic equipment list that should enable students to observe and measure sufficient variables relevant to their own piece of research.

3.2.2 Questionnaire surveys

This section looks at the design and delivery of questionnaires and interview schedules. Questionnaire surveys represent one of the most commonly used means of collecting data from people either in their role as individual members of the public, as living within of families and households, or as representatives and actors in various types of organization (e.g. corporations, public agencies, nongovernment bodies, environmental interest or amenity groups, etc.). The essential purpose of a questionnaire is to ask people questions yielding answers or responses that can be analysed in a rigorous and consistent fashion. Questions can seek **factual** information, such as someone's age, period of work with the same employer, level of educational qualification, years at a residential address, previous residential address, participation in community social activities, or number of times visited a doctor in the last six months, etc., the list is almost endless. Questions can also seek **attitudinal** information or people's **opinions** about various matters, such as about whether they agree with government health policies, whether congestion charging should be introduced, whether the interval between domestic waste collection should be altered, whether 4 × 4 wheel drive vehicles should pay higher or lower licence fees, or whether they anticipate changing their residential address in the next five years, etc., again there is an almost infinite number of things about which people might have an opinion. When carrying out an academic research investigating using a questionnaire the key aim of the technique is to ensure that the questions are delivered in an unambiguous, consistent and reliable fashion without introducing any bias into the responses.

One of the issues to be decided when carrying out a questionnaire survey is how to administer or deliver the questions to the potential respondents. The main methods of delivery are telephone, face-to-face interview, drop-and-collect, postal service and the Internet, including via one of the available survey websites (e.g. www.hostedsurvey.com or www.surveymonkey.com). Face-to-face and telephone interviewing involve an interviewer reading the questions orally and recording the interviewees' answers (interviewer-completion), whereas the latter two require respondents to read and answer the questions unaided (self-completion). In general, telephone interviewing is less successful than face-to-face because the respondent can terminate the interview at any time. It is also more difficult to build a rapport with the interviewee, who might be more inclined to give misleading, untruthful answers, although a higher **response rate** might be achieved. In the past there was a problem with some people not having access to a telephone, which could bias a survey towards certain sections of society, however this difficulty has now been superseded by people have different types of telephone or having one whose number is not listed through a directory enquiry service. Delivery of questionnaires by the postal service, the Internet or some other form of intermediary requires a different approach and the opportunity to identify the postal or email addresses of those selected for inclusion in the survey. Some surveys are distributed on a more speculative basis and involve distribution to a large number of potential respondents, although they are rarely used in academic research investigations. Drop-and-collect represents an intermediate form of delivery, allowing members of the survey team to clarify questions and check completion of answers when collecting the questionnaire. In all cases the questionnaire needs to be delivered to those individuals (persons, households, farmers, tourists, etc.) that have been selected for inclusion in the survey by reference to the sampling strategy (see Chapter 2).

Whatever method of delivery is employed, it is important to introduce potential respondents to the survey, for example by means of a letter and/or introductory section. This should explain why you are seeking the information, the value of the survey, give them the opportunity to withdraw from the process at any stage and confirm the confidentiality of the data obtained. It is also important to stress that it will not be possible to identify individual respondents in the data that are held on computer and that these will be held under secure conditions. Universities, professional institutes, learned societies and other research organizations will invariably have ethical codes that will enable you to ensure that you carry out a survey in an ethical manner that will not cause offence or bring the practice into disrepute. These ethical codes often require investigators to complete an ethics approval form, especially if there research involves contact with human or animal subjects. If research involves using data from secondary sources, such as population censuses or official surveys then ethical approval may not be necessary because the investigator is not having direct contact with human subjects and the statistical data from these sources are increasingly already in the public domain.

Table 3.1 summarizes the advantages and disadvantages of interviewer and self-completion forms of delivery. Some of these counterbalance each other, for example self-completion questionnaires are relatively inexpensive and less time consuming, whereas face-to-face interviews are more costly of time and financial resources. However, the latter allow for more sophisticated and searching forms of questions, whereas self-completion requires standard, easy to understand questions capable of completion in a short time period and without any assistance. Self-completion questionnaires need to be unambiguous, whereas interview questionnaires allow questions to be explained, although this raises the problem of variation in phrasing of questions by interviewers, which may be particularly problematic if a number of interviewers are being used. Surveys completed by an interviewer often produce a higher response

Table 3.1 Advantages and disadvantages of interview and self-completion surveys.

Interviewer-completion		Self-completion	
Advantages	Disadvantages	Advantages	Disadvantages
Opportunity for clarification	Difficult to standardize phrasing	Standardized question format	Impossibility of clarification
Higher response rate	Possibly smaller samples	Potentially larger samples	Lower response rate
Possibility of complex questions	Excessive complexity	Simple question format	Produces only simplistic data
Opportunity for follow-up	Longer data-collection time	Shorter data-collection time	Difficulty of follow-up
Possibility of adding questions spontaneously	Potential of introducing inconsistency	Fixed question content	Impossibility of adding questions spontaneously
	Expensive	Inexpensive	
Allows for variable ability of respondents			Assumes standard ability of respondents
Control question structure			Likelihood of more spoiled responses

rate, although conversely might result in smaller sample size, because potential respondents may be less likely to refuse a face-to-face or even telephone request to take part. Self-completion questionnaires generally achieve lower responses rates, typically less than 40%, but overall sample size may be larger because of the relatively ease of contacting larger numbers of potential respondents.

There are some aspects of questionnaire design that vary according to the mode of delivery: for example, the layout of questions in a postal survey and the instructions for completion need to be clear and completely unambiguous. However, irrespective of the mode of delivery there is a general need to ensure consistency in responses so that the answers obtained from one respondent may be validly compared with all the others. Alternative forms of questions seeking information on what is essentially the same topic can produce very different responses. Consider the following questions and possible answers about where someone lives:

- Where do you live?   In a bungalow

- Where is your home?   England

- Where is your address?   On a main road

All of the answers given to these questions are entirely legitimate, but they do not necessarily provide the information that the investigator was seeking. A questionnaire is a form of communication between one person and another and provided that the investigator (interviewer) and respondent are communicating 'on the same wavelength' as far as understanding what a question is getting at is concerned there should be no problem. Unfortunately, we cannot realistically make this assumption given the variability in people's oral and written communication skills and it is therefore important to eliminate poorly defined questions.

The issues of questionnaire design will be examined with reference to the example project outlined in Box 3.3. The project is complicated by the intention of collecting information about the changing composition of households in a regenerated area. The project's main research question is concerned with whether such areas are suitable for raising children or are only really attractive to single adults or couples without children. There is a basic need for the survey to separate the households at the sampled addresses into those where at least one person has lived there since the accommodation first became habitable and those where all members are at least 'second generation' occupants (i.e. none of the present residents were there at the outset). Sometimes the drop-and-collect method simply involves a member of the survey team dropping the questionnaire through the letter box at an address and then returning to collect it later. However, the use of the drop-and-collect delivery method in this example allows the two types of household to be established quite easily, since the person delivering the questionnaires can make contact with someone at the address and ask this information, which enables the appropriate version to be left at the address.

Box 3.3a: Example project illustrating issues of questionnaire design.

In recent years many cities and towns in Britain and other countries have experienced regeneration when areas that were no longer required for industry or transporation have been redeveloped as offices or for residential, retail and leisure purposes. This process has been associated with industrial restructuring and economic globalization, and has led in some cases to changes in the social and economic character of the areas and contrasts have emerged between residents of the areas surrounding the redevelopment sites and those occupying the new residential accommodation.

The London Docklands Development started in the 1980s was the first major regeneration project of this type in the UK. Some 25 years on, a researcher has decided to investigate 'population turnover' in this area and whether the early residents have remained or moved on. The researcher has decided to select a random sample of residential addresses and to distribute a drop-and-collect questionnaire to each of these. Some of the addresses are likely to be occupied by the same residents who moved in at the start, whilst others will house a completely different group of people. At those addresses where at least one member of the original household is still present, the researcher wants to find out details of the changing composition of the household over the intervening period.

Suppose the researcher had decided to use the postal service to deliver the questionnaire in this project. What differences to the design of the questionnaire might be needed to obtain the sort of information required from the two different types of address (i.e. those where no present residents were there at the start and those where at least one original resident is still there)?

Box 3.3b shows only a selection of the questions that might be included in the two versions of the drop-and-collect questionnaires. Some of the questions are identical on both, whereas others are either phrased differently or included in only one. Some of the questions are concerned with eliciting factual information from the respondents, for example when the last person joined or departed from the household (Box 3.3b). These questions appear on both versions of the questionnaire, although their context is slightly different. It is not necessary to ask those households that moved to the address when it became habitable how long they have lived there, although this information is required for second and subsequent generation households. A predefined set of categories has been used to seek views on the likelihood that either the whole household or any one individual member will leave within the next five years (Box 3.3c). Such a series of categories should not be too excessive in number but nevertheless as exhaustive as possible, although an 'Other' category has been included for completeness in this instance. Asking questions about what might happen in the future is inevitably an inexact science, since people cannot know for certain how their circumstances will alter. However, this project is interested in both past and possible future turnover of residents in the London Docklands and therefore some consideration of possible changes was necessary. If an investigator is interested in possible future occurrences, then the questions should indicate that they are seeking

---

Box 3.3b: Turnover of residents at address.

Asked of households with at least one original resident

1. How many people have **JOINED** your
   household since you first arrived including
   those who do not live here now?
2. In what year did the last new member of this
   household come to live at this address?
3. How many people have **LEFT** your household
   since you first arrived including those who do
   not live here now?
4. In what year did the last person leaving this
   household cease to live at this address?

Enter number of additional adults	
Enter number of additional	
children	
Enter number of departed adults	
Enter number of departed	
children	

Asked of households without any original residents

1. How many years has your household lived at
   this address?
   If there is more than one adult in the
   household, otherwise go to Question 3:
2. Did the adult(s) in your household live
   together at their previous address?

   Yes, please go to next set of   1
       questions in Box 3.3c
   No, please go to question 3   2

3. How many people have **JOINED** your
   household since moving here?
   Enter number of adults
   Enter number of children
4. In what year did the last new member of this
   household come to live at this address?
5. How many people have **LEFT** your household
   since moving here?
   Enter number of adults
   Enter number of children
6. In what year did the last person leaving this
   household cease to live at this address?

---

information about the likelihood of some event and the answers obtained should be treated with due care.

Both of these categorized questions have a second part that attempts to probe a little further into the reasons for the indicated categorical response. Here, the first of these probing questions seeks to understand the reason why the whole household would consider moving within the next five years. There are some obvious problems with this form of questioning. The respondent might only have stated one of the categories (e.g. move to another London Borough) because there is a realistic possibility that this might occur, but the follow-up question asks for the person to justify or rationalize this response. The reason given for moving might therefore be fairly imprecise, misleading and people might be tempted to give a standard response, perhaps even based on how they imagine other people might respond. Although this

---

Box 3.3c: Future turnover of residents at address.

Asked of ALL households

1. Is it likely that your complete household will move from this address within the next 5 years?

Unlikely that the household will move together. (Go to Question 3)	1
Yes, likely to move within London Docklands Development area. (Go to Question 2)	2
Yes, likely to move to another London Borough. (Go to Question 2)	3
Yes, likely to move elsewhere in the South East region. (Go to Question 2)	4
Yes, likely to move elsewhere in the UK or overseas. (Go to Question 2)	5
Yes, location unknown. (Go to Question 2)	6
Don't know. (Go to Question 3)	7

2. If YES, what would be the main reason for your household deciding to move within the next 5 years? ....

3. Is anyone currently living at this address likely to leave within the next 5 years?

Unlikely that anyone will leave.	1
Yes, child(ren) likely to go away to university/college.	2
Yes, adult(s) or child(ren) likely to set up household within London Docklands Development area.	3
Yes, adult(s) or child(ren) likely to set up household in another London Borough.	4
Yes, adult(s) or child(ren) likely to set up household elsewhere in South East region.	5
Yes, adult(s) or child(ren) likely to set up household elsewhere in UK or overseas.	6
Yes, location unknown.	7
Don't know.	8

Box 3.3d: Attitudes to living in the area.

1. Please rank each of the following statements on the scale: 1 = Strongly agree; 2 = Agree; 3 = Neutral; 4 = Disagree; 5 = Strongly disagree; 6 = Don't know.

a) There are not enough facilities for families in this area.	1	2	3	4	5	6
b) You don't need to be wealthy to live in this area.	1	2	3	4	5	6
c) This area should not be developed any further.	1	2	3	4	5	6
d) This area has a strong sense of community.	1	2	3	4	5	6
e) It is not necessary to have a car to live in this area.	1	2	3	4	5	6

---

type of question is more suited to an interviewer completed survey, it can be used on in self-completion questionnaires.

Box 3.3d illustrates another way of obtaining people's opinions and attitudes. It involves presenting respondents with series of statements and asking them to express an opinion, in this case on a five point scale from 'Strongly Agree' to 'Strongly Disagree' together with a 'Don't know' option. This type of attitude battery question can elicit opinions quite successfully in both interviewer and self-completion surveys, they are designed to provoke a response, although some people are unwilling to express opinions at the extremes, which might exaggerate the number of responses in the middle of the range. However, there are issues over how many categories to use and that these questions force people to answer with one of the options. Some people's reaction to such questions is to probe the statements and conclude that 'on the one hand they agree and on the other hand they disagree', strongly or otherwise. This might precipitate people opting for the 'Neutral' choice or 'Don't know', when they actually have a split opinion on the subject. Another problem with attitudinal and to some extent factual questions, particularly in self-completion surveys, relates to whose opinion is obtained. In general, they assume that the person completing the questionnaire is capable of answering on behalf of everyone to whom the survey is addressed (e.g. all residents in a household or all directors of a company). In the case of interviewer-completed questionnaires, then it might be possible to determine the accuracy and comprehensiveness of the answer, for example, by interviewing all members of a household, although this might be quite difficult to achieve.

There is a further aspect of questionnaire design that needs to be examined before moving on to other ways of collecting geographical data. The overall structure of a questionnaire or interview schedule will normally use numbers to identify and separate the different questions, and may also include sections in order to group questions on a topic together and to help respondents and interviewers to progress. In both respects there is also often a need to include 'routing' to guide the passage through completing the questionnaire. Some of the questions asked of households where no original resident is present in Box 3.3b and others in Box 3.3c include examples of routing in questionnaire design. These are instructions addressed to the respondents in the case of self-completion questionnaires or to the interviewers, where this person is responsible for asking the questions and completing the schedule. The most common use of these instructions is to route or direct the respondent or interviewer to a subsequent question or to another section of the questionnaire contingent upon the response given. In the case of Question 2 in the second part of Box 3.3b, it is not appropriate to ask someone who lived alone at the previous address whether they came to the sampled address as a completed household unit. Conversely, if the respondent did live with other people previously, it is relevant to know whether they moved together. The instructions to 'Go to Question 2' in some of the options in Question 1 and 'Go to Question 3' in others in Box 3.3c serve a similar purpose of guiding the respondent or interviewer to the next question to be answered, but here it is contingent upon the response given rather to a characteristic of the respondent. Dichotomous questions, such as those expecting the answer 'Yes' or 'No' can conveniently be used to route respondents through a questionnaire or interviewers through a schedule.

There is often a tension between what information is sought in order to answer the research questions and the methodology of data collection. Some forms of questioning are simply ill-suited to certain modes of delivery. For example, if information about people's motivations for their actions is important, these are unlikely to be adequately sourced by means of a self-completed postal or Internet delivered questionnaire, no matter how sophisticated the design. In some cases these tensions can only satisfactorily be resolved by redesigning the methodology, for example by including use of questionnaires accompanied by indepth interviewing, although in some cases an interviewer-completed questionnaire can allow sufficient probing to obtain the required depth of information.

3.2.3 *Administrative records and documents*

Investigations in Geography and the Earth Sciences focus on the world around us and therefore it is not surprising that data collected as part of the administrative and operational procedures of commercial, public and voluntary sector organizations should be relevant to undertaking research in these fields. However, it may not always be possible to obtain access to these data sources at the required level of detail or indeed at all, although it is invariably worth exploring what might be available at the start of an investigation. By definition these data will have been collected, processed and analysed by another organization and therefore they clearly fall within the type of data known as secondary or borrowed sources, although the means by which the raw data are collected in the first case is variable. It is generally acknowledged that the demand for information increases as governmental and commercial organizations become more complex and the extent of their operations expands into new areas. This process has historically fuelled the demand for more data and particularly in the digital era has raised questions about the compatibility, consistency and confidentiality of data held by different organizations and the various parts of the state bureaucracies. It is not sufficient for data to be made accessible; they need to become usable to be useful.

Many of the obvious examples of such data relate to people rather than to features of the physical environment, although the latter are by no means absent. The former includes population registration systems concerned with births, marriages and deaths, which provide basic data about vital events in people's lives, and procedures for welfare assistance (e.g. free school meals, mobility allowance, housing benefit, etc.), which in its various forms potentially act as indicators of relative deprivation and disadvantage. In many countries a periodic population census constitutes a major source of information about the demographic, economic and social condition of the population, and provides an example of how what is generally conducted as a statistically comprehensive questionnaire-based survey of a population at one point in time by government can become a secondary source of data for researchers across a wide spectrum of subject areas, not only embracing the social sciences, but also the humanities and natural sciences. Population censuses have for many years been a fundamental source of information in geographical studies, since they provide one of the few long-term, reasonably consistent and authoritative statistical summaries of the condition of national, regional and in some cases smaller area populations, and a population represents both a key resource and constituent of a country'scharacter.

Population censuses and other surveys conducted by government are in many respects different from other types of administrative data since they are carried out with the explicit purpose, at least in part, of answering research and policy questions. One of the definitive characteristics of administrative data is that their use for research investigations is essentially a byproduct of the process for which the information was originally collected. This has a number of implications. The data may not be ideally suited for the research questions that the investigator seeks to address. This can result in compromise either in respect of the research aims and objectives, or the analysis being conceptually flawed. The essential problem is the difficulty or impossibility of fitting data obtained and defined for administrative purposes to the requirements of an academic research project. Many of the datasets collected as a result or administrative or statutory processes involve holding names, addresses and other personal details of individuals, who can reasonably expect their interests and privacy to be safeguarded. Thus, a fundamental requirement for treating such information as a secondary source of data for researchers is that records are stripped of details that could be used to identify individuals and organizations.

Administrative and statutory data sources may be held in software systems that do not easily allow information to be extracted in a format suitable for use in statistical analysis software, although greater openness in database structures now makes this less of an issue than formerly. Rather more problematic is that many administrative databases are designed to operate in 'real time', in other words to be subject to continual update and amendment. For example, as people's financial circumstances change they might no longer be eligible for certain types of welfare payment and thus their status on the 'system' would need to be altered, and similarly other individuals might need to be added. The essence of this issue is that things change and in the case of changes involving people and places the connections can easily be severed and rearranged. For example, most people during their lives will reside at several addresses and, in the current climate of high urban regeneration, some of the addresses that people move to will be new and not have previously existed as residential addresses.

Suppose one residential address representing a detached house on a parcel of land with a large garden is redeveloped as three new, smaller houses on the same plot when the elderly resident of the house has moved into a care home because of ill health. What changes would need to be recorded in the address databases held by the local authority, utilities companies (gas, electricity, water, etc.) and by the Royal Mail, which is responsible for assigning postcodes? What operational issues might arise for these organizations in respect of the three new houses?

Administrative databases are often dynamic in their nature, which raises issues from an academic research perspective. A typical research investigation involving the collection of original or new data, especially if carried out by a degree-level student with limited time and financial resources, can generally only relate to a restricted period of time. For example, one intensive period of data collection during the vacation between the second and third years of study, although occasionally one or two repeat visits to an area may be possible. The resultant data will therefore be cross-sectional in nature and provide limited opportunities for examining change. There are examples of longitudinal and panel surveys in many countries involving repeat questioning of the same set of participants often with additions made according to strict criteria to compensate for losses, such as the British Longitudinal Study linking a sample of Population Census records since 1971 with other vital events data. In contrast, the records in an administrative database are potentially changing all the time, which raises the twin questions of when and how often to take a 'snapshot' of the data so that the analysis can be carried out. Academic research is rarely concerned with continual monitoring of phenomena, although clearly issues and locations are revisited as time goes by. There comes a time in all academic research when the data-collection phase comes to an end and the analysis begins. Irrespective of whether we are referring to administrative or other sources of data, additional data could have been collected, which might have altered the results. This is a more worrisome issue when analysing data from administrative sources, because the researchers know that someone is capturing the more up-to-date information while they are carrying out their analysis.

There are some similarities between administrative data sources and those examined in other sections of this chapter. In particular, the issues of accuracy and reliability of the data and the design of the data-collection document are common and merit some attention. The majority of datasets collected by researchers will employ sampling (see Chapter 2) and thus the investigator accepts that some observations will have been omitted. Although this will inevitably introduce the possibility of sampling error into the analysis, the application of a well-designed sampling strategy, ideally as a result of introducing some element of randomness into the selection process, should minimize this source of error. However, when carrying out analysis on data obtained from administrative sources, the investigator usually has little way of knowing how complete or partial the dataset is, and most importantly the reasons why some observations are omitted. Potentially, the important difference between sample surveys and administrative datasets is that in the former case omissions occur as a result of a random element in the sampling process, whereas in administrative data missing information is more likely to arise because certain categories of people have chosen to exclude themselves from the process. In the UK the annual electoral register is created by a process administered by local authorities that involves householders completing a form online or for return by post. Some people may chose for the names to be omitted from the published register or avoid being included altogether, perhaps because they are disenchanted with the political system or because they wish their presence to go unrecorded. Omissions arising from such practices are clearly far from unbiased and random. Although datasets generated from administrative processes can fill an important gap in respect of certain types of information or in obtaining data in sufficient quantity on some topics, their use in research investigations should always be treated with a certain amount of caution. Just because the data exist and are accessible does not mean they are necessarily reliable, even if the source seems reputable.

Most administrative data are obtained compulsorily or voluntarily using a printed or online form. The design of such a document in many respects raises the same issues as those involved in producing a clear, intelligible and easy-to-complete questionnaire. Clarity and simplicity are important features of administrative data-collection forms and this may result in the information sought being even more streamlined and simplified than in a questionnaire survey. Although this facilitates the process of collecting the dataset, it does mean that its utility for answering research questions is likely to be diminished. Governments in most countries in the developed and developing world publish statistical abstracts and digests, which are often based on data obtained through administrative processes, and increasingly these are available over the Internet from the different government websites or from those of international organizations, such as the World Bank or United Nations. The usefulness of these information sources for investigations in Geography and the Earth Sciences is often hampered by their lack of geographical detail.

3.2.4   Interviewing, focus groups and audio recording

The section concerned with questionnaires emphasized the importance of clarity in designing and administering a survey instrument, since it is a form of communication between the investigation team and members of the public. Although questionnaires that are completed with an investigator present can avoid many of the difficulties associated with respondents not understanding the purpose of a question and what sort of information is being sought, there is an alternative approach that may be used as a follow-up to a subset of respondents who have completed a questionnaire or as the main means of collecting data. This involves a more conversational methodology, talking to people so as to obtain a richer, fuller set of information. One of the main problems with questionnaires is that they are designed under the assumption that it is known in advance how people will answer the questions presented and that the answers given will be both truthful and meaningful. Even open-ended or verbatim-response questions, which are often included in a questionnaire, may not be phrased in a way that allows the full range of interpretations and responses to be elicited. In other words, even the most carefully designed questionnaires will have a standardizing effect on data collection.

Consider the three seemingly clear and unambiguous questions contain in Box 3.4.

Box 3.4: Examples of seemingly clear and unambiguous questions.

1. How often do you see a film?

   a. Once a week
   b. Once a month
   c. Once a year
   d. Less often

2. What type of film do you prefer?

   a. Horror
   b. Comedy
   c. Western
   d. Science Fiction
   e. Other

3. Are films less entertaining now?

   a. Strongly agree
   b. Agree
   c. Neutral
   d. Disagree
   e. Strongly disagree

So how did you respond to these questions? Well if your answer to 'How often do you see a film (movie)?' was do you mean on TV, over the Internet or in a cinema, then you already appreciate the problems associated with asking people simple questions to which there are limited set of fixed answers. What if you cannot distinguish a preference for Horror and Science Fiction films? When asked your opinion about whether you think films are more or less entertaining than they used to be, how far back are you supposed to go? If a question prompts respondents to reply with a question of their own, this often indicates that it is not well designed or simply that the standardized rigidity of a questionnaire is not allowing people to answer in a way that is meaningful to them, even if it is convenient for the investigator. Structured or semistructured interviews are more conversational and seek to tease out how people make sense of their lives and the world around them. In many respects, this approach is the antithesis of the statistical analyses examined in later chapters, since interviews are based on a two-way dialogue between interviewer and interviewee that are contextualized not only by their own past but also their present and possibly their future experiences, particularly if they are anxious to be somewhere else. In contrast, rigidly adhering to a prescribed questionnaire reflects only the investigator's past perception and does not usually admit the adaptation and flexibility that is possible as an interview progresses.

It has become increasingly common in recent years for social scientists to adopt a mixed-method approach in their investigations combining a larger-scale questionnaire survey with more indepth interviewing of a smaller subset of respondents. Because the intention with qualitative forms of data collection is to be illustrative rather than representative and to seek corroboration rather than replication of information (Valentine and Cooper, 2003), there is little need to use a random process for selecting interviewees, who might be selected from respondents to a quantitative survey or be chosen entirely separately. Fundamentally, qualitative interviews are based on the principle that interviewees can articulate their 'subjective values, beliefs and thoughts' (Layder, 1994: 125) and communicate these to the researcher without loss or distortion of understanding. Using qualitative methods is as much about investigators getting to know themselves as about trawling data from other people. Researchers should therefore be aware of how they might be perceived by the people they are interviewing from a range of perspectives, such as gender, age, ethnic background, sexuality and wealth.

There are alternative approaches to selecting people to interview. Cold calling entails identifying potential interviewees from a list of 'candidates', such as the electoral register, which lists people aged 18 and over at each address. Careful use of this list can help to identify certain social groups, males and females, people living alone and even people of different ages by following fashions in first names. However, increasing concern over 'identity theft' and unwanted contact from salespeople has resulted in people choosing not to appear in the public register. Combining information from the electoral register with telephone directories may allow researchers to undertake a 'cold call' telephone interview. Disadvantages associated with 'cold calling', its intrusiveness, indeterminate accuracy and potential to cause anxiety amongst vulnerable groups (e.g. elderly people and people living alone), limits its usefulness for research investigations. One way of overcoming these difficulties is to seek an introduction to members of the target social group through an intermediary, sometimes referred to as a 'gatekeeper', such as the leader of community or social organizations. Although this form of approach can result in contact with more interviewees, it is important to be aware that these intermediaries are likely to exercise their own judgement over who to introduce and so the full range of members in the organization may not feature on their list. Thus, the investigator may achieve more seemingly successful interviews, but this may be because of having approached a particularly cooperative or friendly group of people.

The 'snowballing' method of selecting people to interview (see Chapter 2) represents a permutation of approaching potential interviewees through gatekeepers. It is based on the idea of using an initial small number of interviewees to build up a whole series by asking one respondent to suggest other people who might be willing to be interviewed, and for this second tier of interviewees to introduce other people, and so on until sufficient numbers have been obtained. Clearly, the accretion of interviewees through 'snowballing' can suffer from the same problem of the investigator only receiving recommendations towards those who are friends with the initial contacts, are of an amenable disposition or share similar opinions to the introducing interviewee, although this difficulty can be ameliorated by starting with a broad, perhaps random, selection of opening interviewees. One of the main strengths of this method is that the investigator can mention to new contacts that they have been given a person's name as a potential interviewee by an acquaintance, which can help to build trust between the different parties.

Whatever approach is used to recruit interviewees, the outcome is likely to be more successful if an initial contact can be made by post, telephone or email. This is especially the case when seeking interviews with people in organizations. Careful preparation is essential in this case, especially if the investigator seeks to interview people with different levels of responsibility or function in an organization, rather than someone in 'customer relations' or external affairs. Initial contact with a named individual is usually more likely to be successful than simply a letter to the Chief Executive of a company or Director of a public organization. It is often necessary to be flexible when making an initial approach to an organization and the example of trying to interview farmers provides some useful insights into the pitfalls. Farm businesses can be run in a number of different ways. In some cases there might be a single farmer running the business as a sole proprietor, a husband and wife partnership, a group of family members (e.g. two brothers and their sons), a manager working in close conjunction with the owner(s) or a manager working as an employee of a limited company. It will be important when making the initial approach to someone in the 'farm office' to establish how the business is run so that some measure of consistency can be achieved when conducting the interviews. This type of initial approach requires a considerable amount of planning, but almost invariably increases the response rate, although possibly at the expense of only having discussed the questions with cooperative individuals.

The conduct of interviews requires careful planning and it is in their nature that each interview will be different, even if the investigator uses the same script with a series of interviewees. One of the most basic issues to decide is where the interviews are to take place. In many cases the type of respondent will dictate location, thus an interview with someone in a company will often take place in their office, with a member of the public in their home and with a child in their school, most probably in the presence of a responsible adult. The issues associated with locating a series of interviews relate to establishing a comfortable and relaxing setting for the interviewee and a safe and secure environment for the interviewer. It is important for the interviewer to establish a rapport with the interviewee at an early stage in the process and this can usually be achieved by courteous behaviour, by allowing the interviewee time to answer your questions and by being willing to explain the purpose of the investigation. Respondents are more likely to be cooperative if they can see where the questions and the research are leading. Even experienced interviewers find it useful to prepare a set of topics, sometimes called 'lines of enquiry', which may in some circumstances be sent to interviewees in advance. While these are not intended to be as rigid and uniform as the questions on a questionnaire, they are intended to ensure that the issues under scrutiny are covered with each respondent. However, it is in the nature of an interview that this may be achieved by a somewhat rambling, spontaneous, circuitous process. A useful opening gambit in many interviews is to ask the interviewee to 'Tell me about ...', which provides a topic but does not constrain the response. Recording interviews, with respondents' permission, is a very useful way of enabling an interviewer to remain focused on the tasks of keeping the interview on track and of listening to what the interviewee is saying so that the range of issues to be explored is covered and allows unexpected themes to be explored. Recording allows the researcher to revisit the interview on several occasions and to pick up the nuances of what occurred in a way that is impossible if simply relying on written notes. Modern digital recorders help to avoid some of the difficulties associated with the older-style tape recorders, such as the tape running out, and being smaller are generally less intrusive.

Even after completing a recorded interview, it is important to make some brief notes about issues that were discussed and the conduct of the interview itself. These will be useful should there be problems with the recording and will help an interviewer learn from the process so that subsequent interviews are even more effective. Analysis of the qualitative data obtained by means of interviews, whether written verbatim or recorded, involves the researcher teasing out the themes and making connections between the discussions conducted with different interviewees. This requires that the interviews are transcribed so that they can conveniently be compared with each other. Although modern digital recordings can be passed through voice-recognition software to assist with the process of transcription, such software is still being perfected and some notes about the points discussed will always be useful. Ideally, interviews should be transcribed as soon as possible after they have taken place in order to clarify any points that are uncertain and to avoid a backlog of interviews potentially stretching back over several weeks or even months building up.

Interviewing on a one-to-one basis (i.e. interviewer and interviewee) is not the only way of engaging in qualitative methods of data gathering. Focus groups have become not only a fashionable but also a productive way of eliciting information, since if working well they allow an interchange of opinions and views between participants. They have emerged as a popular means of gauging public opinion about political issues, but they are also well suited to generating data to help with addressing certain types of research question. Although it is important for the researcher to be present, opinion is divided over whether it is advisable to employ an independent facilitator to chair the event and to keep the conversation focused on the topic being investigated or whether the researcher can take on this role without biasing the outcome of the discussion. It might be tempting to treat a focus group as an encounter between people with differing or even opposing views on a topic. However, there are significant ethical issues associated with this strategy and from a practical perspective it might be difficult to control if the discussion becomes heated and gets out of hand.

Many of the points raised previously in relation to one-to-one interviewing apply to focus groups. Additionally, it is necessary to organize a neutral location where the event will take place, which might involve hiring a venue that is conveniently accessible to all participants. At the start the facilitator should state clearly the 'ground rules' for the discussion, for example that people should avoid interrupting others when they are speaking and should introduce themselves. It is probably even more important to record a focus group session, since there may be instances when different people are trying to speak and taking written notes would be almost impossible. It may be useful to organize some refreshments for participants at the start of the event, although consumption of these should not be allowed to interfere with the progress of the discussion. Transcription of the discussion should not only seek to provide an accurate record of what was said from start to finish, but also to enable the contributions of individual participants to be tracked through the conversation. This has the potential to allow researchers to explore whether some people maintain or change their opinions in the light of contributions from others.

3.2.5   *Remotely sensed collection methods*

Remote sensing is defined as 'the observation of, or gathering information about, a target by a device separated from it by some distance' (Cracknell and Hayes, 1993, p. 1). Although generally accepted as including such methods as aerial photography and satellite imagery, thus defined remote sensing can include other approaches such as observation and recording without the target's knowledge. However, such approaches in respect of human subjects clearly raise some ethical issues. This section will concentrate on aerial photography and satellite imagery, since they have in common the capture of electromagnetic radiation as a way of measuring characteristics of the Earth's surface together with its atmosphere and oceans. The main advantages of such remote-sensing devices are the rapidity and regularity of collecting large quantities of data and, although the setup costs might be high, their ability to collect data for large areas in some instances over many years results in the unit cost of the data being relatively low in comparison with alternative methods of obtaining the same quantity of information.

The history of remote sensing of the Earth's environment from airborne sensors dates back some 150 years to early experiments with cameras used to take photographs from balloons. During World War I cameras were used in aircraft to take photographs for reconnaissance purposes and between the World Wars such aerial photographic surveys were being used to supplement traditional land-based survey for map making. Although further development occurred with different types of film during WWII, it was the launch of the first Earth Observation satellite in 1960 and the subsequent images of the Earth from space that set the trend in remote sensing for the closing decades of the 20th century. These images confirmed for the public and scientists alike that 'The Earth was … an entity and its larger surface features were rendered visible in a way that captivated people's imagination' (Crackell and Hayes, 1993: p. 4). It is surely no coincidence that the arrival of data from satellite-borne remote sensors was accompanied by an interest in modelling global environmental systems and incipient recognition of the impact of man's activities on the Earth.

It has already been mentioned that the data captured by many remote sensors is electromagnetic radiation (EMR), although others record ultrasonic waves, and before examining the characteristics of the data obtained by this means it is helpful to review the nature of this phenomenon. The principal source of EMR reflected from the Earth's surface and its features is the Sun, which is detected by sensing devices mounted on platforms directed towards the surface. EMR travels at the speed of light in a wave-like fashion creating the electromagnetic spectrum, which provides a scale or metric for its measurement. The length of the EMR waves varies along this scale, which is typically categorized into ranges, such as the near-infrared or visible sections respectively 0.7–$3.0\,\mu m$ and 0.4–$0.7\,\mu m$. A proportion of the EMR reaching the Earth's atmosphere is attenuated by absorption, reflection and scattering in certain parts of the spectrum. The EMR reaching the Earth's surface is reflected back differentially by different features and characteristics, for example there is a different pattern of reflectance by healthy and drought-stressed (desiccated) vegetation. Sensors directed towards the Earth passively record the level and intensity of EMR in one or more sections of the spectrum producing a collection of digital data with characteristic highs and lows. Processing of these data allows the pattern to be determined and the terrestrial features and characteristics to be interpreted. The following reflectance pattern indicates healthy green leaves:

- high level of near-infrared 0.75–1.35 μm;

- medium level of green light 0.5 μm;

- low levels of red and blue 0.4–0.6 μm.

The high level of near-infrared indicates the presence of photosynthetic activity and the medium level for green light the presence of chlorophyll. Variability in the ability of the Earth's atmosphere to absorb EMR along the spectrum means that certain sections are favoured for remote sensing.

- visible or optical (wavelength 0.4–0.7 μm);

- near (reflected) infrared (wavelength 0.7–3.0 μm);

- thermal infrared (wavelength 3–14 μm);

- microwave (wavelength 5–500 mm).

Therefore, rather than attempting to detect EMR along the full range of the spectrum, most sensors are set to capture emitted EMR in combinations of these ranges.

One of the main differences between traditional aerial photographs and satellite images is the format in which the data are stored. In the case of aerial photographs the data are held in an analogue format, corresponding to what the human eye can see, so that a photograph of a tree looks like a tree. The EMR is captured by focusing light onto a photosensitive target, such as a film emulsion, and is then processed with chemicals to produce the visible image. Digital aerial photography has obviously changed this situation. In contrast, sensors on satellites collect data in a digital format and an image of a tree is simply a series of digital numbers representing the EMR reflected by the tree. The digital image comprises a series of cells, called pixels, arranged in a regular grid that each possess an intensity value in proportion to the EMR detected by the sensor. It is possible to convert between analogue and digital data by scanning a photograph, where the scanner is akin to the remote-sensing device on an aircraft or satellite except that it is obtaining digital data from a previously processed photographic image rather than from EMR emitted by the terrestrial features themselves. An important characteristic of remote sensing is that the process does not automatically record either the location or the attributes of the geographical features. In the case of analogue data it is a case of interpreting a photograph to determine what is present, whereas digital data requires processing in order to represent recognizable features.

3.2.5.1 Satellite imagery

Satellite-mounted sensors generally collect digital rather than analogue (photographic) data and these, alongside other spacecraft, have become increasingly important for Earth observation since the 1960s. Such is the importance and detail contained in these data that the Ordnance Survey, Britain's national mapping agency, now uses this source rather than digitizing from aerial photographs and field survey as the primary means of obtaining data for cartographic purposes. The organization believes that 'as the accuracy increases, this method [remote sensing] could replace ground survey and aerial photos' (Ordnance Survey, 2007). There are several texts available that examine in detail the sensors on different satellite series and explain the complexities of image processing (e.g. Curran, 1985; Mather, 1987; Lillesand, Kiefer and Chipman, 2004). In the present context we are concerned with reviewing the essential characteristics of digital satellite imagery as a means of obtaining information about geographical features for statistical analysis.

The first environmental remote-sensing satellite launched in April 1960 has been followed by many others, some of which form part of a series (e.g. the National Aeronautics and Space Administration's (NASA) LANDSAT and France's SPOT series), others are 'one-offs'. The three main distinguishing characteristics of a satellite are its orbit, the spectral resolution and the spatial resolution and viewing area. The orbits of satellites vary in their relationship with the Earth and its position in the solar system in respect of their nature and speed. Geostationary satellites are located in a fixed position in relation to the Earth usually over the equator. Polar-orbiting satellites trace a near-circular orbit passing over the North and South Poles or these regions at different altitudes and speeds according to satellite series. These characteristics define the temporal or repeat frequency at which data for a given portion of the Earth's surface are captured. The combination of a satellite's orbit and the Earth's rotation shifts the swath or scanning track of the sensing device's perpendicular view westwards so that the entirety of the surface is covered over a relatively short period of time. The substantial number of satellites now operating in this fashion means that there is likely to be at least one sensor in position to capture unexpected environmental events, such as the Asian tsunami on 26 December 2004 or the fire at the Buntsfield Oil Storage depot Hemel Hempstead in 12 December 2005.

Our discussion of EMR referred to the fact that some sections of the spectrum are preferentially absorbed by the atmosphere, including by clouds, gases and particulate matter, leaving certain wavelengths to reached the Earth's surface and be reflected back for detection by satellite-mounted sensors. Most systems are multispectral and the sensors are 'tuned' to capture data in more than one section of the EMR spectrum or bandwidth and to avoid those ranges that tend to be absorbed. The clear advantage of multispectral as opposed to single-band systems is that the interplay between the radiance data values for the different bandwidths for each pixel can be processed in a combinatorial fashion, which aids with interpreting the features located on the ground surface. However, a high spectral resolution can result in an increased signal-to-noise ratio, which signifies reduced accuracy. Satellite-mounted sensors, as with most types of measuring instrument, are subject to a certain amount of variation in the signal they receive on account of technical imperfections. This variability can be quantified by means of the signal-to-noise ratio, which can be reduced by increasing the bandwidth or by using a series of individual sensors to detect radiance in each scan line element, the so-called 'push broom' system.

Spatial resolution, although important, is rather more difficult to define and in its simplest form refers to the circular area on the ground, assuming no rectangular framing system is in place, viewed by the sensor at a given altitude and point in time. This is commonly referred to as the instantaneous field of view (IFOV), although alternative definitions of spatial resolution exist that take account of the sensor's ability to detect and record radiance at discrete points and for features to be identified irrespective of their relative size (for a full discussion see Mather, 1987; Lillesand, Kiefer and Chipman, 2004; Campbell, 2007). Spatial resolution should not be confused with pixel size, which refers to a rectangular or more typically square area on the ground surface that is covered by one cell in the digital image. Each of the pixels in the regular grid stores the radiance values emitted in certain sections of the EMR spectrum by the terrestrial features wholly or more commonly partially covered by each cell on the ground. Thus, if a given cell (pixel) straddles two contrasting features, for example bare rock and healthy vegetation the radiance values will reflect both these characteristics, whereas if a cell lies over a single, spectrally undifferentiated surface, such as concrete or bare soil, then the data will be more closely aligned with the spectral imprint of this feature. Reducing the size of the pixels will potentially enable greater surface detail to be determined, as the smaller pixels approach the scale of local variations in the ground surface. However, even seemingly uniform features, such as bare soil or concrete, have small-scale variations in the radiance values that might confound the spectral message. The spatial resolution of satellite imagery has reduced over the years, and careful selection from the range of available sources is required to ensure that the data employed are fit for purpose. Small is not necessarily beautiful when choosing satellite imagery for certain types of application, notably where the feature under investigation possesses little variation over a relatively sizeable extent of the surface. A large number of small pixels in this situation will result in a high degree of data redundancy, which is wasteful of data storage and processing time, and give rises to a spurious sense of detailed variation in something that is inherently relatively constant over a large area.

There have been numerous remote-sensing satellites launched since the start of the Earth-observation era in 1960 and a growing number of countries have become involved individually or collaboratively in operating 'space programmes'. Some notable examples are the LANDSAT (USA), SPOT (FRANCE) and NOAA (USA) (formerly TIROS) series. Conventionally, the digital imagery collected by these systems was expensive and likely to lie beyond the resources of student projects and this remains the case for some of the very recent, high spatial and spectral resolution data. However, a number of the organizations responsible for these satellite series have now started to allow free access to some of the imagery. For example, it is possible to obtain from www.landsat.org orthorectified (i.e. corrected for distortions caused by terrain and sensor characteristics) imagery from LANDSAT together with Landsat 4/5 Thematic Mapper (TM) and Landsat 7 Enhanced Thematic Mapper (ETM) data. The orthorectified imagery can be obtained free of charge for three epochs: 1970s from the Landsat Multispectral Scanner (MSS), 1990s Landsat 4/5 TM and 2000s from Landsat 7 ETM+. The Global Land Cover Facility (http://glcf.umiacs.umd.edu/index.shtml) also provides download access to a selection of data from different satellite series with a particular focus on datasets relating to vegetation and land cover. These include selected recent scenes relating to parts of the Earth's surface where topical events have occurred (e.g. Hurricane Katrina) and data from the Advanced Very High Resolution Radiometer (AVHRR) sensor launched by the National Oceanic and Atmospheric Administration (NOAA) and NASA's Moderate-resolution Imaging Spectroradiometer (MODIS) on the Terra (EOS AM) satellite. Within UK Higher Education sector the LandMap service hosted at MIMAS (University of Manchester) enables institutions to subscribe to a licence enabling image processed and raw satellite data from a similar range of satellite series to be downloaded for teaching and research purposes.

3.2.5.2 *Aerial photography*
The arrival of Earth observation by means of satellite-mounted sensors in the 1960s started a slow process of decline in the importance of aerial photography, although a resurgence of interest arose in the 1990s when these were seen as providing greater detail than the data available from satellite systems at the time. However, as the earlier quotation from the Ordnance Survey indicates aerial photography remains challenged as a means of collecting data about geographical features. Two of the clear advantages of an analogue aerial photograph are that it can be held in the hand and viewed optically, and pairs of overlapping photographs for areas of the surface can be examined stereoscopically in order to determine the relative elevation of terrestrial features, although the latter can now also be achieved with digital aerial imagery.

The three main distinguishing characteristics of aerial photographic images relate to the type of camera and film, and the position of the camera in relation to the horizontal. Typically, a single-lens reflex (SLR) camera is used with either 35 mm or 70 mm format, although 23 cm format is also available, where format refers to the size and shape of the negative images produced. The focal length of the lens, the distance between its centre when focused on infinity and the film plane, influences the field of view of the camera, which in turn affects the scale of the photograph. Increasing the focal length of a camera at a fixed altitude results in an increase in the distance on the photograph between any given pair of points at fixed locations on the ground, this changes the scale of the image. For example, two electricity pylons located 100 m apart on the ground might be measured at 1 cm apart on a photograph with one focal length (i.e. a scale of 1:10 000), whereas with a different focal length the distance on the photograph increases to 1.25 cm (i.e. a scale of 1:8000). However, the question of scale in relation to aerial photography is not as simple as it might seem from this example. Suppose we extend it to consider the situation of there being a series of pylons stretching diagonally across the photograph. Each pylon is exactly 100 m from its neighbour on the ground, but on the aerial photograph if the pair in the middle is 1 cm apart those near the corners will be separated by a greater distance. Thus, the scale of the photograph is not uniform across the entire area. This characteristic arises because the photograph has been produced from a camera with a perspective or central rather than orthographic projection. Techniques are available to apply geometrical correction to aerial photographs, but those using scanned images of the aerial photographs to capture digital data representing the geographical features (see below) should be alert to this issue. The implication of this variation in scale is that aerial photographs have two scales, one relating to the average for each photograph or the whole mission, and the other to each individual point at a particular elevation.

The second differentiating characteristic of aerial photographs relates to the types of film available. In essence, this is similar to distinguishing between satellite-mounted sensors that are 'tuned' to detect radiance in one or more discrete bandwidths. The main types of film used in small, and medium-format aerial photography are pan-chromatic (black and white), near-infrared black and white, colour and false-colour near-infrared (Curran, 1981). The position of the film's plane in the camera relative to the horizontal is generally divided into vertical and oblique. The former is generally preferred since it does not introduce distortion or displacement provided pitching, rolling and height changes of the aircraft are avoided. Photographs above 5° from the vertical are referred to as oblique, which are subdivided according to whether the horizon is included or excluded from the field of view, respectively known as high and low.

The process of converting between the multiscale environment of an aerial photograph and the single scale, orthographic projection of a map is known as photogrammetry. This involves integration of the geometrical properties of the photograph (focal length, displacement, etc.) in relation to its centre. Photogrammetric processing of successive, regularly spaced, partially overlapping aerial photographs can produce three-dimensional data and enable the creation of a digital elevation model (DEM) of the surface. Such models include both X and Y coordinates and Z values, which represent genuine measurements of altitude from which interpolated figures can be obtained over the entire surface. Aerial photographs can also be scanned and information about geographical features digitized from the images, although it is important to be aware of displacement, distortion and scale issues. The success of this process depends on transforming locations on the photograph to match the ground coordinate system using a minimum of three control points that can be precisely located on the photograph and on an existing map or in the field by using a Global Positioning System (GPS) device. The digital datasets created by scanning aerial photographs are in some respects similar to the digital images obtained from satellite-mounted sensors. The scanning process creates an image of pixels of a certain density, usually measured in dots per inch (dpi), and with each pixel holding a data value representing the radiance (EMR) detected by the scanner. The number of grey-scale or colour shades that can be stored is related to the storage capacity of the scanner (e.g. an 8-bit scanner stores 256 colours).

Box 3.5 compares a scanned aerial photograph and satellite image covering part of Colchester in East Anglia. The left section of Box 3.5a shows part of the whole aerial photograph and the right part zooms into a small area after the photo has been

Box 3.5:   Comparison of remotely sensed satellite image and scanned aerial photograph.

(a) Scanned aerial photograph

(b) Satellite image

Source: Copyright University of Manchester/ University College London Year 2001

scanned. It is clear that the image has become pixelated and the reasonably clear impression of buildings, roads, gardens, green open space, etc. that can be interpreted from the nonzoomed photo appear as blurred pixels. There is not only variation in the shade of the pixels but also in the sharpness of the image, which illustrates the problem discussed earlier of variations in the surface cover of across individual pixels producing an indistinct spectral signature because of small-scale variations. The classified panchromatic satellite image in Box 3.5b also covers the Colchester area of East Anglia, but this time at a smaller scale. Thus, the level of detail is not as great and when zooming into part of the image it again pixelates and becomes fuzzy.

3.2.6 Experimental and synthetic data

The previous sections have focused on collecting or obtaining supposedly genuine data about geographical features and phenomena from the 'real' world, albeit in the case of remote sensing without direct contact with the target. However, it is possible to generate experimental or synthetic data in a physical/environmental sciences laboratory or by means of mathematical modelling using computer software that purport to represent processes in the 'real world'. In many respects this offers a completely different approach to investigating geographical phenomena, since it works on the basis of thinking about how the process works as a conceptual model and then turning this model into something that can be reproduced in the artificial setting of a 'laboratory'. However, such models will often be calibrated using 'real-world' data and their operation and results may also be tested in the field. It is beyond the scope of this book to provide a detailed account of modelling in Geography and the Earth Sciences, which has been covered in a number of accessible texts (Hardisty, Taylor and Metcalfe, 1995; Kirkby *et al.*, 1992; Lane, 2003).

Lane (2003) identifies empirical and physical approaches to the modelling of environmental processes, although they are less common in respect of modelling human behaviour in contemporary Human Geography than formerly. However they were a feature of the quantification that occurred in the 1960s and early 1970s, when, for example, models were developed to explain patterns of land use in cities. The empirical approach draws on data measurements obtained in the field or laboratory to construct a model of how the process works. Statistical methods are a key part of the empirical modelling approach, since these allow investigators to test the robustness and predictive power of their models. Empirical models are based on the principle that one variable, or more typically a group of variables, exerts a controlling influence on another variable or variables. This influence is played out in respect of the data values measured for each variable, so that high values on one (or more variables) are almost invariably associated with high (or low) values on the other. The evidence for the existence of such relationships is derived from empirical data and not from the representation of theoretical connections by means of mathematical equations. We will return to the application of statistical methods in respect of empirical models in

Section 3.3. The physical approach to numerical modelling involves making a 'formal statement of processes which link the parameters of interest' (Hardisty, Taylor and Metcalfe, 1995, p. 32). So, rather than relying on collecting information about variables that measure different aspects of the process, numerical models comprise statements describing the links between parameters that can then be expressed as mathematical equations that are solved by simulation of parameter values. The predictions produced by 'running' the model can be compared with observed values, which is usually followed by calibration to optimize the model using a different set of data, which may be obtained by splitting the observed data into two sets, one used for validation and the other for calibration. Finally, an assessment is made as to whether the model and predictions are acceptable.

3.3 Locating phenomena in geographical space

The main thing that distinguishes geographical from other types of data is that the information can be located in relation to the Earth's surface. Therefore, irrespective of whether we are referring to analogue or digital data, there needs to be some way of tying the features to which the data relate and the attributes and variables in the data to some location in respect of the Earth's surface. How this is achieved depends to some extent on the way the data are collected and in another way on how the geographical component of the data is to be analysed. Chapter 2 introduced the idea that as Geographers and Earth Scientists we are interested in the things, variously referred to as entities, observations and phenomena that exist at some location in relation to the Earth's surface and the relationships between these features. Hence, it is important to be able to record where one observation in a given population is located with respect to the others, and where all members of that population are in comparison with other similar or different groups of observations. In other words, the individuals in a population possess variation in their spatial location (i.e. they do not occur on top of each other) and populations of similar observations may be compared and contrasted with each other. This focus on locations requires that some means of recording where populations and their constituent observations are located needs to be considered when planning an investigation.

The essence of the problem is to have some way of unambiguously fixing the location of populations of observations and their individual constituent members in the short, medium or long term. Geographical entities exist at a fixed location at a given point in time, although on some occasion in the past or the future they may have been (or may be) located elsewhere or have a different form. The car park shown in Figure 3.1 serves London's Stansted Airport: the arrival and departure of cars associated with their drivers' and passengers' air travel illustrate the geodynamic nature of most spatial phenomena. At some times there will be an abundance of empty spaces in the car park and at others the majority will be occupied. This example illustrates not only the transient nature of most geographical entities as a whole, but also that

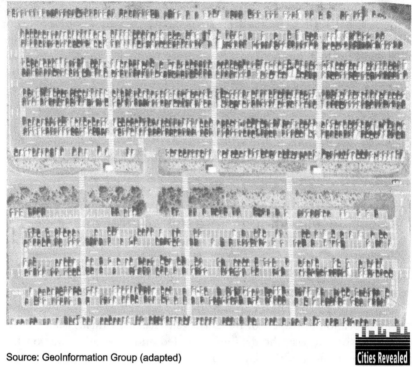

Source: GeoInformation Group (adapted)

Cities Revealed

Figure 3.1 Long Stay Car Park at Stansted Airport, London. Source: GeoInformation Group (adapted).

their individual components (i.e. parking spaces) can change over time. In this case, having been built and assuming no enlargement or internal restructuring has occurred (e.g. changing the layout of roads or parking spaces), the car park may be regarded as relatively fixed. The spatial changes that occur are related to the arrival and departure of vehicles, and essentially the transient occupation of parking spaces. However, it is perfectly feasible for the car park's overall size and internal layout to be altered, thus potentially resulting in its outer boundary being extended, or possibly reduced, and the network of roads connecting blocks of parking spaces being rearranged. Incidentally, the change in the outer boundary impacts upon the adjacent geographical feature, which will necessarily be reduced in physical area, unless the loss of land to the expanded car park is compensated by a gain elsewhere.

The importance of these issues from the perspective of locating geographical features relates to how the data associated with them can be connected with locations in relation to the Earth's surface. The location of a spatial feature is often specified by means of a **georeferencing system**. This usually takes the form of a regular grid with a system of numerical coordinates, which can define global, national or local space (geography). At a global scale, coordinates of longitude (meridian) and latitude (par-allel) normally locate the position of and thereby determine the distance between phenomena. The geometry of space is quantified by a regular, usually square, grid of X and Y coordinates, a Cartesian grid, that enables the position of phenomena to be determined. In the case of longitude and latitude the units of measurement are degrees, minutes and seconds that are defined in relation to the geometrical properties of the Earth (see Figure 3.2a), whereas the British National Grid, introduced by the Ordnance Survey in 1936, is based on the Airy 1830 ellipsoid. It uses the metric system and therefore enables distances to be determined in kilometres or metres directly rather than involving transformation from degrees and minutes. National grids can be directly linked with the global longitude/latitude system as illustrated in Figure 3.2b. An important feature of regular square grids is that distances can be calculated between phenomena not only in the directions of the horizontal and vertical planes of the X and Y axes, but also diagonally. Any pair of points within the grid can form the endpoints of the hypotenuse of a right-angled triangle and its length can be obtained by using Pythagoras' theorem (Figure 3.2c).

Chapter 2 introduced the notion that geographical features could be regarded as constituting points, lines or areas, and that these could be conceptualized in two main ways: for example in the case of areas by means of the boundary drawn around a portion of the Earth's surface or by identifying variation in the concentration or density of some characteristic. These approaches to defining features result in two main ways of georeferencing where they are located. The former case lends itself to capturing a series of lines, themselves formed of smaller segments or arcs, containing vectors of pairs of X and Y coordinates, as examined in Section 2.3.1. Except in the particular case of there being just one area feature of interest, the nodes at the end of each line represent the points where lines bounding two or more areas meet. In con-trast if area features are conceived of as being formed by variations in the values of one or more characteristics, the surface can be divided into a series of regular or irregular units (e.g. squares or triangles) and values of the characteristic(s) recorded for each of these units indicate where a particular feature is located. Population density provides an example of how this system works: each regular or irregular unit possesses a value representing the concentration of people per unit area. Where a number of very high values occur in a group of units that are close together it is reasonable to infer that there is a city or town, whereas at the other extreme many units with low values and the occasional high figure would indicate a sparsely popu-lated rural area of villages and market towns.

The referencing of geographical features to the Earth's surface by means of num-bered vertical and horizontal lines in a regular grid involves using pairs of coordinates to locate where they are to be found and to define their spatial or geometrical form. If geographical features are regarded as being constituted or composed of points, lines and areas, then their location and geometry can be constructed from sets of coordi-nates. However, if they are considered as being built up from variations in the con-centration or density of a particular characteristic, then it is the regular or irregular

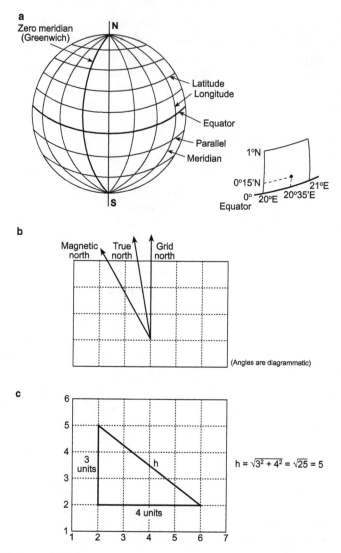

Figure 3.2   Grid-based georeferencing systems. (a) Longitude and latitude system; (b) Relationship between longitude/latitude and British National Grid; (c) Calculation of distance between points.

spatial units holding the data values that are located by means of coordinates. The metric of the grid provides a means of relating distances on the ground to those on a digital or printed map and thus define the scale of representation.

Both of these approaches represent relatively formal ways of recording the location of data about geographical features and phenomena. However, there are alternatives to this quantitative approach that particularly apply when the investigation is less concerned with spatial characteristics of the features and interested in employing a more qualitative methodology with respect to the geographical component of the data. A research question seeking to explore the lifestyles of adolescents in urban areas is likely to be less concerned with determining the precise location of these individuals and more focused on the types and characteristics of places where they 'hang out'. These provide a spatial or geographical context for an adolescent lifestyle that might vary in accordance with different ethnic groupings and with gang membership. In these situations it is less relevant to know the precise boundaries of where such groups congregate as the types of environment where their behaviours can flourish. Despite these contrasts between quantitative and qualitative approaches to locating geographical phenomena, both are concerned with similar spatial questions, such as the juxtaposition of individual and group occurrences of phenomena. The difference is that the former conceives these spatial relationships as capable of direct analysis, whereas the latter argues that the fuzziness of location and uniqueness of places makes this problematic. A focus of where phenomena occur is a distinguishing characteristic of investigations in Geography and the Earth Sciences, even including such abstract, virtual locations as cyberspace. Later chapters exploring the application of statistical techniques in these fields will recognize the variety of ways in which location can be incorporated.

Georeferencing by means of coordinates provides a potentially accurate and relatively precise way of locating and describing the form of geographical features. Ready access to the Global Position System (GPS) via mobile geospatial technologies (e.g. mobile telephone and satellite navigation) allows people to easily determine where they are or to refer to a location on the Earth's surface by means of grid references. However, the notion of people using grid coordinates in everyday conversations seems some way off and both formal and informal languages exist enabling people to say where they are and to describe geographical phenomena. The simplest way of referring to a location is to use its name, which has the advantage of familiarity and the disadvantage of imprecision. Unless an investigator's research questions are concerned with place names, their main function in most studies is to indicate the locale to which the study relates. Possibly more useful than a name is some form of index comprising letters and/or numbers (alphanumeric), particularly where this forms a commonly used system, such as in the case of postal or zip codes, or the standard indexation applied to administrative areas (counties, districts and wards) by the British governmental statistical agencies. A useful feature of these coding systems is that they may be hierarchical, thus enabling a group of smaller or lower areas to be associated with a common larger or higher level area.

These spatial coding systems provide a means of precisely and uniquely identifying locations in space and the things they represent can be regarded as geographical entities, nevertheless, other than in a very simple way, they are unimportant from the perspective of statistical analysis. However, it is possible to capture their location as coordinates, thus enabling the features to be represented according to one of the alternative conceptualizations of spatial entities outlined earlier. The extent to which spatial form and location are important aspects of an investigation will vary according to the research questions asked.

4

Statistical measures (or quantities)

Chapter 4 concentrates on the basic descriptive measures or quantities that can be derived to compare and contrast sets of numbers in their role as measuring quantifiable differences between attributes and variables. Attention is directed towards summary measures referring to numbers, such as the median, mean, range, and standard deviation, and to the location of geographical phenomena, such as mean centre and standard distance. This chapter establishes the characteristics of the basic measures and quantities that may be subject to statistical testing.

Learning outcomes

This chapter will enable readers to:

- explain the purpose of spatial and nonspatial descriptive statistics for comparing and contrasting numerical data;

- identify the appropriate summary measures to use with variables recorded on different measurement scales;

- start to plan how to use descriptive statistics in an independent research investigation in Geography, Earth Science and related disciplines.

4.1 Descriptive statistics

Descriptive statistics are one of two main groups of statistical techniques, the other is known as inferential statistics or hypothesis testing. Descriptive statistics are about obtaining a numerical measure (or quantity), or a frequency count to describe the

Practical Statistics for Geographers and Earth Scientists   Nigel Walford
© 2011 John Wiley & Sons, Ltd

characteristics of attributes and variables of observations in a dataset. We will look at frequency distributions, the principles of hypothesis testing and inferential statistics in Chapter 5, but for the present the latter can be defined as a group of analytical techniques that are used to find out if the descriptive statistics for the variables and attributes in a dataset can be considered as important or significant. Descriptive statistics are useful in their own right and in most geographical investigations they provide an initial entry point for exploring the complex mass of numbers and text that makes up a dataset. Some research questions can satisfactorily be answered using descriptive statistics on their own. However, in many cases this proves insufficient because of an unfortunate little problem known as **sampling error**, to which we will return a number of times in this text.

Descriptive statistics can be produced for both population and sample data, when they relate to a population they are known as **parameters**, whereas in the case of a sample they are simply, but perhaps slightly confusingly, called **statistics**. The purpose of descriptive statistics is to describe individual attributes and variables in numerical terms. We are all familiar with descriptions of people, places and events in books, Box 4.1a reproduces the first paragraph from Jane Austen's novel *Pride and Prejudice*, but try to follow the narrative by reading the words when they have been jumbled (Box 4.1b). It is worse than looking at a foreign language, since the individual words are familiar, but their order makes little sense. Even if the words were reorganized into a sensible order, this may not convey the meaning intended by the original author. We can make a similar comparison with a set of numbers. In Box 4.1c the numbers generated by the UK's National Lottery™ over 26 Saturdays in 2003 have been listed in the order in which they appeared. The range of possible numbers for each draw runs from 1 to 49. Were some numbers selected more often than others over this period? Were some numbers omitted? Were numbers above 25 picked more or less often than those below this midway point between 1 and 49? It is difficult to answer these questions until the numbers are presented in numerical order and Box 4.1d enables us easily to see that 10 and 12 appeared more often than any other number (7 times each), 40 is the only number not to appear in the draw during this 26-week period and that 86 of the numbers were below 25 and 90 were above. This example illustrates that the simple process of reorganizing the numbers into the sequence from lowest to highest allows us to derive some potentially useful information from the raw data. Nevertheless, the sorted sequence still includes 182 numbers (26 weeks × 7 numbers) and there is scope for reducing this detail further to obtain more summary information from the raw data. Indeed, an alternative name for the process of producing descriptive statistics is **data reduction**. Sorting the words of Jane Austen's first paragraph alphabetically is less helpful in conveying a sensible meaning (see lower part of Box 4.1b), although it does allow us to see that the word 'A' was used four times and both 'in' and 'of' twice each.

Which combination of six main numbers would seem to offer the best chance of winning the jackpot prize? Why would using this set of numbers fail to guarantee winning the jackpot?

Box 4.1: Comparison of data reduction in respect of textual and numerical data.

(a) Ordinary paragraph
'It is a truth universally
acknowledged that a single man in
possession of a good fortune must
be in want of a wife'.
Pride and Prejudice by Jane Austen

(b) Jumbled paragraphs
Of man a in single a fortune be good universally
wife that acknowledged want it truth is of
possession a in that a.
A a a a acknowledged be fortune good in in is it
man must of of possession single that truth
universally want wife

(c) 182 Lottery numbers
08/02/03: 43, 32, 23, 37, 27, 41, **35,**
22/02/03: 30, 19, 42, 33, 38, 44, **31,**
01/03/03: 29, 31, 45, 44, 22, 24, **35,**
15/03/03: 47, 28, 04, 25, 38, 24, **11,**
22/03/03: 45, 10, 47, 49, 08, 02, **09,**
29/03/03: 33, 42, 05, 15, 35, 21, **26,**
12/04/03: 45, 25, 03, 16, 05, 43, **23,**
19/04/03: 48, 08, 38, 31, 13, 10, **14,**
26/04/03: 35, 27, 21, 09, 48, 33, **18,**
03/05/03: 07, 03, 49, 46, 06, 29, **37,**
10/05/03: 26, 32, 18, 08, 38, 24, **31,**
17/05/03: 48, 15, 36, 12, 08, 22, **37,**
31/05/03: 22, 25, 02, 09, 26, 12, **21,**
14/06/03: 46, 43, 37, 03, 12, 29, **11,**
21/06/03: 36, 20, 45, 12, 39, 44, **49,**
28/06/03: 23, 17, 35, 10, 01, 29, **36,**
05/07/03: 10, 18, 06, 19, 22, 43, **08,**
12/07/03: 15, 13, 21, 12, 09, 33, **43,**
19/07/03: 09, 34, 03, 17, 10, 48, **36,**
26/07/03: 27, 49, 16, 01, 12, 26, **23,**
02/08/03: 45, 28, 49, 47, 05, 26, **08,**
09/08/03: 41, 25, 14, 30, 19, 11, **38,**
16/08/03: 01, 09, 42, 45, 12, 10, **27,**
23/08/03: 42, 25, 15, 10, 18, 48, **11,**
30/08/03: 18, 38, 43, 44, 26, 33, **02,**
06/09/03: 21, 25, 39, 01, 27, 42, **17**

(d) Sorted lottery numbers

01	01	01	01	**02**	02	02	03
03	03	03	04	05	05	05	06
06	07	**08**	**08**	08	08	08	08
09	**09**	09	09	09	09	10	10
10	10	10	10	10	**11**	**11**	**11**
11	12	12	12	12	12	12	12
13	13	**14**	14	15	15	15	15
16	16	**17**	17	17	**18**	18	18
18	18	19	19	19	20	**21**	21
21	21	21	22	22	22	22	**23**
23	23	23	24	24	24	25	25
25	25	25	25	**26**	26	26	26
26	26	**27**	27	27	27	27	28
28	29	29	29	29	30	30	**31**
31	31	31	32	32	33	33	33
33	33	34	**35**	**35**	35	35	**35**
36	**36**	36	36	**37**	**37**	37	37
38	38	38	38	38	38	39	39
41	41	42	42	42	42	42	**43**
43	43	43	43	43	44	44	44
44	45	45	45	45	45	45	46
46	47	47	47	48	48	48	48
48	49	49	**49**	49	49		

Note: 'Bonus Ball' numbers are shown in bold.

There are two main, commonly used groups of summary measures or quantities known as **measures of central tendency** and **measures of dispersion**. These are examined in detail later in this chapter. The first group includes the mode, median and mean (commonly called the average). The main measures of dispersion are the range, interquartile range, variance and standard deviation. The two sets of measures provide complementary information about numerical measurements, in the first case a central or typical value, and in the second how 'spread out' (dispersed) the data values are. The choice of which measure to use from each group depends largely upon the measurement scale of the attribute or variable in question. Most analyses will involve using two or three of the measures from each group according to the nature of the data. In addition to summarizing the central tendency and dispersion of a numerical distribution, it may be important to examine two further characteristics, namely its **skewness** and its **kurtosis** ('peakedness'). Measures of skewness describe whether the individual values in a distribution are symmetrical or asymmetrical with respect to its mean. Measures of kurtosis indicate the extent to which the overall distribution of data values is relatively flat or peaked.

4.2 Spatial descriptive statistics

One thing setting geography and other geo (earth) sciences apart from many other subjects is an interest in investigating variations in the **spatial location** of geographically distributed phenomena as well as their attributes and variables. In some respects the spatial location of geographical phenomena can be considered as 'just another attribute or variable', but in practice where phenomena are located in respect of each other and the patterns thus produced are often of interest in their own right. When measuring differences between nonspatial phenomena we concentrate on the values of one or more attributes and variables, but usually ignore the position of each observation in the set. Thus, a nonspatial analysis of sampled households is likely to have little interest in distinguishing between, for example, whether any of the households live next to each other or how far apart any pair of households lives. Spatially distributed phenomena have the extra dynamic that their position can also be quantified. Of course, most measurable entities, even if infinitely small or large, are spatially located, what distinguishes 'spatially aware' disciplines is that this spatial distribution itself is not dismissed as a chaotic, random occurrence, but as being capable and deserving of interpretation.

Suppose a researcher is concerned about variations in the quality of households' accommodation and wellbeing. Respondents in each sampled household might be interviewed and information obtained about the number of rooms and their size, the availability of various amenities (e.g. central heating, air conditioning, mains gas and electricity), the entrance floor level, the number of people in the household, the presence/absence of a swimming pool, etc. Analysis of these data might reveal certain nonspatial associations or connections, for instance a large living area per person and the presence of a swimming pool may both indicate relative financial and social advantage. However, if analysis of the survey data reveals households possessing these and other similar combinations of characteristics live close together, then perhaps we can also conclude that this is a relatively wealthy suburban area. In other words, the physical distance between phenomena and the patterns they display may add to the analysis of their socioeconomic characteristics. In this situation, we can start to say something about the characteristics of the places people occupy as well as the people themselves.

The underlying reason for geographers investigating spatial distributions is that they are indicative of the outcome of a process of competition for space amongst the phenomena of interest. In some plant communities, for example, there may be a mix of species that flower at different times of the year and so it does not matter too much if there is some variation in height, provided that tall, medium and short plants do not flower at the same time. Similarly, if a gardener tries to grow too many plants in a limited space, some will thrive, whereas others will decline and possibly die. In a commercial context, a farmer producing apples will try to ensure that the fruit trees are planted an equal distance apart in order to maximize each tree's exposure to sunlight, rainfall, soil nutrients and to facilitate harvesting the fruit. However, suppose the farmer decides to grub up the apple trees and sow the field to grass in order to graze cattle. The free movement of these animals within the field is normally unconstrained and their competing demand for fresh herbage is controlled not by locating the animals at different locations in the field, but by limiting the overall number within this bounded portion of space and allowing the animals to roam freely in search of fresh pasture. Competition for space is therefore rarely, if ever, an unrestricted process, since there is normally some form of controlling or managing function in place, which may be either 'natural' (e.g. competing demand for exposure the sun's radiation), or 'human' (e.g. restricting demand for access to fresh grass).

Figure 4.1 follows on from these examples by illustrating the three main ways in which phenomena, in this case cars parked in the long-stay car park at Stansted airport, can be distributed spatially. In some parts of the car park the cars are parked in a regular fashion either in pairs next to each other or singly with one or two spaces in between. In other parts there are cars parked in small clusters with between 4 and 8 cars in each. However, taking an overall view the cars would seem to be parked in a fairly random fashion with some close together and others further apart, some forming small groups and others on their own. Figure 4.1 presents a snapshot of the spatial pattern of cars in this car park, but thinking about the process whereby cars arrive and depart an airport car park helps in understanding how the pattern may have arisen. It provides an interesting example of the dynamic interrelationships between space and time.

Where might drivers arriving at this section of the car park their cars? How will length of stay in the car park influence the spatial pattern? How might the way the operator of the car park influence the spatial pattern by controlling car arrivals and departures?

This interest in geographical patterns has led to the development of a whole series of spatial statistics, sometimes called **geostatistics** that parallel the standard measures used by researchers in other fields. Descriptive spatial statistics aim to describe the distribution and geometrical properties of geographical phenomena. For example, lines can be straight or sinuous; areas may be large or small, regular (e.g. a square) or

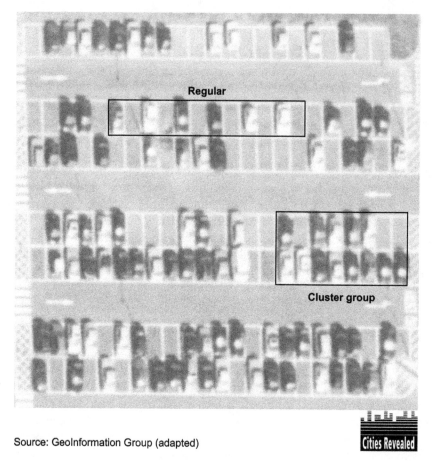

Source: GeoInformation Group (adapted)

Figure 4.1 Spatial distribution patterns. Source: GeoInformation Group (adapted).

irregular (e.g. a polygon). Although such spatial statistics are not relevant in all geographical investigations, it is appropriate to complement explanations of the standard measures with descriptions of their spatial equivalents.

The division of statistical procedures into descriptive and inferential categories also applies to spatial statistics. Numerical measures and frequency counts that can be derived to describe the patterns produced by spatial phenomena either relate to one occasion, as in the snapshot of car parking shown in Figure 4.1, or as they change between different times, which could be illustrated by continuous filming of cars arriving and departing the car park. Geographical phenomena are usually divided into three basic types when examining spatial patterns, namely points, lines and areas. However as we saw in Chapter 2, to the question of scale can sometimes confuse this simple classification, especially between points and areas. Figure 4.2 shows the complete aerial photograph from which the extract in Figure 4.1 was taken. Zoomed out this distance the individual cars in certain parts of the car park can still be identified, but in the large central area (highlighted) they have virtually merged into one

Source: GeoInformation Group (adapted)

Figure 4.2 The blurring of individual points into areas. Source: GeoInformation Group (adapted).

mass and just about transformed into an area, although the circulation routes between the parking spaces enable lines of cars to be identified. Clearly few geographers or Earth scientists spend their time analysing the patterns produced by cars in parking lots, but nevertheless this example illustrates that the scale at which spatial phenomena are considered not only affects how they are visualized but also how their properties can be analysed.

Measures of central tendency and dispersion also form the two main groups of descriptive spatial statistics and include equivalent descriptors to their nonspatial counterparts. The details of these are examined later in this chapter. These measures focus in the first case on identifying a central location within a given collection of spatial phenomena and in the second on whether the features are tightly packed together or dispersed. They are sometimes referred to as **centrographic techniques** and they provide a numerical expression of the types of spatial distribution illustrated in Figure 4.1.

4.3   Central tendency

4.3.1   Measures for nonspatial data

We have already seen that a certain amount of descriptive information about the numerical measurements for a population or sample can be obtained simply by sorting them into ascending (or descending) order. However, calculating a single quantity representing the central tendency of the numerical distribution can reveal more. There are three main measures of central tendency known as the mode, median and mean, and each provides a number that typifies the values for a nonspatial attribute or variable. The procedures for calculating these measures are outlined in Box 4.2 with respect to a variable measured on the ratio/interval scale, water temperature in a fluvioglacial stream recorded at half-hourly intervals during daytime. The lowest of the central tendency measures is the mean (9.08); the mode is highest (11.40) with the median (10.15) in between. Thinking about the variable recorded in this example and the time period over which measurements were taken, this particular sequence is not surprising. Undoubtedly it takes some time for a rise in air temperature to warm up the water melting from the glacier, which can be sustained by the warmth from the midday and early afternoon sunshine, when the peak temperature occurred. Recording of water temperature stopped sooner after this peak than it had started before it in the morning. It took five hours to reach the peak (09.00–14.00 hrs), but recording stopped three and a half hours afterwards (17.30 hrs). In other words, the length of the period over which the data measurements were captured has possibly had an impact on the statistical analysis.

Why are there three measures of central tendency?

Each has its relative strengths and weaknesses and should be used with attributes or variables recorded according to the different scales of measurement. If a large number of values are possible, then the mode is unlikely to provide much useful information. At one extreme it might simply record that each value occurs only once or, slightly more promisingly, that one out of 50 values is repeated twice or perhaps even three times. In other words, the mode is only really helpful when dealing with nominal attributes that have a limited range of values representing the labels attached to the raw data. For example, respondents in a survey may be asked to answer an attitude battery question with 'Strongly Agree', 'Agree', 'Neutral', 'Disagree' or 'Strongly Disagree' as the possible predefined responses. Suppose further that these responses have been assigned numerical codes from 1 to 5 respectively. Although the number of potential values has been limited, it is entirely feasible that the mode will be one of the extreme values (1 or 5). Hence, the notion of central tendency with respect to the mode relates the most common nominal category, since in this example the values 1 to 5 do not imply any order of magnitude in measurement and there is no requirement that 'Strongly Agree'

Box 4.2a: Central tendency measures for nonspatial data: Sample site for the measurement of water temperature in a fluvioglacial stream from Les Bossons Glacier, France.

Mean – population symbol: μ; sample symbol: $\bar{x}$

Sample site at stream exit from sandur (outwash) plain

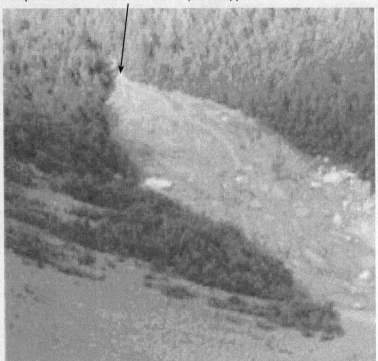

Box 4.2b: Calculation of the mode, median and mean.

The various measures of central tendency are perhaps some of the most intuitive statistics (or parameters) available. Their purpose is to convey something about the typical value in an attribute or variable and allow you to quantify whether the central values in two or more sets of numbers are similar to or different from each other. Each of the main measures (mode, median and mean) can be determined for variables measured on the interval or ratio scale, ordinal variables can yield their mode and median, but only the mode is appropriate for nominal attributes. So, if you have some data for two or more samples of the same category of observations, for instance samples of downtown, suburban and rural households, then you could compare the amounts of time spent travelling to work.

The mode is the most frequently occurring value in a set of numbers and is obtained by counting. If two or more adjacent values appear the same number of times, the question arises as to whether there are two modes (i.e. both values) or one, midway between them. If no value occurs more than once, then a mode cannot be determined. The median is the value that lies at the midpoint of an ordered set of numbers. One way of determining its value is simply to sort all the data values into either ascending or descending order and then identify the middle value. This works perfectly well if there is an odd, as opposed to even, number of values, since the median will necessarily be one of the recorded values. However, if there is an even number, the median lies halfway between the two observations in the middle, and will not be one of the recorded values when the two middle observations are different. Calculation of the arithmetic mean involves adding up or summing all the values for a variable and then dividing by the total number of values (i.e. the number of observations).

The methods used to calculate the mode, median and mean are illustrated below using half-hourly measurements of water temperature in the fluvioglacial stream from Les Bossons Glacier near Chamonix in the European Alps (see Box 4.2a).

	Mode	Median		Mean	$\bar{x} = \dfrac{\sum x}{n}$
	x	x	Sorted x		x
09.00 hrs	4.30	4.30	4.30		4.30
09.30 hrs	4.70	4.70	4.70		4.70
10.00 hrs	4.80	4.80	4.80		4.80
10.30 hrs	5.20	5.20	5.20		5.20
11.00 hrs	6.70	6.70	6.70		6.70
11.30 hrs	10.10	10.10	7.80		10.10
12.00 hrs	10.50	10.50	8.80		10.50
12.30 hrs	11.20	11.20	9.30		11.20
13.00 hrs	**11.40**	11.40	10.10		11.40
13.30 hrs	11.80	11.80	10.20		11.80
14.00 hrs	12.30	12.30	10.50		12.30
14.30 hrs	11.90	11.90	11.10		11.90
15.00 hrs	**11.40**	11.40	11.20		11.40
15.30 hrs	11.10	11.10	11.40		11.10
16.00 hrs	10.20	10.20	11.40		10.20
16.30 hrs	9.30	9.30	11.80		9.30
17.00 hrs	8.80	8.80	11.90		8.80
17.30 hrs	7.80	7.80	12.30		7.80
					$\sum x = 163.50$
			$\dfrac{10.10 + 10.20}{2}$		$\dfrac{163.50}{18}$
	Mode = 11.40	Median = 10.15			$\bar{x} = 9.08$

should have been labelled 1, 'Agree' as 2, etc. It is simply a matter of arbitrary convenience to allocate the code numbers in this way – they could just as easily have been
completely mixed up ('Strongly Agree' as 3, 'Agree' as 5, 'Neutral' as 4, 'Disagree' as 1
and 'Strongly Disagree' as 2). The median is, by definition, more likely to provide a
central measure within a given set of numbers. The main drawback with the median is
its limited focus on either a single central value or on the two values either side of the
midpoint. It says nothing about the values at either extreme.

The arithmetic mean is probably the most useful measure of central tendency,
although its main feature is recognized as both a strength and weakness. All values in
a set are taken into account and given equal importance when calculating the mean,
not just those at the extremes or those in the middle. However, if the values are asymmetrical about the mean, for example there are a few outlying values at either the
lower or upper extreme, then these may 'pull' the mean in that direction away from
the main group of values. Trimming values at both extremes, for example by ignoring
the lowest and highest 5% or 10% of values may reduce this effect, but the decision
to do so represents a subjective judgement on the part of the investigator, which may
or may not be supported from either a theoretical or statistical perspective. Another
advantage of the mean over the median is that it can easily be calculated from totals
(i.e. total count of phenomena and total sum of data values). For example, the 2001
UK Population Census reports that there were a total of 3 862 891 persons living in
1 798 864 privately rented furnished housing units in England, from which it can
readily be determined that the mean number of persons per dwelling unit under this
form of tenure was 0.47 (1 798 864/3 862 891). The equivalent median value cannot
be determined in this fashion. The arithmetic mean possesses two further important
properties. First, if the mean is subtracted from each number in the set, then these
differences will add up to zero; and, secondly, if these positive and negative differences
are squared and then summed, the result will be a minimum (i.e. the sum of the
squared differences of any number apart from the mean would be larger). Therefore,
unless the median and mean are equal, the sum of the squared differences from the
median would be greater than those from the mean. These properties of the mean
might at this point seem somewhat esoteric, although intuitively they suggest that it
really does encapsulate the central tendency of a set of data values.

4.3.2 Measures for spatial data

The equivalent measures of central tendency with respect to spatial phenomena represented graphically as points are known as the **median centre** and the **mean centre**,
both of which can be weighted according to the numerical values associated with
attributes or variables quantified with respect to the phenomena occurring at each
point. The procedures involved with determining the median and mean centres, and
the latter's weighted equivalent, are given in Box 4.3 using the example of the location
of Burger King food outlets in Pittsburgh. Box 4.3a indicates that the distribution of

the companies' outlets is concentrated downtown, but that there are also some outlying enterprises. The question is whether the overall centre of the corporation's presence in the city is in the urban core or elsewhere. The mean and median centres are relatively close together in this example and can be regarded as providing a reasonable summary of the centre of the spatial distribution and as indicating that the Burger King has good coverage across the city as a whole.

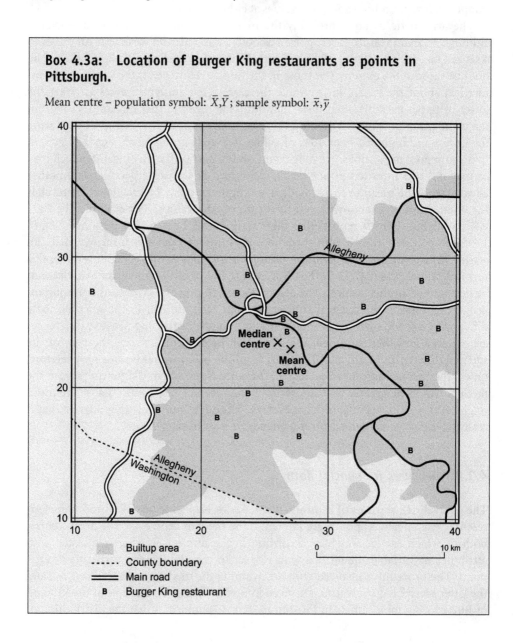

Box 4.3a: Location of Burger King restaurants as points in Pittsburgh.

Mean centre – population symbol: $\bar{X}, \bar{Y}$; sample symbol: $\bar{x}, \bar{y}$

Box 4.3b: Calculation of the mean and median centres of Burger King Restaurants in Pittsburgh.

Measures of central tendency for spatially distributed phenomena serve much the same purpose as their nonspatial counterparts – they capture the typical spatial location rather than a typical value. In the case of spatial phenomena this is a physically central location in the distribution. Calculation of the mean and/or median centre for the distributions of two or more spatial phenomena within a given area enables you to determine if their centres are relatively close or distant. Commercial enterprises competing with each other for customers may seek to open outlets in locations where they can attract each other's clientele. In this case you might expect the mean centres of each corporation's outlets to be reasonably close to each other. In contrast, if enterprises obtained a competitive advantage by operating at a greater distance from rival companies, then their mean centres would be further apart.

The median and mean centres are normally calculated from the X and Y coordinate grid references that locate the phenomena in Euclidean space. The median centre is determined by interpolating coordinate values midway between the middle pairs of X and Y data coordinates. The median centre lies at the intersection of the two lines drawn at right angles to each other (orthogonal lines) through these points and parallel with their corresponding grid lines. This procedure partitions the point phenomena into 4 quadrants (NW, NE, SW and SE). However, this procedure will not necessarily assign equal numbers of points to each quadrant. An alternative method involves dividing the points into four equal-sized groups in each of the quadrants and then determining the median centre as where two orthogonal lines between these groups intersect. The problem is that many such pairs of lines could be drawn and hence the median centre is ambiguous.

The mean centre is far more straightforward to determine and simply involves calculating the arithmetic means of the X and Y coordinates of the data points. The mean centre is located where the two orthogonal lines drawn through these 'average' coordinates intersect. The mean centre is thus unambiguously defined as the grid reference given by the mean of the X and Y coordinates.

The methods used to calculate the mean and median centres are illustrated below using points locating Burger King outlets in Pittsburgh, denoted by subscript B for X and Y coordinates (see Box 4.3a).

	$\bar{x} = \dfrac{\sum x_B}{n}$	$\bar{y} = \dfrac{\sum y_B}{n}$	Median	
	x_B	y_B	Sorted x_B	Sorted y_B
1	11	27	11	10
2	14	10	14	15
3	16	18	16	16
4	19	23	19	16
5	21	17	21	17
6	22	16	22	18
7	22	27	22	19
8	23	19	23	20
9	23	28	23	20

	$\bar{x} = \frac{\sum x_B}{n}$	$\bar{y} = \frac{\sum y_B}{n}$	Median	
	x_B	y_B	Sorted x_B	Sorted y_B
10	26	20	26	22
11	26	24	26	23
12	26	25	26	24
13	27	16	27	24
14	27	25	27	25
15	27	32	27	25
16	32	26	32	26
17	36	15	36	27
18	36	35	36	27
19	37	20	37	28
20	37	22	37	28
21	37	28	37	32
22	38	24	38	35
	$\sum x_B = 583$	$\sum y_B = 497$		
	$\frac{583}{22}$	$\frac{497}{22}$		
	$\bar{x}_B = 27$	$\bar{y}_B = 23$	Median $x_B = 26.0$	Median $y_B = 23.5$
	Mean centre = 27.0, 23.0			

The median centre shares the same problem as the (nonspatial) median, since it focuses on the centre of the spatial distribution and hence ignores possibly important outlying occurrences. Figure 4.3 illustrates how the median centre can occur with a range of *X* and *Y* coordinates in the shaded area at the intersection of the two sets of parallel dashed lines. Nevertheless, it is a useful, intuitively simple location for the centre of a spatial distribution. The mean centre also has some issues. A few observations located at the extremes of a distribution can lead to the mean centre being drawn in their direction. An alternative to the Median centre, especially in American texts is the Centre of Minimum Travel, which is located at the point of minimum total distance between all of the points within a given set. The mean centre lies within a rectangular area with its dimensions defined by boundary lines drawn north–south and west–east through most northerly, southerly, westerly and easterly points in the distribution. This is known as the Minimum Bounding Rectangle and defines the extreme extent of the distribution. Consequently, if a few or even one point lies at some distance from the remainder, its extreme *X* and/or *Y* coordinates will drag the mean centre away from where the majority of phenomena are located. Another seemingly anomalous result can occur if the distribution of points includes two or possibly more distinct clusters each of a similar size. The mean centre could fall between the groups. Similarly,

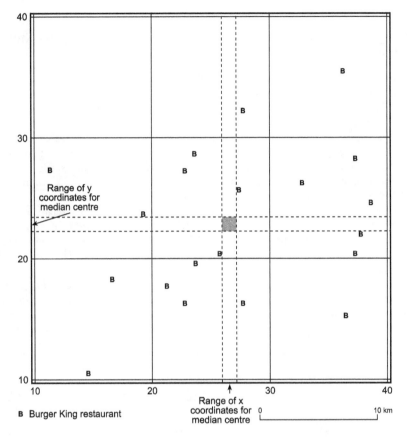

Figure 4.3 Square area containing alternative locations for the median centre.

if the points form a crescent-shaped group, the mean centre is likely to fall in space between the tips of the crescent where the phenomena themselves are not located. These problems do not entirely negate the value of the mean centre as a descriptive measure of the central tendency of spatial distributions, they simply re-emphasize the issue relating to the application of all statistical techniques – *caveat investigator*, let the researcher beware!

Weighting the location of the mean centre according to the values of a measured attribute or variable can enhance its usefulness. The distribution of spatial phenomena may be of interest in its own right, nevertheless, the underlying processes associated with producing the pattern is often equally important. Consequently, we are usually not only interested in the distribution of the phenomena, but also the values with respect to particular variables and attributes associated with each point, line or area. For example, if we were interested in finding out the mean centre of Burger King's customer base in Pittsburgh, a team of researchers could be positioned at each restaurant and count the number of people entering the premises on one day, assuming the corporation were not willing to release these data for reasons of their commercial

Box 4.4:   Calculation of the weighted mean centre of Burger King Restaurants in Pittsburgh.

Weighted mean centre – population symbol: $\bar{X}_w$, $\bar{Y}_w$; sample symbol: $\bar{x}_w$, $\bar{y}_w$

The weighted mean centre takes into account the 'centre of gravity' in a spatial distribution, since it is a measure that combines the location of spatial phenomena with the values of one or more attributes or variables associated with them. If all of the spatial entities had the same data values, then the weighted mean centre would occur at exactly the same location as the 'basic' mean centre. However, if, as an extreme example, one entity in a set dominates the data values, then the weighted mean centre will be drawn towards that single case.

Calculation of the weighted mean centre involves multiplying the X and Y coordinates for each point by the corresponding data value(s) or possibly in more complex cases by the results of a mathematical function. The outcome of these calculations is a weighted X coordinate and a weighted Y coordinate, which are each totalled and divided by the sum of the weights to produce the weighted mean centre. This point is located where two lines at right angles to each other (orthogonal lines) drawn through these 'average' weighted coordinates intersect.

The method used to calculate the weighted mean centre is illustrated below using points locating Burger King outlets in Pittsburgh, denoted by subscript B for X and Y coordinates (see Box 4.3a). Note that the figures for daily customers (the weighting variable w) are hypothetical.

	$\bar{x} = \sum \dfrac{x_B}{n}$	$\bar{y} = \sum \dfrac{y_B}{n}$		$\dfrac{\sum x_B w}{\sum w}$	$\dfrac{\sum y_B w}{\sum w}$
	x_B	y_B	w	$x_B w$	$y_B w$
1	11	27	1200	13 200	32 400
2	14	10	7040	98 560	70 400
3	16	18	1550	24 800	27 900
4	19	23	4670	88 730	107 410
5	21	17	2340	49 140	39 780
6	22	16	8755	192 610	140 080
7	22	27	9430	207 460	254 610
8	23	19	3465	79 695	65 835
9	23	28	8660	199 180	242 480
10	26	20	7255	188 630	145 100
11	26	24	7430	193 180	178 320
12	26	25	5555	144 430	138 875
13	27	16	4330	116 910	69 280
14	27	25	6755	182 385	168 875
15	27	32	1005	27 135	32 160
16	32	26	4760	152 320	123 760
17	36	15	1675	60 300	25 125
18	36	35	1090	39 240	38 150

	$\bar{x} = \sum \dfrac{x_B}{n}$	$\bar{y} = \sum \dfrac{y_B}{n}$		$\dfrac{\sum x_B w}{\sum w}$	$\dfrac{\sum y_B w}{\sum w}$
	x_B	y_B	w	$x_B w$	$y_B w$
19	37	20	1450	53 650	29 000
20	37	22	2500	92 500	55 000
21	37	28	1070	39 590	29 960
22	38	24	1040	39 520	24 960
	$\sum x_B = 583$	$\sum y_B = 497$	93 025	2 283 165	2 039 460
	$\dfrac{583}{22}$	$\dfrac{497}{22}$		$\dfrac{2\,283\,165}{93\,025}$	$\dfrac{2\,039\,460}{93\,025}$
	$\bar{x}_B = 27$	$\bar{y}_B = 23$		$\bar{x}_w = 24.5$	$\bar{y}_w = 21.9$
	Mean centre	27.0, 23.0		Weighted mean centre	24.5, 21.9

sensitivity. Each point in the distribution would then be weighted by multiplying its X and Y coordinates by the corresponding number of customers, then totalling the results and dividing them by the overall total (see Box 4.4). This process takes into account the size of some activity or characteristic occurring at the point locations, which can be thought of as a height measurement. Weighting the mean centre can also be achieved by applying a mathematical function that connects several variables together, for example the customers entering the Burger King premises during one day expressed as a percentage of the total number of people living or working within a 50 m radius.

One of the strengths of the arithmetic mean over the median was that it could be calculated from data where the individual values were unknown and the only information available was the total number of observations and the sum total of a variable. In a similar fashion the mean centre and its weighted equivalent can be estimated even if the coordinates for the individual points are missing. This may be useful if the phenomena can be aggregated into counts per regular grid square. Suppose that the exact X and Y coordinates of the Burger King restaurants in Pittsburgh were unknown, but there was a count of the number per 10×10 km grid square (see Box 4.5a). Conventionally, all the points within a square are assigned grid references representing one of the corners, typically the lower left, or the centre of the square, as in this case. In comparison with Box 4.3a, the detail of exactly where each outlet is located has been lost with their positions generalized to within a series of 10×10 km squares. Both the mean centre and the weighted version relating to such aggregate distributions of data points are clearly less accurate than those pertaining to phenomena whose exact grid coordinates are available. However, they represent useful alternative measures when analysing data obtained from secondary sources.

Box 4.5a: Allocation of Burger King restaurants to an arbitrary grid of squares in Pittsburgh.

Mean centre – population symbol: $\bar{X}$, $\bar{Y}$; sample symbol: $\bar{x}$, $\bar{y}$

Weighted mean centre – population symbol: $\bar{X}_w$, $\bar{Y}_w$; sample symbol: $\bar{x}_w$, $\bar{y}_w$

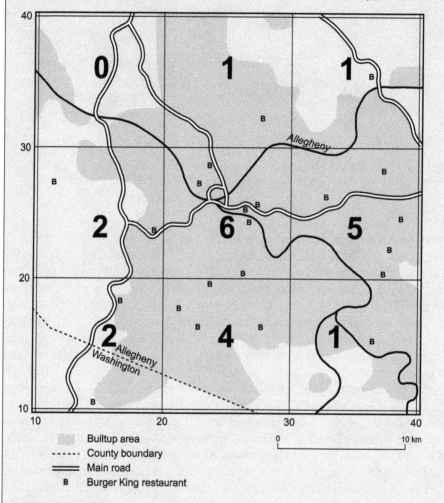

Builtup area
County boundary
Main road
B   Burger King restaurant

0   10 km

Box 4.5b: Calculation of the aggregate mean centre and the aggregate weighted mean centre of Burger King Restaurants in Pittsburgh.

The aggregate mean centre treats the spatial units, usually grid squares, in which entities are located as the primary unit of analysis. This results in some loss of detail, but may be useful where either the exact georeferenced locations of entities cannot be determined or there are so many occurrences that it is not necessary to take each one into account individually. Since the aggregate mean centre generalizes the spatial distribution of entities, local concentrations of points can be obscured and its location will depend on which coordinates (e.g. bottom left corner or midpoint) of the superimposed grid are used. If the number of entities per grid square is fairly similar across the whole study area, then the grid unit representing the aggregate mean centre is likely to contain the 'true' mean centre of the individual point distribution. However, if, as an extreme example, one grid square includes the majority of cases with only a few in some of the others, then the aggregate mean centre will probably be in a different grid square to the one containing the 'true' mean centre.

Calculation of the aggregate mean centre involves using the X and Y grid references for the grid squares in which the individual entities occur. Multiply the chosen grid references by the number of entities within each square, then sum these values and divide the totals by the number of squares covered by the study area. This results in X and Y coordinates to a grid square that constitutes the aggregate mean centre. A weighted version of the aggregate mean centre can be calculated, which involves multiplying the count of points per square by the values of an attribute or variable.

The method used to calculate the aggregate mean centre is illustrated below in Box 4.5c using the number of Burger King restaurants in the six 10×10 km grid squares in Pittsburgh (see Box 4.5a) and grid references for the midpoint of the squares (e.g. 10.5, 20.5). The procedure for calculating the aggregate weighted mean centre is shown in Box 4.5d.

Box 4.5c: Calculation of the aggregate mean centre of Burger King restaurants in Pittsburgh.

$$\bar{x} = \frac{\sum x_B n}{\sum n} \qquad \bar{y} = \frac{\sum y_B n}{\sum n}$$

	Square	Pts/sq n	x_B	$x_B n$	y_B	$y_B n$
1	10,10	2	10.5	21.0	10.5	21
2	10,20	2	20.5	21.0	20.5	41
3	10,30	0	30.5	0.0	30.5	0
4	20,10	4	10.5	82.0	10.5	42
5	20,20	6	20.5	123.0	20.5	123
6	20,30	1	30.5	20.5	30.5	30.5
7	30,10	1	10.5	30.5	10.5	10.5
8	30,20	5	20.5	152.5	20.5	102.5
9	30,30	1	30.5	30.5	30.5	30.5

$$\sum n = 22 \qquad\qquad \sum x_B n = 481 \qquad\qquad \sum y_B n = 401$$

$$\frac{481}{22} \qquad\qquad \frac{401}{22}$$

$$\bar{x} = 21.9 \qquad\qquad \bar{y} = 18.2$$

Aggregate mean centre

Box 4.5d: Calculation of the aggregate weighted mean centre of Burger King restaurants in Pittsburgh.

$$\bar{x} = \frac{\sum x_B w}{\sum w} \qquad \bar{y} = \frac{\sum y_B w}{\sum w}$$

Sq		Pts/sq n	Weight/sq w	x_B	$x_B w$	y_B	$y_B w$
1	10,10	2	8590	10.5	90 195.0	10.5	90 195.0
2	10,20	2	5870	20.5	61 635.0	20.5	120 335.0
3	10,30	0	0	30.5	0	30.5	0
4	20,10	4	18 890	10.5	387 245.0	10.5	198 345.0
5	20,20	6	45 085	20.5	924 242.5	20.5	924 242.5
6	20,30	1	1005	30.5	20 602.5	30.5	30 652.5
7	30,10	1	1675	10.5	51 087.5	10.5	17 587.5
8	30,20	5	10 820	20.5	330 010.0	20.5	221 810.0
9	30,30	1	1090	30.5	33 245.0	30.5	33 245.0
			93 025		$\sum x_B w = 1898\,263$		$\sum y_B w = 1\,636\,413$

$$\frac{1\,898\,263}{93\,025} \qquad\qquad \frac{1\,636\,413}{93\,025}$$

$$\bar{x} = 20.4 \qquad\qquad \bar{y} = 17.6$$

Aggregate weighted mean centre   20.4, 17.6

Unfortunately, mean centres based on aggregate counts of spatially distributed phenomena within areal units such as grid squares suffer from the problem that such units are arbitrarily defined. This arbitrariness arises from three decisions that have been taken about the nature of the grid squares: the size of the units; the position of the grid's origin; and its orientation. Figure 4.4 illustrates the impact of these decisions with respect to placing a regular square grid over the location of Burger King fast-food outlets in Pittsburgh. Three alternative grid square sizes – 500 m, 1000 m and 1500 m – and three different origins – x_1, y_1, x_2, y_2, and x_3, y_3 – are shown. Although calculations of the mean centres in each of these cases is not included, it will be apparent that decisions about the size and origin of regular grids can have a profound impact on aggregate spatial counts and hence on the location of the mean centre. There is, of course, nothing to dictate that a regular square, hexagonal or triangular grid has to be used, and once irregular, variable sized units are admitted, the arbitrariness becomes even more obvious. This example illustrates the **modifiable areal unit**

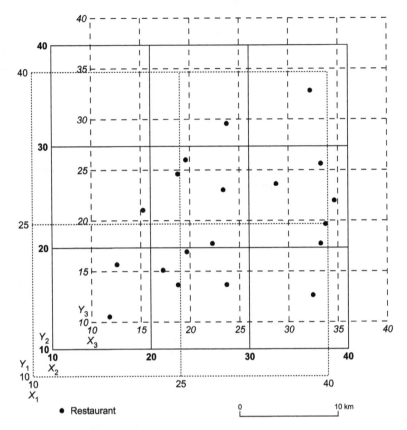

Figure 4.4 Alternative origins and sizes of regular square grid.

problem (MAUP), which is a common difficulty encountered when attempting to analyse spatial patterns.

4.3.3 Distance measures for spatial data

An alternative approach to analysing spatial patterns is to summarize distributions by reference to the **mean distance** between the set of spatial entities. Once this matrix of distances between each pair of points in the set has been determined, these measurements can be treated like any other variable that is quantified on the ratio scale and the overall mean or the **nearest-neighbour mean distance** can be calculated by summation and division. In general terms, a small mean distance implies that the collection of phenomena is more tightly packed together in comparison with a larger mean distance. But, how small is small and how large is large? The answer partly depends on the size of the study area and how it is defined, which again raises issues associated with the MAUP. What might seem a short mean distance within a relatively large area may appear as a long distance in a small area. Box 4.6 illustrates the procedure for calculating the mean distance between the Burger King outlets in Pittsburgh. The overall mean distance is 13.0 km and the nearest-neighbour mean is 4.4 km, with modal and median values of 11.2 km and 12.0 km, respectively. This variation partly reflects the presence of some outlets outside the central downtown area. For comparison, the equivalent measures for the central downtown 10×10 km grid square are a mean of 4.3 km with the median and mode both 5.0 km, which suggests the outlets are much closer together downtown than in the study area overall.

Box 4.6a: Mean distance for Burger King Restaurants in Pittsburgh.

Mean distance – population symbol: $\bar{D}$; sample symbol: $\bar{d}$

The physical distance that separates each pair of entities can be treated as a variable much like 'standard' thematic attributes and variables in order to calculate measures of central tendency. The mean, and for that matter the modal and median, distance between phenomena provide measures of how far apart, or close together, things are. Each of these measures potentially suffers from the same problems as their nonspatial counterparts. Insofar as particular entities lying at the extremes of spatial distributions can distort the results, by suggesting a larger distance between entities is the norm rather than the exception.

The overall mean distance is obtained by determining the distance measurements between every pair of entities in the set, adding them up or summing the distances and then dividing by the total number of distance measurements (not the total number of points). The nearest-neighbour mean distance is calculated by summing the distance between each point and its nearest neighbour and dividing by the total number of points. The distance measurements can be calculated using Pythagoras' theorem, although standard statistical and spreadsheet software is not well suited to this task. Spatial analysis software is more amenable to the task.

The upper part of the table below illustrates how to calculate the distance between a selected subset of the pairs of points locating Burger King restaurants in Pittsburgh (see Box 4.3a) and indicates the complexity of the computational problem. In this example, there are 22 points giving rise to 231 pairs $(1 + 2 + 3 + 4 + \dots + 19 + 20 + 21$, etc.) and, the calculations are shown for the distance between points 1 and 2, 2 and 3, 3 and 4, to 22 and 21. The calculations between 1 and 3, 1 and 4, 1 and 5, etc., or 2 and 4, and 2 and 5, 2 and 6, etc., and so on are not considered. The calculations shown produce the diagonal in the matrix in Box 4.6c, which includes the 231 distance measurements resulting from pairing each Burger King restaurant with all others in the set. The overall modal, median and mean distance between the Burger King restaurants in Pittsburgh can be determined only when all the distance measurements have been calculated and sorted into ascending (or descending) order (see Box 4.6d).

Box 4.6b: Calculation of distances between selected pairs (1 and 2, 2 and 3, 3 and 4, etc.) of Burger King restaurants in Pittsburgh, USA.

Calculation of point-to-point distances using Pythagoras' theorem

Pt	Pt	Start	End	Start	End			Distance =
m	n	x_m	x_n	y_m	y_n	$(x_m - x_n)^2$	$(y_m - y_n)^2$	$\sqrt{(x_m - x_n)^2 + (y_m - y_n)^2}$
1	2	11	14	27	10	9.0	289.0	17.3
2	3	14	16	10	18	4.0	64.0	8.2
3	4	16	19	18	23	9.0	25.0	5.8
4	5	19	21	23	17	4.0	36.0	6.3
5	6	21	22	17	16	1.0	1.0	1.4
6	7	22	22	16	27	0.0	121.0	11.0
7	8	22	23	27	19	1.0	64.0	8.1
8	9	23	23	19	28	0.0	81.0	9.0
9	10	23	26	28	20	9.0	64.0	8.5
10	11	26	26	20	24	0.0	16.0	4.0
11	12	26	26	24	25	0.0	1.0	1.0
12	13	26	27	25	16	1.0	81.0	9.1
13	14	27	27	16	25	0.0	81.0	9.0
14	15	27	27	25	32	0.0	49.0	7.0
15	16	27	32	32	26	25.0	36.0	7.8
16	17	32	36	26	15	16.0	121.0	11.7
17	18	36	36	15	35	0.0	400.0	20.0
18	19	36	37	35	20	1.0	225.0	15.0
19	20	37	37	20	22	0.0	4.0	2.0
20	21	37	37	22	28	0.0	36.0	6.0
21	22	37	38	28	24	1.0	16.0	4.1
22		38		24				

Box 4.6c: Distance matrix for pairs of Burger King restaurants in Pittsburgh, USA.

Pts	Coords	1	2	3	4	5	6	7	8	9	10	11	12	13	14	15	16	17	18	19	20	21	22
Coords		27	10	18	23	17	16	27	19	28	20	24	25	16	25	32	26	15	35	20	22	28	24
1	11																						
2	14	17.3																					
3	16	10.3	8.2																				
4	19	8.9	13.9	5.8																			
5	21	14.1	9.9	5.1	6.3																		
6	22	15.6	10.0	6.3	7.6	1.4																	
7	22	11.0	18.8	10.8	5.0	10.0	11.0																
8	23	14.4	12.7	7.1	5.7	2.8	3.2	8.1															
9	23	12.0	20.1	12.2	6.4	11.2	12.0	1.4	9.0														
10	26	16.6	15.6	10.2	7.6	5.8	5.7	8.1	3.2	8.5													
11	26	15.3	18.4	11.7	7.1	8.6	8.9	5.0	5.8	5.0	4.0												
12	26	15.1	19.2	12.2	7.3	9.4	9.8	4.5	6.7	4.2	5.0	1.0											
13	27	19.4	14.3	11.2	10.6	6.1	5.0	12.1	5.0	12.6	4.1	8.1	9.1										
14	27	16.1	19.8	13.0	8.2	10.0	10.3	5.4	7.2	5.0	5.1	1.4	1.0	9.0									
15	27	16.8	25.6	17.8	12.0	16.2	16.8	7.1	13.6	5.7	12.0	8.1	7.1	16.0	7.0								
16	32	21.0	24.1	17.9	13.3	14.2	14.1	10.0	11.4	9.2	8.5	6.3	6.1	11.2	5.1	7.8							
17	36	27.7	22.6	20.2	18.8	15.1	14.0	18.4	13.6	18.4	11.2	13.5	14.1	9.1	13.5	19.2	11.7						
18	36	26.2	33.3	26.2	20.8	23.4	23.6	16.1	20.6	14.8	18.0	14.9	14.1	21.0	13.5	9.5	9.8	20.0					
29	37	26.9	25.1	21.1	18.2	16.3	15.5	16.6	14.0	16.1	11.0	11.7	12.1	10.8	11.2	15.6	7.8	5.1	15.0				
20	37	26.5	25.9	21.4	18.0	16.8	16.2	15.8	14.3	15.2	11.2	11.2	11.4	11.7	10.4	14.1	6.4	7.1	13.0	2.0			
21	37	26.0	29.2	23.3	18.7	19.4	19.2	15.0	16.6	14.0	13.6	11.7	11.4	15.6	10.4	10.8	5.4	7.1	7.1	8.0	6.0		
22	38	70.7	62.1	22.8	19.0	18.4	17.9	16.3	15.8	15.5	12.6	12.0	12.0	13.6	11.0	13.6	6.3	9.2	11.2	4.1	2.2	4.1	

Box 4.6d: Sorting and summation of all distances to determine mean, modal and median distance measures between Burger King restaurants in Pittsburgh, USA.

Mean distance $\bar{d} = \sum \dfrac{d_B}{n}$

Sorted distances (d)	d	d	d	d	d	d	d	d	d
1.0	5.1	7.1	9.2	11.2	12.6	14.3	16.3	19.4	26.9
1.0	5.1	7.1	9.4	11.2	12.7	14.4	16.6	19.4	27.7
1.4	5.4	7.2	9.5	11.2	13.0	14.8	16.6	19.8	29.2
1.4	5.4	7.3	9.8	11.2	13.0	14.9	16.6	20.0	33.3
1.4	5.7	7.6	9.8	11.2	13.0	15.0	16.8	20.1	62.1
2.0	5.7	7.6	9.9	11.2	13.3	15.0	16.8	20.2	70.7
2.2	5.7	7.8	10.0	11.4	13.5	15.1	16.8	20.6	
2.8	5.8	7.8	10.0	11.4	13.5	15.1	17.3	20.8	
3.2	5.8	8.0	10.0	11.4	13.5	15.2	17.8	21.0	
3.2	5.8	8.1	10.0	11.7	13.6	15.3	17.9	21.0	
4.0	6.0	8.1	10.2	11.7	13.6	15.5	17.9	21.1	
4.1	6.1	8.1	10.3	11.7	13.6	15.5	18.0	21.4	
4.1	6.1	8.1	10.3	11.7	13.6	15.6	18.0	22.6	
4.1	6.3	8.2	10.4	11.7	13.6	15.6	18.2	22.8	
4.2	6.3	8.2	10.4	**12.0**	13.9	15.6	18.4	23.3	
4.5	6.3	8.5	10.6	12.0	14.0	15.6	18.4	23.4	
5.0	6.3	8.5	10.8	12.0	14.0	15.8	18.4	23.6	
5.0	6.4	8.6	10.8	12.0	14.0	15.8	18.4	24.1	
5.0	6.4	8.9	10.8	12.0	14.1	16.0	18.7	25.1	
5.0	6.7	8.9	11.0	12.0	14.1	16.1	18.8	25.6	
5.0	7.0	9.0	11.0	12.1	14.1	16.1	18.8	25.9	
5.0	7.1	9.0	11.0	12.1	14.1	16.1	19.0	26.0	
5.0	7.1	9.1	11.0	12.2	14.1	16.2	19.2	26.2	
5.1	7.1	9.1	11.2	12.2	14.2	16.2	19.2	26.2	
5.1	7.1	9.2	11.2	12.6	14.3	16.3	19.2	26.5	

$$\sum d = 2993.8$$

$$\dfrac{2993.8}{231}$$

Mode = 11.2   Median = 12.0 $\bar{d} = 12.96$

What can we conclude about the distribution of Burger King restaurants in Pittsburgh from this information?

4.4 Dispersion

4.4.1 Measures for nonspatial data

We have seen that there are various ways of quantifying the central tendency of values in a set of data. It is also useful to know the spread of data values. This information can be obtained by calculating a quantity representing the dispersion of the numerical distribution. Imagine sitting in a stadium and looking down at a football match, the 22 players (and the referee) are all located somewhere within the 90 m by 45 m rectangle that defines the pitch (minimum dimensions). At kick off, when the match starts, the players will be fairly evenly distributed around the pitch in their allotted positions. However, if one side is awarded a free kick, the large majority of the players will rush to cluster together in the goal area with those on one team attempting to score and the others to defend. In other words, sometimes the players form a compact distribution and other times they will be more dispersed.

There are three main measures of dispersion known as the range, variance and standard deviation, all of which are numerical quantities indicating the spread of values of a nonspatial attribute or variable. These measures are interpreted in a relative fashion, in the sense that a small value indicates less dispersion than a large one, and vice versa. The range is simply calculated as the difference between the smallest and largest values. Alternatively, various percentile ranges may be determined that omit an equal percentage of values at the upper and lower ends of the distribution. For example, excluding both the largest and smallest 25% of data points and calculating the difference between the values at either end of the remainder produces the interquartile range. The procedures for obtaining the variance and standard deviation measures involve rather more complicated calculations than simply sorting the data values into ascending order and are presented in Box 4.7 with a variable measured on the ratio/interval scale. Nevertheless, the basic purpose in each case is the same, namely to determine whether the data values are close together or spread out.

The reason for there being several measures of dispersion is essentially the same as why there are different measures of central tendency. Each has advantages and disadvantages, and should be used for attributes and variables recorded according to the different scales of measurement. Intuitively, the range seems the most sensible measure, since it reports the size of the numerical gap between the smallest and largest value. Unfortunately, this may be misleading, since extreme outlier values may exaggerate the situation. Discarding the top and bottom 10% or 25% helps to overcome this problem, but the choice of exclusion percentage is entirely arbitrary. A further problem with using the absolute or percentile range is that it fails to take into account the spread in the values between the upper and lower values. We are none the wiser about whether the data values are clustered, regularly or randomly dispersed along the range.

The variance and the standard deviation help to overcome this problem by focusing on the differences between the individual data values and their mean. However, this implies that these dispersion measures are not suitable for use with attributes recorded on the nominal or ordinal scales. If most of the data values are tightly clustered around the mean, then the variance and standard deviation will be relatively

---

Box 4.7a: Variance and standard deviation of water temperature measurements in a fluvioglacial stream from Les Bossons Glacier, France.

Variance – population symbol: σ^2; sample symbol: s^2

Standard deviation – population symbol: σ; sample symbol: s

The variance and its close relation the standard deviation are two of the most useful measures of dispersion when your data are recorded on either the interval or ratio scales of measurement. They measure the spread of a set of numbers around their mean and so allow you to quantify whether two or more sets of numbers are relatively compact or spread out, irrespective of whether they possess similar or contrasting means. The concept of difference (deviation) is important when trying to understand what the variance and standard deviation convey. A small variance or standard deviation denotes that the numbers are tightly packed around their mean, whereas a large value indicates that they are spread out. So, if you have data for two or more samples of the same category of observations, for instance samples of soil from the A, B and C horizons, and then you can see whether or not the spreads (dispersions) of particle sizes are similar.

The standard deviation is closely related to the variance, but one of its main advantages is that its value is measured in the same units as the original set of measurements. So, for example, if these measurements are of water temperature in degrees Celsius or of journey time in minutes, their standard deviation is expressed in these units. The standard deviation is obtained by taking the square root of the variance. There are two main ways of calculating the variance and the standard deviation. The method on the left in the calculations table below involves simpler computation, since it is based on the sum of the X values and the sum of the squared X values. The second method uses the sum of the squared differences (deviations) of the X values about their mean and, although this entails slightly more complicated calculations, it provides a clearer illustration of what the standard deviation measures. These methods are illustrated using sample data that represent measurements of water temperature (X) in the fluvioglacial stream flowing from Les Bossons Glacier, France where the stream leaves the sandur plain taken at half-hourly intervals between 09.00–17.30 hrs (see Box 4.2).

The calculations show that the sample with a mean of 9.08 °C has a variance of 7.78 and a standard deviation of 2.79 °C – thankfully the two methods for deriving the standard deviation produce the same result.

Box 4.7b: Calculation of the variance and standard deviation.

$$s = \sqrt{\frac{n(\sum x^2)-(\sum x)^2}{n(n-1)}} \qquad s = \sqrt{\frac{\sum(x-\bar{x})^2}{(n-1)}}$$

	x	x^2	x	$(x-\bar{x})$	$(x-\bar{x})^2$
09.00 hrs	4.30	18.49	4.30	−4.78	22.88
09.30 hrs	4.70	22.09	4.70	−4.38	19.21
10.00 hrs	4.80	23.04	4.80	−4.28	18.35
10.30 hrs	5.20	27.04	5.20	−3.88	15.08
11.00 hrs	6.70	44.89	6.70	−2.38	5.68
11.30 hrs	10.10	102.01	10.10	1.02	1.03
12.00 hrs	10.50	110.25	10.50	1.42	2.01
12.30 hrs	11.20	125.44	11.20	2.12	4.48
13.00 hrs	11.40	129.96	11.40	2.32	5.37
13.30 hrs	11.80	139.24	11.80	2.72	7.38
14.00 hrs	12.30	151.29	12.30	3.22	10.35
14.30 hrs	11.90	141.61	11.90	2.82	7.93
15.00 hrs	11.40	129.96	11.40	2.32	5.37
15.30 hrs	11.10	123.21	11.10	2.02	4.07
16.00 hrs	10.20	104.04	10.20	1.12	1.25
16.30 hrs	9.30	86.49	9.30	0.22	0.05
17.00 hrs	8.80	77.44	8.80	−0.28	0.08
17.30 hrs	7.80	60.84	7.80	−1.28	1.65
$n = 18$	$\sum x = 163.50$	$\sum x^2 = 1617.33$	$\sum x = 163.50$		$\sum(x-\bar{x})^2 = 132.21$

Variance $\quad s^2 = \dfrac{(18(1617.33)-(163.50)^2}{18(18-1)} \qquad\qquad s^2 = \dfrac{132.21}{18-1}$

$$s^2 = 7.78 \qquad\qquad s^2 = 7.78$$

Standard deviation $\quad s = \sqrt{\dfrac{18(1617.33)-(163.50)^2}{18(18-1)}} \qquad\qquad s = \sqrt{\dfrac{132.21}{18-1}}$

$$s = 2.79 \qquad\qquad s = 2.79$$

Sample Means $\quad \dfrac{\sum x}{n} \quad \dfrac{163.50}{18} = 9.08 \qquad\qquad \dfrac{\sum x}{n} \quad \dfrac{163.50}{18} = 9.08$

small even if there are a small number of extremely low and high values as well. A similar result will occur if the values are bimodally distributed with most lying at **each** end of the range and comparatively few in the middle around the mean. Conversely, the variance and standard deviation will be moderately large, if the data values are regularly spaced across the range. In the highly unlikely event that all data values are identical, then the range, variance and standard deviation measures will all equal zero and hence denote an absence of variation of the variable. In practical terms the main difference between the variance and the standard deviation is that the latter is measured in the same units as the original data values. Thus, Box 4.7 the standard deviation of 2.79 for melt water stream temperature is in °C, whereas the variance of 7.78 is in °C squared, which does make much sense.

4.4.2 Measures for spatial data

When analysing the patterns of spatial phenomena, it is often useful to indicate their dispersion, which may be achieved by calculating the **standard distance**. This is essentially the spatial equivalent of the standard deviation. It provides a measure of the extent to which the individual data points are clustered around the mean centre in two-dimensional space. Calculation of standard distance focuses on the X and Y coordinates of the set of data points and is a measure of how spread out or dispersed they are. Because a spatial distribution of points is located in two dimensions, X and Y, it is feasible for one set of coordinates to be clustered tightly around their mean and for the other to be more spread out. Figure 4.5a illustrates how a distribution of points depicting the positions of former coal pits within a relatively confined approximately north-south oriented valley in South Wales can lead to a narrower spread of X coordinates in comparison with Y coordinates. Figure 4.5b shows how points representing the location of volcanoes associated with the predominantly west–east aligned tectonic plates in the eastern Mediterranean region results in a wide spread of X coordinate values and the Y coordinates in a narrower range. These patterns contrast within a distribution in which the features are more random or tightly clustered.

Box 4.8 shows the calculation of the standard distance for the points representing Burger King Restaurants in Pittsburgh and produces the value 9.73 km. It is difficult to tell just by considering this value whether these restaurants are dispersed or compact, you would want to compare this result with another fast-food chain in Pittsburgh, with Burger King's outlets in other cities, or possibly with some theoretical figure perhaps denoting a perfectly regular distribution. Unfortunately, if comparing the standard distance values for study areas that are substantially different in size, then an adjustment should be introduced to compensate. Neft (1966), for example used the radius of his study area countries to produce a relative measure of population dispersion. Standard distance can be used as the radius of a circle drawn around the mean centre of a distribution of points and thus provide a graphic representation of their dispersion.

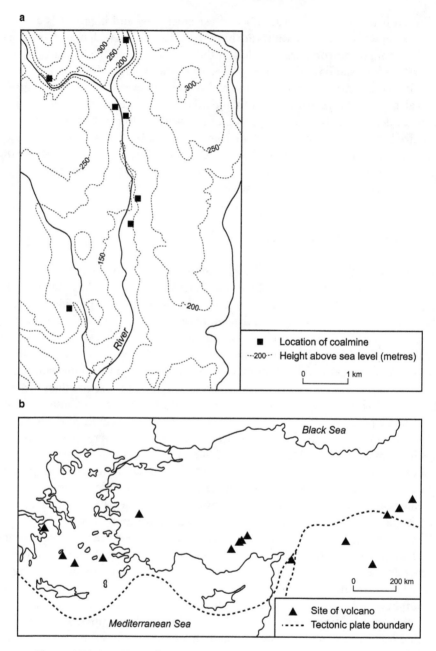

Figure 4.5 *X* or *Y* coordinates constrained within relatively narrow ranges.

Box 4.8a: Standard distance for Burger King restaurants in Pittsburgh.

Standard distance – population symbol: S_d; sample symbol: s_d

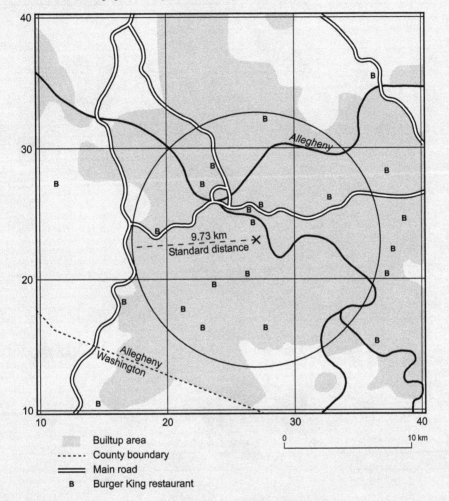

Builtup area
County boundary
Main road
B   Burger King restaurant

0   10 km

The notion that all points representing the location of a set of geographical phenomena will occur exactly on top of each other is just as unrealistic as to imagine that the values of any nonspatial variable for a given collection observations will be the same. The standard distance measure enables us to quantify how spread out the points are, just as the standard deviation measures the dispersion of a set of numerical values. The larger the standard distance then the greater the dispersion amongst the points.

Standard distance is calculated from the X and Y coordinates of the points and does not require you to calculate the distances between all the pairs of points as described in Box 4.7. The procedure given below shows calculation of the variances for the X and Y coordinates, these are summed and the square root of the result produces the standard distance. Examination of the X and Y variances indicates that the X coordinates are more dispersed that the Y coordinates.

Box 4.8b: Calculations for standard distance.

$$s_d = \sqrt{\frac{\sum (x-\bar{x})^2}{(n-1)} + \frac{\sum (y-\bar{y})^2}{(n-1)}}$$

Point	X	$(x-\bar{x})$	$(x-\bar{x})^2$	y	$(y-\bar{y})$	$(y-\bar{y})^2$
1	11	−15.5	240.25	27	4.4	19.36
2	14	−12.5	156.25	10	−12.6	158.76
3	16	−10.5	110.25	18	−4.6	21.16
4	19	−7.5	56.25	23	0.4	0.16
5	21	−5.5	30.25	17	−5.6	31.36
6	22	−4.5	20.25	16	−6.6	43.56
7	22	−4.5	20.25	27	4.4	19.36
8	23	−3.5	12.25	19	−3.6	12.96
9	23	−3.5	12.25	28	5.4	29.16
10	26	−0.5	0.25	20	−2.6	6.76
11	26	−0.5	0.25	24	1.4	1.96
12	26	−0.5	0.25	25	2.4	5.76
13	27	0.5	0.25	16	−6.6	43.56
14	27	0.5	0.25	25	2.4	5.76
15	27	0.5	0.25	32	9.4	88.36
16	32	5.5	30.25	26	3.4	11.56
17	36	9.5	90.25	15	−7.6	57.76
18	36	9.5	90.25	35	12.4	153.76

$$\sum x = 583 \qquad \sum (x-\bar{x})^2 = 1335.50 \qquad \sum y = 497 \qquad \sum (y-\bar{y})^2 = 749.32$$

$$\frac{1335.50}{22} \qquad\qquad\qquad\qquad \frac{749.32}{22}$$

$$60.61 \qquad\qquad\qquad\qquad 34.06$$

$$s_d = \sqrt{60.61 + 34.06}$$

Standard distance $s_d = 9.73$

4.5   Measures of skewness and kurtosis for nonspatial data

Skewness and kurtosis are concerned with the shape of the distribution of data values and so in some respects belong with discussion of frequency distributions in Chapter 5. However, since each is a quantity capable of being calculated for data not expressed in the form of a frequency distribution and they are known, respectively, as the third and fourth moments of descriptive statistics, following on from central tendency, the first, and dispersion, the second, they are discussed here. The principle underlying skewness in respect of statistics is closely allied to the everyday word skewed, which indicates that something is slanted in one direction rather than being balanced. For example, a political party seeking re-election may be selective in the information it chooses relating to its current period in office in order to put a 'spin' or more favourable interpretation on its record. If information or statistics are skewed, this implies that, for whatever reason, there is a lack of symmetry or balance in the set of data values.

Kurtosis may most simply be translated into everyday language as 'peakedness', and therefore provides a method of quantifying whether one set of numbers is more or less peaked than another. Just as a mountainous landscape may be described as undulating if the height of the mountains is low relative to their surface area, in contrast with a deeply dissected upland region with steeper slopes. From the numerical point of view it is a question of there being a difference in the ratio of the maximum height to radius of the mountain, assuming each peak to be perfectly circular when viewed from above (plan view).

Following on from this analogy, Box 4.9a shows two typical shapes for volcanic cones that are approximately circular in plan view, but have rather different cross sections. In one case, Mauna Loa on Hawaii, the cone has been formed from relatively free-flowing lava that was able to spread extensively over the existing surface during successive eruptions and build up what is sometimes called a 'shield cone'. In contrast, Mount Teide on Tenerife in the Canary Islands was created during a more violent series of eruptions of more viscous lava that constructed a conical volcano. Lines have been superimposed along the profiles of the volcanoes to illustrate that one is relatively flat and the other more peaked (see Box 4.9a). The calculations involved with obtaining the skewness and kurtosis measures are illustrated with elevation

---

Box 4.9a: Calculation of skewness and kurtosis using examples of contrasting volcanic cones in Hawaii and Tenerife.

Skewness – population symbol: β_1; sample symbol: b_1.

Kurtosis – population symbol: γ_2; sample symbol: g_2.

Mauna Loa (shield volcano)

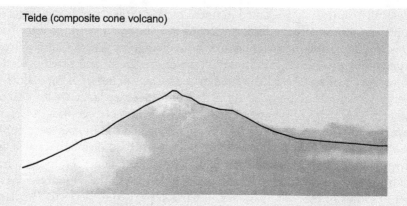

Teide (composite cone volcano)

Skewness is most useful when you are dealing with data recorded on the interval or ratio scales of measurement. It can be viewed as measuring the degree of symmetry and the magnitude of the differences of the values of a variable around its mean. A skewness statistic of zero indicates perfect symmetry and differences of equal magnitude either side of the mean, whereas a negative value denotes that more numbers are greater than the mean (negative skewness) and a positive quantity the reverse. A large skewness statistic signifies that the values at one extreme exert a disproportionate influence. Both symmetrical and severely skewed distributions can display varying degrees of kurtosis depending on whether the values are tightly packed together or are more spread out. A kurtosis statistic of 3 indicates that the distribution is moderately peaked, whereas a set of data values possessing a relatively flat shape will produce a statistic less than 3 and those with a pronounced peak will be greater than 3. As with the other descriptive quantities skewness and kurtosis are commonly used to help with comparing different sets of numbers to see if they are characteristically similar or different. Calculating the skewness and kurtosis statistics for two or more samples of data enables you to see whether or not the distributions of their values have a similar shape.

The sum of the differences between the individual values and their mean always equals zero and the square of the differences is used in calculating the variance (see Box 4.7), so the cube of the deviations provides the basis of the skewness measure. These cubed differences are summed and then divided by the cubed variance of the values multiplied by the number of data points. Following a similar line of argument, calculation of kurtosis involves raising the differences to the fourth power and then dividing the sum of these by the variance raised to the power of four multiplied by the number of observations. These calculations are shown below in Boxes 4.8b and 4.8c with respect to samples of elevation measurements for two contrasting volcanic cones, Mauna Loa in Hawaii and Mount Teide on Tenerife. The classical broad, reasonably symmetrical profile of the Mauna Loa shield volcano results in a skewness and kurtosis statistics, respectively, of −0.578 and 2.009. The profile of Mount Teide is more conical (peaked in statistical terms) and slightly less symmetrical than Mauna Loa, which produces skewness and kurtosis values of −0.029 and 1.673.

Box 4.9b: Calculation of skewness for samples of spot heights on Mauna Loa, Hawaii (x_L) and Mount Teide, Tenerife (x_T).

(Note: standard deviation for Mauna Loa elevations $(s) = 73.984$; and standard deviation for Mount Teide elevations $(s) = 363.354$.)

	Mauna Loa $b_1 = \dfrac{\sum(x-\bar{x})^3}{(n-1)s^3}$				Mount Teide $b_1 = \dfrac{\sum(x-\bar{x})^3}{(n-1)s^3}$		
N	x_L	$(x_L - \bar{x}_L)$	$(x_L - \bar{x}_L)^3$	N	x_T	$(x_T - \bar{x}_T)$	$(x_T - \bar{x}_T)^3$
1	3840	−146.44	−3 140 358.0	1	2600	−572.7	−187 817 497.0
2	3852	−134.44	−2 429 883.8	2	2700	−472.7	−105 609 182.0
3	3864	−122.44	−1 835 565.8	3	2800	−372.7	−51 761 668.0
4	3889	−97.44	−925 149.3	4	2900	−272.7	−20 274 953.0
5	3901	−85.44	−623 711.5	5	3000	−172.7	−5 149 038.3
6	3913	−73.44	−396 093.8	6	3100	−72.7	−383 923.6
7	3938	−48.44	−113 661.2	7	3200	27.3	20 391.2
8	3962	−24.44	−14 598.3	8	3300	127.3	2 063 905.9
9	3974	−12.44	−1 925.1	9	3400	227.3	11 746 621.0
10	3986	−0.44	−0.1	10	3500	327.3	35 068 535.0
11	4011	24.56	14 814.4	11	3600	427.3	78 029 650.0
12	4023	36.56	48 867.3	12	3700	527.3	146 629 965.0
13	4035	48.56	114 508.1	13	3717	544.3	161 273 450.0
14	4047	60.56	222 104.6	14	3700	527.3	146 629 965.0
15	4059	72.56	382 025.0	15	3600	427.3	78 029 650.0
16	4072	85.56	626 343.1	16	3500	327.3	35 068 535.0
17	4088	101.56	1 047 533.9	17	3400	227.3	11 746 621.0
18	4072	85.56	626 343.1	18	3300	127.3	2 063 905.9
19	4059	72.56	382 025.0	19	3200	27.3	20 391.2
20	4047	60.56	222 104.6	20	3100	−72.7	−383 923.6
21	4035	48.56	114 508.1	21	3000	−172.7	−5 149 038.3
22	4023	36.56	48 867.3	22	2900	−272.7	−20 274 953.0
23	4011	24.56	14 814.4	23	2800	−372.7	−51 761 668.0
24	3986	−0.44	−0.1	24	2700	−472.7	−105 609 182.0
25	3974	−12.44	−1 925.1	25	2600	−572.7	−187 817 497.0
	$\sum x_L = 113356$		$\sum(x_L - \bar{x}_L)^3 = -5618013.1$		$\sum x_T = 79317$		$\sum(x_T - \bar{x}_T)^3 = -33600939.0$

$$b_1 = \frac{5\,618\,013.1}{24(404\,965.8)}$$

$$b_1 = -0.578$$

$$b_1 = \frac{33\,600\,939.0}{24(47\,972\,275)}$$

$$b_1 = -0.029$$

Box 4.9c: Calculation of kurtosis for samples of spot heights on Mauna Loa, Hawaii (x_L) and Mount Teide, Tenerife (x_T).

(Note: standard deviation for Mauna Loa elevations (s) = 73.984; and standard deviation for Mount Teide elevations (s) = 363.354).

	Mauna Loa				Mount Teide		
	$g_2 = \dfrac{\sum (x-\bar{x})^4}{ns^4}$				$g_2 = \dfrac{\sum (x-\bar{x})^4}{ns^4}$		
N	x_L	$(x_L - \bar{x}_L)$	$(x_L - \bar{x}_L)^4$		x_T	$(x_T - \bar{x}_T)$	$(x_T - \bar{x}_T)^4$
1	3840	−146.44	459 874 025.8	1	2600	−572.7	107 559 324 269.5
2	3852	−134.44	326 673 582.4	2	2700	−472.7	49 919 348 352.4
3	3864	−122.44	224 746 679.3	3	2800	−372.7	19 290 538 323.2
4	3889	−97.44	90 146 548.1	4	2900	−272.7	5 528 574 182.1
5	3901	−85.44	53 289 906.6	5	3000	−172.7	889 135 929.0
6	3913	−73.44	29 089 126.0	6	3100	−72.7	27 903 563.8
7	3938	−48.44	5 505 750.6	7	3200	27.3	557 086.7
8	3962	−24.44	356 783.5	8	3300	127.3	262 776 497.6
9	3974	−12.44	23 948.7	9	3400	227.3	2 670 241 796.4
10	3986	−0.44	0.0	10	3500	327.3	11 478 632 983.3
11	4011	24.56	363 842.5	11	3600	427.3	33 343 630 058.2
12	4023	36.56	1 786 589.4	12	3700	527.3	77 320 913 021.0
13	4035	48.56	5 560 511.1	13	3717	544.3	87 784 364 145.9
14	4047	60.56	13 450 656.0	14	3700	527.3	77 320 913 021.0
15	4059	72.56	27 719 736.4	15	3600	427.3	33 343 630 058.2
16	4072	85.56	53 589 919.4	16	3500	327.3	11 478 632 983.3
17	4088	101.56	106 387 540.5	17	3400	227.3	2 670 241 796.4
18	4072	85.56	53 589 919.4	18	3300	127.3	262 776 497.6
19	4059	72.56	27 719 736.4	19	3200	27.3	557 086.7
20	4047	60.56	13 450 656.0	20	3100	−72.7	27 903 563.8
21	4035	48.56	5 560 511.1	21	3000	−172.7	889 135 929.0
22	4023	36.56	1 786 589.4	22	2900	−272.7	5 528 574 182.1
23	4011	24.56	363 842.5	23	2800	−372.7	19 290 538 323.2
24	3986	−0.44	0.0	24	2700	−472.7	49 919 348 352.4
25	3974	−12.44	23 948.7	25	2600	−572.7	107 559 324 269.5
$\sum x_L = 99661$		$\sum (x_L - \bar{x}_L)^4 = 1 501 060 350.0$		$\sum x_T = 79317$		$\sum (x_T - \bar{x}_T)^4 = 704 367 516 272.0$	

$$g_2 = \frac{1 501 060 350.1}{24(29 961 099.8)}$$

$$g_2 = \frac{704 367 516 272.0}{24(17 430 924 527.8)}$$

$$g_2 = 2.009$$

$$g_2 = 1.683$$

measurements for these differing shapes in respect of 5 km long cross sections whose centres correspond with the peaks of the cones. Both cones are reasonably symmetrical across this distance, although the measures of kurtosis reflects the difference in profile.

From a computational perspective, there is no reason why measures of skewness and kurtosis should not be calculated for spatial as much for nonspatial data. For example, the skewness statistic could be calculated for all the pairs of distance measurements between Burger King restaurants in Pittsburgh and some commentary supplied about whether there are more or less smaller than the mean distance, and whether the magnitude of the differences are larger in the case of the latter or former. However, it is less common for analysis to consider the skewness or kurtosis of spatial distributions.

4.6 Closing comments

In some respects this chapter has been all about 'playing with numbers', seeing how we can describe the essential characteristics of a set of numbers. The four main elements of this statistical description are central tendency, dispersion, skewness and kurtosis. Most of these elements can be quantified in several different ways, for example by taking the most typical or frequently occurring (mode), the middle (median) or mean (average) value in the case of central tendency. In the case of measuring dispersion, the different types of range (absolute and percentile) provide an intuitive impression of how spread out the values of a variable are, but in practice the variance and standard deviation are more useful quantities. In most cases there are equivalent spatial statistics to these more 'standard' versions. Although this chapter has tended to focus on spatial phenomena as points, similar measures exist for lines and areas and in some instances linear and areal feature may be located spatially as a point by means of a pair of X and Y coordinates. Such points may relate to the midpoint of a line or the spatial or weighted centroids of an area (polygon). In these cases then the measures outlined in this chapter can reasonably be used to describe, for example the mean centre of a collection of areas.

Of course when carrying out quantitative data analysis the intention is not just to observe how different statistical quantities vary between one set of numbers and another. There needs to be a transition from regarding the numerical values as not just a set of digits but as having importance as far as the variable or attribute in question is concerned. So, for example, when looking at the values for the temperature of Les Bossons Glacier melt water stream, the purpose is not simply to regard them as 'any old set of numbers' that happen to have a mean of 9.08 and a standard deviation of 2.79, there are many other sets of numbers that can produce these results. The important point is that the numbers are measurements or values of a variable relating to a particular geographical and temporal context. In these circumstances the statistical measures are not simply the outcome of performing certain mathematical computations, but are tools enabling us to say something potentially of interest about the diurnal pattern of temperature change associated with the melting of ice from this specific glacier. When planning research investigations and specifying the attributes variables that will be relevant, it is as well to think ahead to the types of analysis that will be carried out, and without knowing what the values will be in advance, thinking about how to interpret the means, variances and standard distances that will emerge from the data.

5

Frequency distributions, probability and hypotheses

Frequency distributions are an alternative way of unlocking the information contained in a set of data in the form of absolute and relative frequency counts. These are linked to the summary measures or quantities in the underlying distribution of data values examined previously. Examining the characteristics of frequencies provides a starting point for understanding how mathematically defined probability distributions form the basis inferential statistical techniques. Three probability distributions are covered relating to situations where the outcome of events arising from chance is either dichotomous (e.g. Yes/No, Male/Female, etc.), discrete or integer (e.g. the 10 rock hardness categories, travel mode (train, car, bus, bicycle, etc.)) or continuous (e.g. commuter travel time, water temperature, nitrate concentration, etc.).The chapter introduces the broad principles of statistical inference and hypothesis testing.

Learning outcomes

This chapter will enable readers to:

- recognize the relationships between frequency counts and summary measures;

- identify the three main probability distributions and their association with attributes and variables recorded on different measurement scales;

- connect the concepts of summary measures, frequency distributions and probability in the context of statistical analysis;

- formulate simple Null and Alternative Hypotheses;

- continue planning how to use frequency distributions to summarize data in an independent research investigation in Geography, Earth Science and related disciplines.

Practical Statistics for Geographers and Earth Scientists   Nigel Walford
© 2011 John Wiley & Sons, Ltd

5.1 Frequency distributions

The different types of descriptive statistic examined in Chapter 4 comprise a single number that provides some information about the whole set of data values from which they are obtained. The mean and its fellows represents a central or typical value, the range, variance and standard deviation indicate the spread of values, whilst skewness and kurtosis, respectively, signifying their symmetry and 'peakedness'. However, it is also useful to consider all of the values in a set by means of a frequency distribution. Univariate frequency distributions focus on one set of values, whereas bivariate and multivariate frequencies, respectively, include two or three plus variables, although in the latter case it can be difficult to visualize and interpret the information. Sometimes, particularly when using secondary data from official, governmental sources, these can only be obtained in the form of a frequency distribution.

In simple terms a frequency distribution is a count of the data values allocated between classes or categories that span their full range. These counts can be expressed as absolute counts showing the number of observations in the classes or relative counts usually expressed as a percentage or proportion of the total, respectively, totaling either 100 per cent or 1.0. Often, it is also useful to examine cumulative relative frequencies, in which case the percentage or proportion of observations in successive classes are added together or accumulated. This is especially useful with classes that follow some type of sequence, for example from lower to higher sets of values.

Absolute and relative frequency distributions can be presented in the form of tables or as various types of graph, although multivariate distributions can be difficult to interpret in both tabular and graphical format. Tables showing univariate distributions can include the absolute, relative and cumulative frequencies. The main graphical alternative is the histogram in which the heights of the columns projecting from the horizontal X-axis are scaled according to the absolute or relative frequency on the Y-axis. An alternative method used in respect of a continuous variable is a frequency polygon that is produced by a line connecting the midvalues of each class of data values and sometimes extended beyond the lowest and highest classes containing observations to meet the horizontal axis at the midpoints of next lower and upper class.

The frequency distributions for nominal attributes can be produced without any further processing of the data values, since they are held in the form of categories. However, when dealing with ordinal or continuous (interval or ratio) data the values need to be separated into classes before the distribution can be examined. Box 5.1 illustrates these aspects of frequency distribution tables and histograms with reference to data examined previously, namely the numbers of fast-food restaurants operated by two companies (Burger King and McDonald's) in Pittsburgh and the half-hourly temperature of the fluvioglacial stream at the exit from the outwash plain of the Bossons Glacier near Chamonix in the French Alps. The continuous variable has been divided into three mutually exclusive classes using two decimal places since the original data were recorded to one decimal place so that there is no question about which

Box 5.1a: Frequency distributions and histograms for nominal attributes and data values of continuous variables (interval scale measurements).

The allocation of nominal scale attributes into categories means that it is often simpler to produce frequency distributions from this type of data. The categorization of the fast-food restaurants operated by two companies (Burger King and McDonald's) in Pittsburgh provides an illustration below. Since the two types of restaurant have been defined as part of the data-collection process, it is relatively simple to determine their absolute and relative frequencies and to visualize these as a histogram.

The data values of ordinal, interval and ratio scale variables need to be grouped into classes in order to produce frequency distributions. The process is illustrated using the data values for the temperature of the water in the fluvioglacial stream at the exit from the outwash plain of the Bossons Glacier near Chamonix in the French Alps that was introduced previously. The 18 data values have been sorted in ascending order and then classified into data ranges or classes, sometimes referred to as 'bins', 0.00–4.99, 5.00–9.99 and 10.00–14.99. These are equal range classes (i.e. each is 5 °C wide).

The methods used to produce frequency distributions and histograms for types of fast-food restaurant and water temperature are shown below, respectively, in Boxes 5.1b and 5.1c.

Box 5.1b: Frequency distributions and histograms for nominal attributes and data values of continuous variables (interval scale measurements).

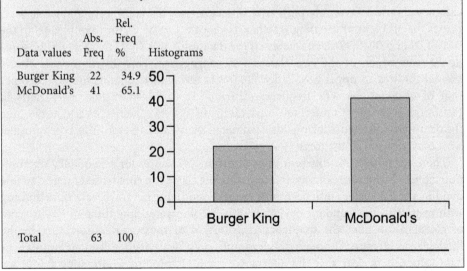

Data values	Abs. Freq	Rel. Freq %	Histograms
Burger King	22	34.9	
McDonald's	41	65.1	
Total	63	100	

Box 5.1c: Classification of data values of continuous variables (interval scale measurements).

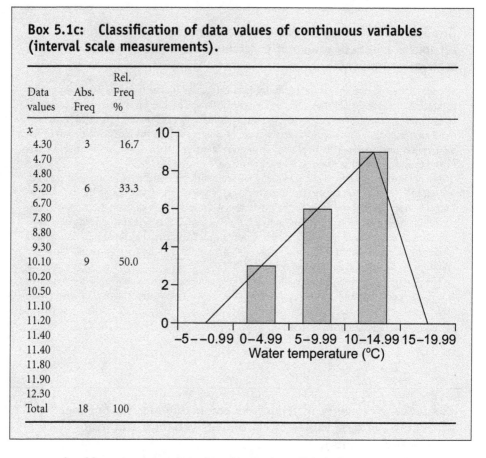

Data values	Abs. Freq	Rel. Freq %
x		
4.30	3	16.7
4.70		
4.80		
5.20	6	33.3
6.70		
7.80		
8.80		
9.30		
10.10	9	50.0
10.20		
10.50		
11.10		
11.20		
11.40		
11.40		
11.80		
11.90		
12.30		
Total	18	100

category should receive any particular observation. This is important since it over-comes the problem of deciding whether to allocate a value of, for example, 4.9 to the 0.00–4.99 or 5.00–9.99 class; whereas if the data value was 4.997 there would be some doubt over which of the two classes was appropriate. This distribution is clearly skewed towards to uppermost range (10.00–14.99) which contains nine or 50.0 per cent of observations. The frequency distribution for the fast-food restaurants in Pittsburgh is simpler to interpret, since there are only two categories and at the time the data were collected, McDonalds dominated the local market for this type of outlet with 65.1 per cent of the total.

There are number of different ways of producing classes for a frequency distribu-tion table or histogram. Alternatives to basing the classes on equal-sized ranges include dividing the data values into percentile ranges, identifying natural breaks or delimiting with respect to the standard deviation. Box 5.2 illustrates how these different types of classification alter the frequency distribution of the water temperature in the fluvioglacial stream. Although this example only deals with a limited number of data values, it is clear that the alternative methods produce contrasting frequency

Box 5.2a: Alternative methods for classifying data values of continuous variables (interval scale measurements).

It is sometimes useful to change the categories of nominal scale attributes, for example to combine detailed types of housing tenure into a smaller number of summary groups. For instance owning a property outright and purchasing with a mortgage might be collapsed into an owner occupier category. The example of two categories of fast-food restaurant does not lend itself to this type of adjustment, although if the location and food type of all restaurants in Pittsburgh had been recorded then if would have been possible to collapse the detailed types into categories representing the cultural background of the food available (e.g. European, Latin American, Oriental, etc.).

There are a number of alternative ways of classifying ordinal, interval and ratio scale variables so that they can be examined as frequency distributions. Classification of the data values into quartiles relates to the interquartile range measure of dispersion and separates the values into four groups each containing 25 per cent of the observations. If, as in the example here, the total does not divide into classes with equal numbers of cases, then some will contain the extras. Classification according to where 'natural breaks' occur in the set of data values represents a rather more subjective method. It essentially locates the class boundaries where there are abrupt steps in the ranked sequence of values. Definition of classes in terms of units of the standard deviation about the average of the data values provides a more rigorous approach to the procedure, since it takes account of dispersion about the mean.

These classification methods are illustrated with respect to the variable recording half-hourly water temperature in the fluvioglacial stream at the exit from the outwash plain of the Bossons Glacier in Boxes 5.2b, 5.2c and 5.2d. The classes based on units of the standard deviation on either side of the mean are calculated by subtracting and adding the standard deviation (2.79) to the mean value (9.08). The next pair of classes moving outwards are obtained by subtracting and adding twice the standard deviation (5.56) and so on.

Box 5.2b: Quartile (percentile) classes for data values of continuous variables (interval scale measurements).

4.30–6.97	6.98–10.14	10.15–11.34	11.35–12.30
4.30	7.80	10.20	11.40
4.70	8.80	10.50	11.40
4.80	9.30	11.10	11.80
5.20	10.10	11.20	11.90
6.70			12.30

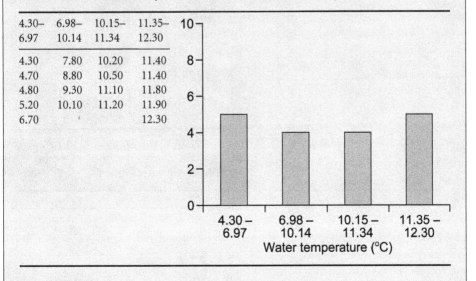

Box 5.2c: Natural classes for data values of continuous variables (interval scale measurements).

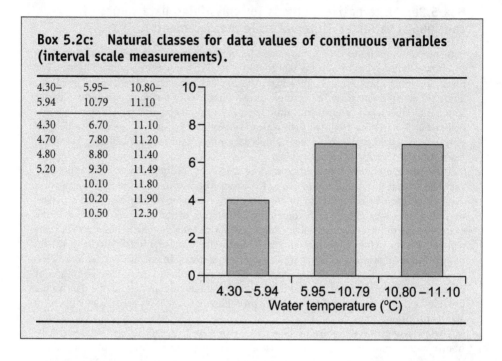

| 4.30– | 5.95– | 10.80– |
5.94	10.79	11.10
4.30	6.70	11.10
4.70	7.80	11.20
4.80	8.80	11.40
5.20	9.30	11.49
	10.10	11.80
	10.20	11.90
	10.50	12.30

Box 5.2d: Standard deviation-based classes for data values of continuous variables (interval scale measurements).

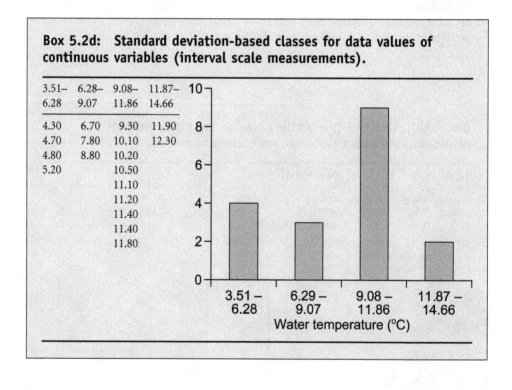

| 3.51– | 6.28– | 9.08– | 11.87– |
6.28	9.07	11.86	14.66
4.30	6.70	9.30	11.90
4.70	7.80	10.10	12.30
4.80	8.80	10.20	
5.20		10.50	
		11.10	
		11.20	
		11.40	
		11.40	
		11.80	

histograms. Not surprisingly, the quartile classification produced a rather flat distribution with some 25 per cent of values in each of the four classes. The slightly higher numbers in the lowest and uppermost classes in simply an artefact of the procedure where the total observations do not divide exactly into four groups with an equal number. The clearest natural break in the set of data values occurred between observations four and five (5.20 and 6.70 – difference 1.50) with a slightly less obvious step between 12 and 13 (10.50 and 11.40 – difference 0.90). Classes based on units of the standard deviation from the mean produced a rather more distinctive distribution with a clear modal range of 9.08–11.86 (i.e. the mean plus one standard deviation). In each case the class boundaries do not occur at simple integers when compared with the regular, equal 'user-defined' ranges used in Box 5.1c. Examination of the four histograms reveals how the same set of data values can be allocated to classes in various ways to produce quite different frequency distributions and thus provide alternative information about the temperature of the water in this particular fluvioglacial stream.

5.2 Bivariate and multivariate frequency distributions

The univariate frequency distributions examined the previous section focus on one attribute or variable. Therefore, we have no idea whether the Burger King and McDonald's restaurants share the same pattern in relation to distance from the centre of Pittsburgh. Similarly, we have no way of telling whether water temperature changes in relation to the time of day. Examination of bivariate frequency distributions will help to answer these questions. In some cases it might be useful to explore how three or even more variables are associated with each other by viewing their multivariate frequency distribution. For example, is the occurrence of the two types of fast-food restaurant in different zones moving outwards from the central business district of Pittsburgh also connected with variations in land values? Water temperature in the stream might be linked not only to the time of day but also the amount of cloud cover. Before considering the characteristics of bivariate frequency distributions, it is worth noting that often the aggregate data used in geographical research from secondary sources exist in the form of bivariate or multivariate frequency distributions, possibly with location acting as one of the variables or dimensions. For example, many of the statistics from the UK national Population Census take the form of tables of aggregate absolute counts of people or households with different demographic and socioeconomic characteristics. The presentation of data in the form of aggregate statistics helps to preserve the confidentiality of the raw observations, since the identity of particular individuals becomes subsumed with others in the geographical unit in question.

The creation of a bivariate frequency distribution is similar to dealing a standard, shuffled pack of 52 playing cards according to the suit (spades, clubs, diamonds or hearts) and whether a picture (Ace, King, Queen and Jack) or a number (2–10).

Dealing the cards in this way distributes the cards into piles according to two criteria (suit and high or low value), but assuming no cards are missing from the pack, the outcome can be confidently predicted in advance: there will be eight piles four each containing nine numbered cards and another four with the four picture cards. However, when the 'observations' are not cards in pack but geographical phenomena in a population or sample the outcome is less certain, although the total number to be distributed will be known. There are three main ways of representing bivariate or multivariate frequency distributions: either as a (contingency) table or cross-tabulation, as a graph or chart, or as a thematic map.

Contingency tables can be produced from two or more nominal attributes without any further processing, since the data are held in the form of categories. However, when producing a table from ordinal or continuous (interval or ratio) variables or from one or more of these in combination with a nominal attribute(s), it will be necessary to classify the raw data values using one of the methods already examined (equal classes, percentiles, units of the standard deviation, etc.). Box 5.3 illustrates key features of bivariate frequency distribution tables with reference to the data with which we are already becoming familiar. It shows that there are a number of different types of table that can be produced each allowing a slightly different interpretation of the information obtained from the same raw data. Relative frequency cross tabulations can contain row, column or overall percentages and the absolute counts (Box 5.3b) and may be accompanied by the expected frequency counts, which represent the number of observations that would occur in each cell of the table if they were distributed in proportion to the row and column totals. As in the case of Box 5.3c, expected frequencies will often be calculated as decimal values (e.g. 4.8 in the Burger King 0.00–4.9 km cell). The observed absolute frequency count for this cell was five, very close to the expected value. However, the McDonald's and 15.00–19.9 km shows an observed count of 9 and an expected one of 11.2, indicating there were '2.2' fewer restaurants than expected. The importance of the differences between these two types of frequency will become apparent when we examine how to test to see whether the observed counts are significantly different from the expected ones in Chapter 7. Box 5.3 illustrates two-way or two-dimensional contingency tables, however, multivariate versions (i.e. with three or more dimensions) can also be produced. However, these can become quite complex to summarize and visualization of multivariate distributions is more often undertaken graphically.

Describe some other interpretations of the information contained in the absolute and relative cross-tabulations apart from those referred to in Box 5.3a. There are at least three more things you could mention.

There are many different ways of representing bivariate and multivariate frequency distributions graphically and the choice of method depends upon the nature of the data to be presented rather more than the visual appearance of the result. Again, the

Box 5.3a: Bivariate contingency tables from nominal attributes and classified data values of continuous variables (ordinal and interval scale measurements).

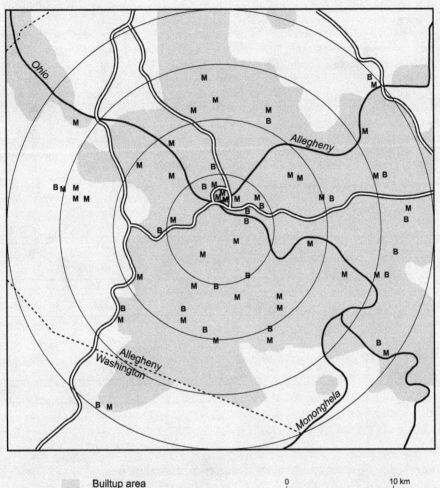

Builtup area

County boundary

Main road

B   Burger King restaurant

M   McDonald's restaurant

Location of Burger King and McDonald's restaurants within four concentric zones of downtown Pittsburgh.

The diagram indicates that Burger King and McDonald's restaurants are distributed spatially throughout Pittsburgh and that both companies have outlets in the four concentric zones moving outwards from the downtown area. However, it is not immediately obvious whether one type of restaurant dominates any particular zone, and if so to what extent. Trying to answer these questions involves interpreting the bivariate absolute and relative frequency distributions.

Absolute and relative frequency cross tabulations provide a way of finding how many and what percentage (or proportion) of observations possess different combinations of characteristics. Thus, in the boxes below it is possible to see that there were very similar numbers of Burger King and McDonald's restaurants (eight and nine, respectively) in the outermost zone (15.0–19.9 km). However, these absolute figures accounted for 36.4 per cent of the total in the case of Burger King outlets, but only 21.4 per cent of the McDonald's. Perhaps this result indicates Burger King had a greater presence in the suburban areas of the city. As far as the two types of fast-food restaurant enumerated McDonald's seems to dominate the sector and overall outlets of both types are reasonably evenly distributed across the four zones with almost a quarter of the total, between 21.9 and 28.1 per cent, in each.

Absolute frequency and expected counts, and relative percentage frequencies together with the calculations for these are given in Boxes 5.3b, 5.3c, 5.3d, 5.3e and 5.3f.

Box 5.3b: Absolute frequency distribution.

	0.0–4.9 km	5.0–9.9 km	10.0–14.9 km	15.0–19.9 km	Marginal Column Total
Burger King	5	5	4	8	22
McDonald's	9	13	11	9	42
Marginal Row Total	14	18	15	17	64

Box 5.3c: Expected absolute frequency distribution.

	0.0–4.9 km	5.0–9.9 km	10.0–14.9 km	15.0–19.9 km	Marginal Column Total
Burger King	$(14 \times 22)/64$ = 4.8	$(18 \times 22)/64$ = 6.2	$(15 \times 22)/64$ = 5.2	$(17 \times 22)/64$ = 5.8	22
McDonald's	$(14 \times 42)/64$ = 9.2	$(18 \times 42)/64$ = 11.8	$(15 \times 42)/64$ = 9.8	$(17 \times 42)/64$ = 11.2	42
Marginal Row Total	14	18	15	17	64

Box 5.3d: Row percentage relative frequency distribution.

	0.0–4.9 km	5.0–9.9 km	10.0–14.9 km	15.0–19.9 km	Marginal Column Total
Burger King	$(5 \times 100)/22$ $= 22.5$	$(5 \times 100)/22$ $= 22.7$	$(4 \times 100)/22$ $= 18.2$	$(8 \times 100)/22$ $= 36.4$	**100.0**
McDonald's	$(9 \times 100)/42$ $= 21.4$	$(13 \times 100)/42$ $= 31.0$	$(11 \times 100)/42$ $= 26.2$	$(9 \times 100)/42$ $= 21.4$	**100.0**
Marginal Row Total	$(14 \times 100)/64$ $= 21.9$	$(18 \times 100)/64$ $= 28.1$	$(15 \times 100)/64$ $= 23.4$	$(17 \times 100)/64$ $= 26.6$	**100.0**

Box 5.3e: Column percentage relative frequency distribution.

	0.0–4.9 km	5.0–9.9 km	10.0–14.9 km	15.0–19.9 km	Marginal Column Total
Burger King	$(5 \times 100)/14$ $= 35.7$	$(5 \times 100)/18$ $= 27.8$	$(4 \times 100)/15$ $= 26.7$	$(8 \times 100)/17$ $= 47.1$	$(22 \times 100)/64$ $= 34.4$
McDonald's	$(9 \times 100)/14$ $= 64.3$	$(13 \times 100)/18$ $= 72.2$	$(11 \times 100)/15$ $= 73.3$	$(9 \times 100)/17$ $= 52.9$	$(41 \times 100)/64$ $= 65.6$
Marginal Row Total	**100.0**	**100.0**	**100.0**	**100.0**	**100.0**

Box 5.3f: Overall percentage relative frequency distribution.

	0.0–4.9 km	5.0–9.9 km	10.0–14.9 km	15.0–19.9 km	Marginal Column Total
Burger King	$(5 \times 100)/64$ $= 7.8$	$(5 \times 100)/64$ $= 7.8$	$(4 \times 100)/64$ $= 6.3$	$(8 \times 100)/64$ $= 12.5$	$(22 \times 100)/64$ $= 34.4$
McDonald's	$(9 \times 100)/64$ $= 14.1$	$(13 \times 100)/64$ $= 20.3$	$(11 \times 100)/64$ $= 17.2$	$(9 \times 100)/64$ $= 14.1$	$(41 \times 100)/64$ $= 65.6$
Marginal Row Total	$(14 \times 100)/64$ $= 21.9$	$(18 \times 100)/64$ $= 28.1$	$(15 \times 100)/64$ $= 23.4$	$(17 \times 100)/64$ $= 26.6$	**100.0**

focus here will be on using graphs to illustrate the characteristics of bivariate and multivariate frequency distributions from the perspective of helping to understand statistical methods rather than as devices for communicating results visually. The choice of graph depends upon the nature of the data being examined, in other words whether the data values refer to nominal attributes or continuous measurements on the ordinal, interval or ratio scales. Box 5.4 illustrates bivariate graphing with a clustered column chart or histogram in respect of the fast-food restaurants in Pittsburgh that is equivalent to the contingency table in Box 5.3b, which has been duplicated to emphasize the point. A clustered column graph is an extension of the univariate histogram and may be used with nominal attributes or categorized continuous variables. It allows the two or more frequency distributions to be viewed side by side and their shapes to be compared. In this example it seems that the number of McDonald's is nearly double the number of Burger King restaurants in the first three zones, whereas the two types are nearly equal in the outermost zone.

One of the most common methods for displaying a bivariate distribution of unclassified continuous variables is as a scatter-graph (also called a scatterplot and scattergram). Box 5.4c includes a scatter-graph for water temperature at two sampling points on the fluvioglacial stream flowing from Les Bossons Glacier in France. The variables, usually assigned the shorthand labels X and Y, are plotted, respectively, on the horizontal and vertical axes of the graph. These are scaled according to the measurement units of the variables, degrees Celsius in this case, and provide a visual impression of the spread of values and how the data values for one variable are connected with those of the other. The water-temperature measurements on the fluvioglacial stream have been recorded at exactly the same time of day at each location. Thus, there is a 'hidden' variable in the data providing a temporal dimension to the information. If they had been made randomly through the day and the time recorded independently, then insufficient data might have been available for certain periods that would have limited the opportunity to examine whether temperature varies between the locations at the same time of day. This, together with differences in the ranges of data values, has helped to produce the slightly curious shape of the scatter-graph. Strictly speaking scatter-graphs do not provide a count of the number of observations possessing certain values. However, it is feasible to overlay the graph with a grid to represent the cells in a contingency table, which is equivalent to grouping the raw data values into classes. Such a grid has been superimposed on the scatter graph in the lower part of Box 5.4c and the corresponding absolute frequency cross-tabulation appears on Box 5.4d. The data values for both sets of temperature measurements have been allocated to groups using the classification scheme originally applied to one of the sample locations (see Box 5.1b) and because those recorded near the glacier snout have a narrower range, there are some empty cells in the table.

How could the classification schemes for the water-temperature data values at the two locations be altered to avoid so many empty cells in the table?

Box 5.4a: Bivariate graphical display with nominal attributes and classified continuous variables (interval and ratio scale measurements).

Sample site at stream exit from outwash plain

Sample site on stream near glacial snout

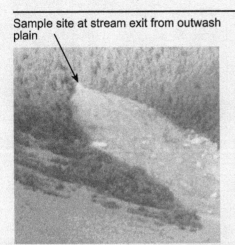

The clustered histogram or column graph for fast-food restaurants in Pittsburgh has been created with the frequency counts of the restaurants on the vertical axis and the four distance zones on the horizontal. The graph indicates whether the two companies' outlets possess a similar or different frequency distribution. The McDonald's restaurants show a rising then falling trend over the set of distance zones, whereas Burger King dips in the middle two zones and reaches a peak furthest from the city centre.

The scatter-graph plots water temperature for the two locations on the fluvioglacial stream at half-hourly intervals throughout the day when the field work was carried out. The X and Y axes have been scaled to cover the same spread of units (0.0 to 14.0 °C); although data values for the measurements made near the stream's exit from the outwash plain have a wider dispersion than those from near the snout. The range of values on the X-axis range between 4.0 and 8.0, which is much narrower than those on Y (4.0–12.0). Lower temperature values tend to occur together at both locations, although the high values near the stream's exit were accompanied by only moderate figures at the glacier snout. The overlain grid used to indicate classification of the temperature recordings further emphasizes this point.

A clustered column chart for type and zone of fast-food restaurant is given in Box 5.4b and a scatter-graph of water temperature together with a bivariate classification of the values are in Boxes 5.4c and 5.4d.

Box 5.4b: Clustered column graph of fast-food restaurants by distance from central Pittsburgh.

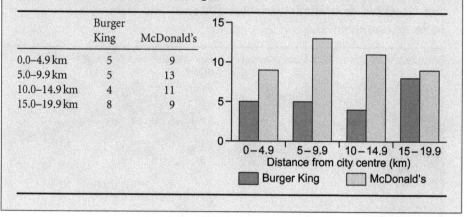

	Burger King	McDonald's
0.0–4.9 km	5	9
5.0–9.9 km	5	13
10.0–14.9 km	4	11
15.0–19.9 km	8	9

Box 5.4c: Scatter graph of raw data values of water temperature at glacier snout and outwash exit of Les Bossons Glacier, France.

	Snout	Exit
09.00 hrs	4.10	4.30
09.30 hrs	4.60	4.70
10.00 hrs	4.60	4.80
10.30 hrs	4.70	5.20
11.30 hrs	4.80	6.70
12.00 hrs	5.20	10.10
12.30 hrs	5.40	10.50
13.00 hrs	6.30	11.20
13.30 hrs	6.90	11.40
14.00 hrs	7.40	11.80
14.30 hrs	7.10	12.30
15.00 hrs	6.90	11.90
15.30 hrs	6.30	11.40
16.00 hrs	6.10	11.10
16.30 hrs	5.50	10.20
17.00 hrs	5.50	9.30
17.30 hrs	5.30	8.80
18.00 hrs	5.10	7.80

Box 5.4d: Contingency table of water temperature at glacier snout and outwash exit.

	0.00–4.99	5.00–9.99	10.00–14.99	Exit
0.00–4.99	3	0	0	5
5.00–9.99	2	3	0	11
10.00–14.99	0	10	0	2
Snout	5	13	0	18

5.3   Estimation of statistics from frequency distributions

This chapter has so far considered frequency distributions from the perspective that it is possible to access the individual raw data values for the full set of (sampled) observations (i.e. details of fast-food outlets and the water-temperature measurements). These values are either nominal categorical attributes or have been classified into such groups for the purpose of producing an absolute or relative frequency distribution. This process of classification entails discarding some of the detail in the raw data to produce an overview of how the observations are distributed. When analysing data from secondary sources this detail has often already been removed by the 'publisher' of the information and so the summary measures of central tendency and dispersion examined in Chapter 4 cannot be obtained from the raw data values to provide a statistical description. There are occasions when it is useful to estimate the summary measures from the frequencies. However, it is highly likely that such estimated measures (e.g. a mean or standard deviation) will differ from the 'true' figures that could have been calculated, if the raw data were available.

The main descriptive measures used in respect of continuous data values are the mean, variance and standard deviation. Each of these can be estimated for variables in respect of observations distributed as frequencies in discrete classes. Box 5.5 illustrates these calculations in respect of the frequency distribution of farm sizes in Government Office Regions covering London and South East England for 2002. This frequency distribution has been obtained from a secondary source and the classes are those employed by the Department of the Environment, Food and Rural Affairs. The data show that farm area is highly skewed to the left of the distribution indicating a very large number of small holdings, especially under 2.0 ha, and very few in the larger size categories (3.3 per cent over 300.0 ha). Notice that the unequal class widths (e.g. 10, 50 and 100 ha) produce a distribution with decreasing numbers of farms in the classes up to 50 ha followed by a rise that leads into a second series of decreasing columns from 100 ha onwards. There is likely to be a difference between the values of the true and estimated mean because the estimation calculations (see Box 5.5b)

Box 5.5a: Estimation of summary descriptive statistical measures from frequency distribution of farm sizes in London and South East England.

Mean – population symbol: μ; sample symbol: $\bar{x}$

Variance – population symbol: σ^2; sample symbol: s^2

Standard deviation – population symbol: σ; sample symbol: s

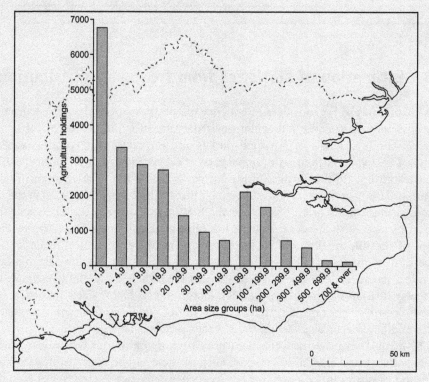

Frequency distribution of farm area in London and South East England 2002.

The mean as a measure of central tendency conveys something about the typical value of a continuous variable whereas the variance and standard deviation indicate how dispersed all the data values are around the mean. The usual ways of calculating these statistics (adding up the values and dividing by the number of observations for the mean and summing the squared differences from the mean and dividing by the number of observations for the variance) are not possible when the data are held in the form of a frequency distribution. An estimate of the mean can be calculated by multiplying the midvalue of each class by the corresponding number of observations, then adding up the results and dividing this sum by the total number of observations. Class midvalues are obtained by dividing the width of each class by 2 and adding the result to the lower class limit (e.g. 4.9–2.0/2 + 2.0 = 3.45). The variance may be estimated by subtracting the estimate of the overall mean from the midvalue of each class, squaring the results and summing before dividing by n (or $n–1$ in the case of a sample data). The square root of the variance provides the standard deviation.

The methods used to calculate estimates of the mean, variance and standard deviation for the frequency distribution of farm (holding) areas in South East England are illustrated below in Box 5.5b, where the letters f and m denote the frequency count and class midvalue, respectively, for the farm area variable (x). Note that the uppermost open ended class (700 ha and over) has been omitted from the calculations to enable comparison with the published statistics.

Box 5.5b: Calculation of estimates for the mean, variance and standard deviation from data in a frequency distribution.

Size class	Class midvalue m	Count f	mf	$\sqrt{\dfrac{\sum(f(m-\bar{x})^2)}{n}}$ $(m-\bar{x})$	$f(m-\bar{x})^2$
0.0–1.9	0.95	6759	6421.1	−43.75	12 937 148.44
2.0–4.9	3.45	3356	11 578.2	−41.25	5 710 443.75
5.0–9.9	7.45	2872	21 396.4	−37.25	3 985 079.50
10.0–19.9	14.95	2715	40 589.3	−29.75	2 402 944.69
20.0–29.9	24.95	1433	35 753.4	−19.75	558 959.56
30.0–39.9	34.95	961	33 587.0	−9.75	91 355.06
40.0–49.9	44.95	724	32 543.8	0.25	45.25
50.0–99.9	74.95	2103	157 619.9	30.25	1 924 376.44
100.0–199.9	149.95	1668	250 116.6	105.25	18 477 374.25
200.0–299.9	249.95	721	180 214.0	205.25	30 373 972.56
300.0–499.9	399.95	523	209 173.9	355.25	66 003 940.19
500.0–699.9	599.95	155	92 992.3	555.25	47 786 897.19
		23 990	$\sum x = 1071985.5$	$\sum(f(x-\bar{x})^2) = 190252536.98$	

Mean
$$\frac{1071985.5}{23\,990}$$
$$\bar{x} = 44.7$$

Variance
$$s^2 = \frac{190\,252\,536.98}{23\,990}$$
$$s^2 = 7930.5$$

Standard deviation
$$s = \sqrt{\frac{190\,252\,536.98}{23\,990}}$$
$$s = 89.1$$

Box 5.5c: Calculation of estimates for the mean, variance and standard deviation from data in a frequency distribution with approximately equal size classes.

Size class	Class midvalue m	Count F	$\bar{x}=\dfrac{\sum fm}{n}$ mf	$s^2=\sqrt{\dfrac{\sum(f(m-\bar{x})^2)}{n}}$ $(m-\bar{x})$	$f(m-\bar{x})^2$
0.0–99.9	49.95	20 923	1 045 103.9	−26.15	576 690.19
100.0–299.9	199.95	2 389	477 680.6	123.85	57 581 021.81
300.0–499.9	399.95	523	209 173.9	323.85	66 003 940.19
500.0–699.9	599.95	155	92 992.3	523.85	47 786 897.19
		23 990	$\sum x = 1824950.5$	$\sum(f(x-\bar{x})^2)=148338606.78$	

$$\dfrac{1824\,950.5}{23\,990}$$

Mean $\bar{x}=76.1$

Variance

$$s^2=\dfrac{148\,338\,606.78}{23\,990}$$

$$s^2=6183.4$$

Standard deviation

$$s=\sqrt{\dfrac{148\,338\,606.78}{23\,990}}$$

$$s=78.6$$

Source: Department of the Environment, Food and Rural Affairs (2003).

assume that the data values within each class are evenly distributed across the entire class interval. However, they may in fact be unevenly distributed and clustered towards one or other end of some or all of the class ranges. The data source in this instance has provided not only the frequency count of farms in each class, but also the total area of farmland that these represent. Therefore, it is possible to calculate the true mean, which turns out to be 43.2 ha compared with an estimate of 44.7 ha (excluding 700 and over ha class).

In this example, the skewness of the distribution and the use of variable class intervals across the entire spread of data values are likely to have limited the difference between the estimated and true means. The calculations used to estimate the variance and standard deviation in Box 5.5b assume that there is no dispersion amongst the observations (farms) in each class and that the area of each one equals the class mid-value (i.e. the within class variance is zero). The effect of the highly skewed distribution of farm areas can be discovered by collapsing the classes so that they are equal (as far as the original classes allow) (see Box 5.5c), with the exception of the lowest category that is 100 ha. The effect is to inflate the estimated mean to 76.1 ha with the variance and standard deviation reducing slightly. Since the original estimated mean is relatively close to the known true mean in this example, it may be concluded that the unequal class interval employed in assembling the frequency distribution help to overcome the underlying skewness in the raw data values.

5.4 Probability

People encounter probability throughout their everyday lives without necessarily thinking about it and how it affects them. Our experience of probability is so commonplace that we often have some difficulty when it comes to conceptualizing what is meant by the term in a more formal fashion. Texts dealing with statistical techniques often resort to explaining the concept with reference to tossing coins and other similar devices where there is clearly an element of chance involved in the action. At an early age many children encounter chance events when playing board games that involve throwing dice. They soon learn that when a number between 1 and 6 lands uppermost then it is possible to move forward by that number of spaces and at each throw of the die the same number does not appear. With experience they learn that there is an equal chance they could move forwards by one, two, three, four, five or six spaces. In other words, they learn by practice the probabilities associated with the outcome of the activity.

One such board game with a long pedigree is Snakes and Ladders (or Chutes and Ladders) as represented in a slightly simplified format in Figure 5.1. If the number on the die results in the player's token landing on the head of a snake, then it has to 'fall down the snake' to its tail and resume its progress towards the Finish from that space at the next turn. However, a player landing on a square at the foot of a ladder can 'climb' to the space at its top and continue from there. The number of squares on a board of Snakes and Ladders is normally 8×8, 10×10 or 12×12 and the ladders and snakes are of various lengths, some of which will cascade or promote the player a small number of rows, whereas others will cause a more precipitous fall or dramatic ascent. The probabilities associated with this game are therefore more complex than one player simply having the luck of throwing more high numbers on the die than his/her opponents and thereby overtake them in the race from Start to Finish. Nevertheless, it is the probabilities associated with each throw of the die that are of interest here.

This description of playing Snakes and Ladders begs one important question. How does a player of this or any other game that involves throwing a die **know** that on each throw it can fall with one and only one numbered side uppermost, and that each of the six numbers has an equal probability of facing upwards. There are two ways of answering this question: either by 'thinking about it' or theorizing the characteristics of the die, the action of throwing it and the possible range of outcomes when it lands; or by spending a lot of time experimentally throwing a die

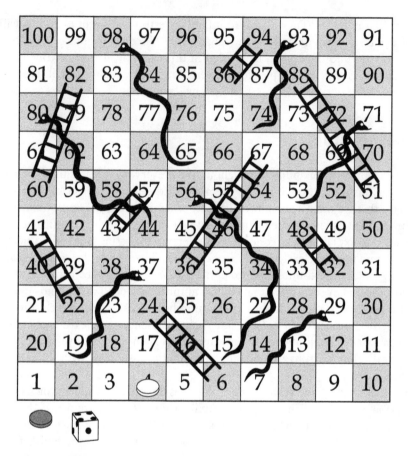

Figure 5.1   Typical layout of Snakes and Ladders Board.

and carefully recording the number of times each digit faces upwards. The first approach entails theorizing about the event and, assuming the die is unweighted and unbiased, reaching the conclusion that it is extremely unlikely to land on one of its edges or corners. Therefore, it must land with one face hidden from view lying on the surface where it has fallen four facing outwards at right angles to each other on the sides and the sixth visible appearing uppermost. There is an equal chance (probability) of any one of the numbered faces in this upwards facing position. This is known as *a priori* probability, since the probability of a particular outcome is known or assessed in advance of ever actually throwing the die. The alternative way of determining probability involves experimentally throwing the die over and over again and recording the number facing upwards on each occasion, which will produce a frequency distribution of the range of possible outcomes. In the long run, if this experiment is carried a sufficient number of times, the frequency distribution will show each of the six possible outcomes having occurred an equal or very nearly equal number of times. This type of probability is referred to as *a posteriori* because it is determined empirically after the (throwing) events have taken place.

In this example, both types of probability should produce the same answer. So, if you were to take time off from reading this chapter and to throw a die, what is the probability that a two will appear on the uppermost surface? How many different ways can you write the answer to this question?

The action of throwing a die produces discrete or integer outcomes, since it is possible to obtain a 1, 2, 3, 4, 5 or 6 only a whole number of times. For example, a six cannot be thrown on 0.45, 2.36 or any other decimal or fractional number of occasions. In other words, the type of event or observation represented by throwing a die is a nominal attribute with more than two categories. It would be quite feasible to produce a die where the faces were numbered with the values 0.75, 1.25, 2.50, 3.75, 4.25 and 5.50 rather than as conventionally with 1, 2, 3, 4, 5 or 6. It would be difficult to use such a die in a game of Snakes and Ladders, after all how can a player's token move 3.75 squares on the board. However, if this die was thrown on many occasions the number of outcomes associated with each fractional value (0.75, 1.25, etc.) would still be an integer or whole number even though the labels of the nominal categories were decimals. Few, if any research projects in Geography and the Earth Sciences study the throwing dice. Nevertheless, investigators are often concerned with analysing events that have a discrete number of possible outcomes and/or with situations where the values of the variables being measured can be decimals.

A project will often involve a mixture of attributes and variables quantified according to these different types of measurement. The probabilities associated with these 'research' attributes and variables can also be conceived theoretically or empirically. Other things being equal, the proportion of males and females in paid employment should be the same as the proportion of the two genders in the population of normal working age (currently 16–65 years in the UK). Assuming these proportions are 0.55 for males and 0.45 for females, a random sample of 600 people of working age people should contain 270 women and 330 men (see Table 5.1). This probability is based on the known proportions of the two genders in the parent population. If a die was to be thrown 600 times you would expect a 1 to face upwards 100 times, a 2 100 times and also the same frequency count (100) for the 3, 4, 5 and 6 faces of the die. However, in both cases the observed or empirical results could be different from the expected or theorized outcomes. Table 5.1 also shows the hypothetical cases of slightly more males and less females appearing in the sample, and more or less than 100 observed occurrences of some of the numbered faces of the die. The expected outcomes are based on *a priori* probabilities since they can be asserted without the empirical results (the sample and the die throwing) having been obtained. In principle, similar arguments can be advanced in respect of the data values of continuous variables, although these relate to decimal measurements rather than integer outcomes. Thus, other things being equal, the temperature just below the surface of a water body on a day when the sun is shining will be higher than on days when it is raining. In other words,

Table 5.1 Observed and expected frequency distributions for dichotomous and discrete outcomes.

Gender split in working age population	Expected outcome	Observed outcome	Throwing a die	Expected outcome	Observed outcome
	Count (%)	Count (%)		Count (%)	Count (%)
Male	330 (55.0)	338 (56.3)	1	100 (16.6)	96 (16.0)
Female	270 (45.0)	262 (43.7)	2	100 (16.7)	97 (16.2)
			3	100 (16.7)	107 (17.8)
			4	100 (16.6)	103 (17.2)
			5	100 (16.7)	100 (16.6)
			6	100 (16.7)	97 (16.2)
Total	600 (100.0)	600 (100.0)		600 (100.0)	600 (100.0)

the temperature measurements in these two situations can be put in order by reference to theoretical prior probability about the effect of thermal incoming electromagnetic radiation from the sun (i.e. asserting in advance they should be higher on a sunny day and lower on a cloudy day). Unfortunately, it is more difficult to say by how much the set of measurements on the sunny day are likely to be higher than those on the cloudy day, although this may be determined from past empirical results (i.e. *a posteriori* probability).

In summary, from a statistical perspective it is possible to refer to three types of outcome for different events:

- dichotomous, where there are two possibilities (e.g. male or female gender; yes or no answer);

- integer or discrete, where there are more than two possibilities, but each can only occur completely (e.g. rock hardness category 10, age group of 20–24 years);

- continuous, where there are decimal or fractional possibilities resulting from measurement (e.g. annual gross income of £23 046.35, temperature of 15.6 °C.

The results of these different types of event can be expressed in the form of frequency distributions and the likelihood of obtaining a particular distribution can be determined by reference to mathematically defined probabilities. Statistical testing relies on such probability distributions to act as a 'sounding post' or 'benchmark' against which to test results obtained from data collected empirically. The three main probability distributions are the Binomial, the Poisson and the Normal. The frequency distribution associated with each of these corresponds to what would be produced if the values of the variables and attributes occurred randomly. Before examining the tests that are available to help with deciding whether distributions produced from empirical data are significantly different from those expected according to a probability distribution, we will examine the characteristics of these distributions.

5.4.1 Binomial distribution

Words starting with the letters 'bi' usually indicate that their meaning has something to do with two or twice. For example, bisect means to cut into two pieces or sections and bilateral, as in bilateral talks, refers to discussions between two parties or countries. The second part of the word binomial ('nomial') looks a little like nominal, which we have seen relates to categorical attributes where the values correspond to names or labels. Thus, it is not difficult to see that the Binomial Distribution is associated with those situations in which there are two nominal categories or outcomes to an experiment or event. The Binomial Distribution was devised by James Benouilli in the 17th century and applies to a series of similar events each of which has two (dichotomous) outcomes possible. Box 5.6 outlines the how the Binomial Distribution is calculated using the now familiar example of Burger King restaurants in Pittsburgh, whose presence or absence in the nine 10×10 km grid squares covering the city is shown in Box 5.6a.

The form and shape of the frequency distribution varies according to the length or number of events and the probability of each possible outcome. While it might be relatively easy to determine the number of events, determination of the probabilities of each outcome can be a little more problematic. However, as discussed previously we can usually resort to either theoretical reflection or empirical precedent. The frequency distribution produced from binomial probabilities has a symmetrical shape if the probabilities of each outcome are identical (50/50 or 0.5/0.5), or very nearly so. A skewed or asymmetrical distribution will result if there is distinct difference in the two probabilities, for example 75/25 or 0.75/0.25. The mean, variance, standard deviation and skewness statistics can be calculated for a frequency distribution produced from binomial probabilities. In this case, the mean refers to the average number of outcomes of one or other of the two types from a certain number of events. The variance and standard deviation relate to the dispersion of the distribution around the mean and skewness to the degree of symmetry.

5.4.2 Poisson distribution

There is nothing in the word 'Poisson' to suggest why the Poisson Distribution should refer to the probabilities associated with events that have a discrete number of integer (complete) outcomes and a diminishing probability of occurrence as the number of possible outcomes increases. However, the distribution is useful as a means of determining the probabilities associated with each of an ordered or ranked series of

Box 5.6a: Binomial Distribution.

Binomial Distribution probability: $p(X) = \dfrac{N!\,p^X q^{N-X}}{X!(N-X)!}$

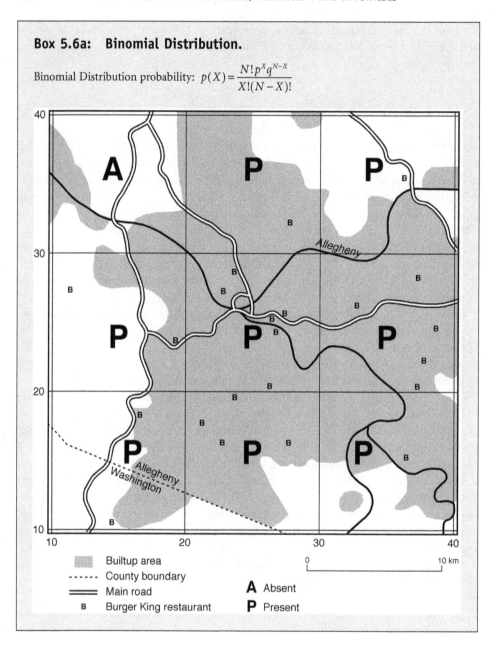

Box 5.6b: Calculation of Binomial probabilities.

The Binomial Distribution is used to determine the probability of each of two possible outcomes in a series of events. In this example the outcomes are the presence or the absence of a Burger King restaurant in a set of 10 x 10 km grid squares superimposed over the built-up area of Pittsburgh. The probability of the occurrence of the event under investigation is denoted by p in the formula and probability of non-occurrence by q. In this example the probabilities of occurrence and non-occurrence of a Burger King in any one grid square are equal (0.5). Across the complete set of n (9) grid squares, at one extreme they might each include at least one of these restaurants whereas at the other there is not a restaurant in any square. The empirical result of counting the number of squares with and without these fast food outlets is shown as a frequency table below. It reveals that 88.9% or 8 squares include a Burger King and 11.1% does not.

The Binomial Distribution in this example determines the probability that a certain number of squares, referred to as X in the Binomial formula, will contain a Burger King. In other words the number of squares in which the event occurs. The calculations in Box 5.5c provide the Binomial probability for the occurrence of at least one Burger King restaurant in 2, 3 and 5 squares. Now you might ask why is it useful to find out these probabilities when the empirical data on their own provide the answer – there are eight squares with a restaurant. The answer is that the Binomial Distribution has indicated the probable number of squares with a restaurant if they were distributed randomly. The probability of there being eight squares with a Burger King is quite small (0.08 or 8 times in 100), certainly compared the probability of only five squares with a restaurant (0.25).

Box 5.6c: Calculation of Binomial probabilities.

$$p(X) = \frac{N!\,p^X q^{N-X}}{X!(N-X)!}$$

Burger King	Count	$p(2) =$	$p(3) =$	$p(5) =$
0 (absent)	1	$\dfrac{9!0.5^2 0.5^{9-2}}{2!(9-2)!}$	$\dfrac{9!0.5^3 0.5^{9-3}}{3!(9-3)!}$	$\dfrac{9!0.5^5 0.5^{9-5}}{5!(9-5)!}$
1 (present)	8	$\dfrac{362880(0.25)(0.0078)}{2(5040)}$	$\dfrac{362880(0.13)(0.016)}{6(720)}$	$\dfrac{362880(0.031)(0.063)}{120(24)}$
	9	$p(2) = 0.07$	$p(3) = 0.16$	$p(5) = 0.25$

'counted' outcomes. Two important applications of the Poisson Distribution relate to counting the number of times a particular outcome occurs within a given collection of spatial units or in a set of time periods. These units or periods are treated as 'events' in such situations. Whereas the Binomial Distribution requires a fixed, finite number of outcomes to an event (e.g. Yes or No answers to a question), the Poisson does not impose an upper limit and theoretically an outcome could occur a very large number of times within any individual spatial or temporal unit. However, one common feature of both distributions is that they deal with variables measured according to the ordinal scale, in other words a ranked or sorted sequence of outcomes.

Box 5.7 illustrates that the Poisson distribution is defined mathematically as a series of discrete terms, each representing an integer outcome in the sequence. Thus, the first provides the probability of a zero outcome (i.e. none of the phenomena being counted),

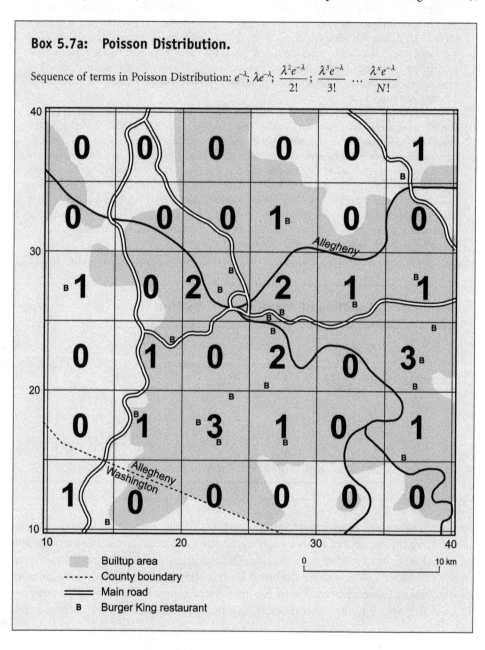

Box 5.7a: Poisson Distribution.

Sequence of terms in Poisson Distribution: $e^{-\lambda}$; $\lambda e^{-\lambda}$; $\dfrac{\lambda^2 e^{-\lambda}}{2!}$; $\dfrac{\lambda^3 e^{-\lambda}}{3!}$ ... $\dfrac{\lambda^x e^{-\lambda}}{N!}$

Builtup area
County boundary
Main road
B   Burger King restaurant

Box 5.7b: Calculation of Poisson probabilities.

The Poisson Distribution is used to discover the probability of the number of times that a discrete outcome will be expected to occur in a sequence. The present example complements the application of the Binomial Distribution in Box 5.6 by providing Poisson probabilities for the number of Burger King restaurants in a set of 5 km grid squares superimposed over the built-up area of Pittsburgh. Each square might contain 0, 1, 2, 3, 4, ... N restaurants and the Poisson Distribution uses a series of terms to represent each of these integer outcomes. The terms include a common element ($e^{-\lambda}$) where the constant e represents the base of natural logarithms (2.718 281 8) and λ is the average number of outcomes per unit. This equals 0.611 in this example (22/36). The lower part or denominators of the terms in the Poisson Distribution are the factorials of the integer number of outcomes that are of interest ($3! = 3 \times 2 \times 1 = 6$). These integers have been limited to 0 through 5 in the calculations in Box 5.7.

The probability calculated for each term is multiplied by the total number of outcomes, restaurants in this example, to produce the frequency distribution that should have occurred if they were distributed at random across the 36 grid squares. These are shown to 1 decimal place below, but can easily be rounded to facilitate comparison with the empirical results. The Poisson Distribution indicates that there should be 20 squares with zero Burger King restaurants and 12 with one, whereas the field survey reveals frequency counts of 21 and 10, respectively. Looking further at the frequency distribution suggests that it is quite surprising that two squares with three should have occurred, when the Poisson probability indicates there should have been one.

Box 5.7c: Calculation of Poisson probabilities.

Burger King	Count of squares	Count of B Kings	Poisson term		P	Est. count
0	21	0	$e^{-\lambda}$	$2.718^{-0.611}$	0.54	19.5
1	10	10	$\lambda e^{-\lambda}$	$(0.611)2.718^{-0.611}$	0.332	11.9
2	3	6	$\dfrac{\lambda^2 e^{-\lambda}}{2!}$	$\dfrac{(0.611^2)2.718^{-0.611}}{2}$	0.101	3.7
3	2	6	$\dfrac{\lambda^3 e^{-\lambda}}{3!}$	$\dfrac{(0.611^3)2.718^{-0.611}}{6}$	0.021	0.7
4	0	0	$\dfrac{\lambda^3 e^{-\lambda}}{4!}$	$\dfrac{(0.611^4)2.718^{-0.611}}{24}$	0.003	0.1
5	0	0	$\dfrac{\lambda^3 e^{-\lambda}}{5!}$	$\dfrac{(0.611^5)2.718^{-0.611}}{144}$	0.000	0.0
	36	22			1.000	36.0

the second one, the third two and so on to the maximum considered necessary. The calculations show that probabilities become smaller as the sequence progresses. Thus, there is a relatively high probability that zero or one outcome will occur in a given unit of space or time, but there is a comparatively low chance that four, five or more will occur. Standard descriptive statistics (mean, standard deviation and skewness) can be calculated for a frequency distribution produced from Poisson probabilities. These are derived from one of the key elements in the series of Poisson terms (λ), which equals the mean and variance. The standard deviation is the square root of λ and skewness is the reciprocal of this statistic. The value of λ is usually obtained from empirical data as in this example and varies from one application to another.

There are two issues that should be noted. First, when determining the Poisson probabilities for a particular set of empirical data it is necessary to calculate λ, the average number of outcomes that have occurred across the group of spatial or temporal units, or other types of event and use this value in the calculation of the Poisson probabilities. However, this average density of outcomes could easily vary across the overall study area. Suppose the research was interested in the numbers of independent retail outlets in a county and decides to superimpose a regular grid of 1 km squares over the area. The count of shops is likely to vary considerably between sparsely and densely populated areas: urban neighbourhoods might have a few shops competing with each other for business, whereas villages, if they are lucky, might have one facility. Thus, the mean or average number of shops per 1 km square is likely to be higher or lower in different localities in the county. It would be possible to calculate separate means for high- and low-density areas and then combine these by weighting. However, this introduces the problem of deciding how to define the physical extents of these localities in terms of individual grid squares.

Another way of dealing with this issue would be to change the size of the spatial units, for example from 1 km squares to 2.5 km, 5.0 km or even larger squares. Each increase in area would mean that the number of squares would decrease, or conversely their number would increase if their size was reduced. Given that the number of phenomena being counted is fixed, these changes in size of unit will impact on the frequency distribution. It is likely that a reduction in size (e.g. to 0.25 km squares) would increase the number of units without any or only one shop and the count with three, four or more would probably decrease. Changing the size of the spatial or temporal units is still likely to produce a skewed frequency distribution, but the average number of outcomes per unit will alter, which results in a different value for λ.

5.4.3   Normal distribution

The Normal Distribution, also referred to as the Gaussian distribution, was first specified by the German mathematician Carl Gauss. Unlike the previous Binomial and Poisson Distributions, it applies in situations where variables are measured on the interval or ratio scales. It is relatively easy to imagine how a frequency distribution can be produced when counting the number of discrete outcomes from a series of events. Each additional occurrence increases the number of outcomes in a particular category by one unit each time. However, when dealing with variables measured on a continuous scale, such as the concentration of chemical pollutants in water, it is perhaps less obvious how these constitute a frequency distribution. At one extreme the pollutant's concentration could have a different value for each water sample that is measured, whereas conversely they could all be the same. Neither of these extreme situations is likely to occur. Each additional measurement might be the same as one already recorded, very similar or dissimilar. The Normal Distribution is based a recognition that individual measurements in a set when taken together will tend to cluster around the mean or average rather than occur at the extremes of the range. In other words, while measured variables will normally include some relatively high and low values, the majority will tend towards the mean. In its ideal form the Normal Distribution traces a symmetrical, 'bell-shaped' curve with a convex peak and tails extending outwards towards plus and minus infinity (Figure 5.2). The points of inflection on the curve where it changes from convex to concave occur at one standard deviation from the mean in either direction (i.e. mean minus 1 standard deviation and mean plus 1 standard deviation). Since the standard deviation quantifies dispersion in terms of the original measurement scale units of the variable, unlike the variance, these points can be marked on the horizontal axis of X.

Application of the Binomial and Poisson Distributions involved calculating the probabilities associated with discrete outcomes (e.g. the number of grid squares with 0, 1, 2, etc. Burger King restaurants in Box 5.7). Application of the Normal Distribution seeks to achieve a similar purpose, namely to calculate the probability of obtaining a value for the variable X within a certain range. Since values of X are likely to cluster around the mean as a measure of central tendency, the probability of obtaining a value close to the mean will be higher than for one further away. The heights of the normal

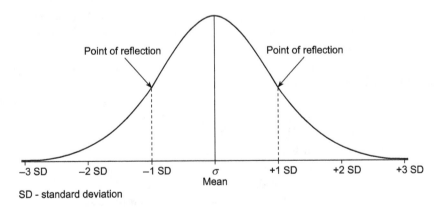

Figure 5.2 The Normal Distribution Curve.

Box 5.8:   Normal Distribution.

Normal Distribution: $Y = \dfrac{1}{\sqrt{2\pi}\sigma} e^{-((X-\mu)^2/2\sigma^2)}$

The Normal Distribution provides a way of determining the probability of obtaining values of a continuous variable measured on the interval or ratio scale. The equation for the Normal Distribution calculates the Y frequencies or ordinates of a collection of data values for a variable (X) whose mean (μ) and standard deviation (σ^2) are known. If the values of the variable follow the Normal Distribution, plotting the ordinates will produce a symmetrical, normal curve. The equation can be broken down into three groups of terms: the constants π(3.1416...) and e(2.7128...); the mean (μ) and standard deviation (σ^2); and the known data values of the variable (X) for which ordinates are required.

---

curve for different values of the variable, X, known as **ordinates**, represent the expected frequency of occurrence. While from a theoretical perspective the symmetrical, bell-shaped form of the normal curve can be created by inserting numerous different values of X into the equation given in Box 5.8, provided that the mean and standard deviation are known, although there is usually a more restricted number of X values available when working with empirical data. It is rarely necessary to carry out the computations associated with the normal curve in order to make use of its important properties. However, it is still important to check whether the known data values indicate that the variable follows the Normal Distribution. Several of the statistical tests examined in later chapters strictly speaking should only be applied under the assumption that the sampled data values are drawn from a population that follows the Normal Distribution.

The mean of the data values corresponds to the point on the horizontal axis of X that lies under the central peak of the normal curve. A different set of values for the same variable might result in a new mean, larger or smaller than the first, but that still forms the centre of a symmetrical, bell-shaped normal curve. In other words, the position of the mean on the horizontal axis for a particular measured variable can be thought of as determined by the individual data values that have been recorded. The standard deviation, as has been explained previously, measures the spread of the data values. It is possible to imagine a series of symmetrical curves above a single mean value that are themselves relatively more or less spread out. The mean and standard deviation are thus of crucial importance in determining the position and spread of the normal curve. This variability in the form of the normal curve produced from different sets of data for the same variable would limit the usefulness of the Normal Distribution were it not for the possibility of standardizing the raw data.

Standardization expresses the differences between each X value and the overall mean as a proportion of the standard deviation. The resulting quantities are known as Z values. In some ways this procedure is similar to expressing raw data values as a percentage of the total. Percentages are useful for making comparisons between different pieces of information, for example it is more informative to know that there are 15 per cent unemployed working age adults in area A and 4 per cent in area B, rather than that Area A has 7500 and Area B has 12 000. It might be implied from the absolute counts that unemployment is more problematic in Area B, whereas the lower percentage indicates it is possibly of less importance.

What do you conclude about the numbers of working age adults in areas A and B from this information?

Standardization of X data values results in positive and negative Z scores, respectively indicating that the original value was less than or greater than the mean. These Z values are measured in units of the standard deviation, which can be plotted on the horizontal axis. Thus, a Z value of say +1.5 indicates that it is one-and-a-half times the standard deviation. Provided the frequency distribution of the original values of X (see Section 5.1 for classification of raw data values to produce frequency distributions) appears to show that they follow a normal curve, the equation for the Normal Distribution can be simplified by substituting Z for a term on the right-hand side, which allows all other terms involving the standard deviation to equal 1.0 (see Box 5.9). Box 5.9 shows the calculations involved in producing the standardized normal curve for the sampled water-temperature measurements taken in the fluvioglacial stream from Les Bossons Glacier near Chamonix.

Probability statements in respect of the Normal Distribution refer to the probabilities of obtaining X values, or their Z score equivalents, in particular ranges or sections of the horizontal axis. The two types of probability statement that can be made are best illustrated by means of a pair of questions:

Box 5.9a: Z Scores and the Standardized Normal Distribution.

Z scores: $Z = \dfrac{X-\mu}{\sigma^2}$; Z standardized normal distribution: $y = \dfrac{1}{\sqrt{2\pi}}e^{-(Z^2/2)}$

There are two steps involved in producing a standardized normal curve: first, standardization of X values to Z scores; and then insertion of these scores into the standardized normal distribution equation. Z scores are calculated relatively easily by subtracting each value in turn from the overall mean and dividing by the standard deviation. Calculations for the ordinates (Y values) are a little more complicated but involve the constants π(3.1416...) and e(2.7128...) and the Z scores.

The methods used to calculate Z scores and Y ordinates for the water-temperature measurements at the snout Les Bossons Glacier near Chamonix are shown below in Box 5.9b and the plot of the standardized normal curve for these data is included in Box 5.9c. Although this does not exactly conform to the classical symmetrical, bell-shaped curve, its shape provides a reasonable approximation.

Box 5.9b: Calculations for ordinates of standardized normal curve.

X (sorted)	$Z = \dfrac{X-\mu}{\sigma^2}$	$Z =$	$Y = \dfrac{1}{\sqrt{2\pi}}e^{-(Z^2/2)}$	$Y =$
09.00 hrs	4.10	$\dfrac{4.10-5.66}{0.98}$	-1.67	$\dfrac{1}{\sqrt{2(3.1418)}}2.7128^{-(-1.67^2/2)}$	0.140
09.30 hrs	4.60	$\dfrac{4.60-5.66}{0.98}$	-1.17	$\dfrac{1}{\sqrt{2(3.1418)}}2.7128^{-(-1.17^2/2)}$	0.285
10.00 hrs	4.60	$\dfrac{4.60-5.66}{0.98}$	-1.17	$\dfrac{1}{\sqrt{2(3.1418)}}2.7128^{-(-1.17^2/2)}$	0.285
10.30 hrs	4.70	$\dfrac{4.70-5.66}{0.98}$	-1.07	$\dfrac{1}{\sqrt{2(3.1418)}}2.7128^{-(-1.07^2/2)}$	0.318
11.00 hrs	4.80	$\dfrac{4.80-5.66}{0.98}$	-0.97	$\dfrac{1}{\sqrt{2(3.1418)}}2.7128^{-(-0.97^2/2)}$	0.352
17.30 hrs	5.10	$\dfrac{5.10-5.66}{0.98}$	-0.67	$\dfrac{1}{\sqrt{2(3.1418)}}2.7128^{-(-0.67^2/2)}$	0.451
11.30 hrs	5.20	$\dfrac{5.20-5.66}{0.98}$	-0.57	$\dfrac{1}{\sqrt{2(3.1418)}}2.7128^{-(-0.57^2/2)}$	0.479
17.00 hrs	5.30	$\dfrac{5.30-5.66}{0.98}$	-0.47	$\dfrac{1}{\sqrt{2(3.1418)}}2.7128^{-(-0.47^2/2)}$	0.505
12.00 hrs	5.40	$\dfrac{5.40-5.66}{0.98}$	-0.37	$\dfrac{1}{\sqrt{2(3.1418)}}2.7128^{-(-0.37^2/2)}$	0.527
16.00 hrs	5.50	$\dfrac{5.50-5.66}{0.98}$	-0.27	$\dfrac{1}{\sqrt{2(3.1418)}}2.7128^{-(-0.27^2/2)}$	0.544
16.30 hrs	5.50	$\dfrac{5.50-5.66}{0.98}$	-0.27	$\dfrac{1}{\sqrt{2(3.1418)}}2.7128^{-(-0.27^2/2)}$	0.544
15.30 hrs	6.10	$\dfrac{6.10-5.66}{0.98}$	0.33	$\dfrac{1}{\sqrt{2(3.1418)}}2.7128^{-(0.33^2/2)}$	0.535
12.30 hrs	6.30	$\dfrac{6.30-5.66}{0.98}$	0.53	$\dfrac{1}{\sqrt{2(3.1418)}}2.7128^{-(0.53^2/2)}$	0.491
15.00 hrs	6.30	$\dfrac{6.30-5.66}{0.98}$	0.53	$\dfrac{1}{\sqrt{2(3.1418)}}2.7128^{-(0.53^2/2)}$	0.491
13.00 hrs	6.90	$\dfrac{6.90-5.66}{0.98}$	1.13	$\dfrac{1}{\sqrt{2(3.1418)}}2.7128^{-(1.13^2/2)}$	0.299
14.30 hrs	6.90	$\dfrac{6.90-5.66}{0.98}$	1.13	$\dfrac{1}{\sqrt{2(3.1418)}}2.7128^{-(1.13^2/2)}$	0.299
14.00 hrs	7.10	$\dfrac{7.10-5.66}{0.98}$	1.33	$\dfrac{1}{\sqrt{2(3.1418)}}2.7128^{-(1.33^2/2)}$	0.234
13.30 hrs	7.40	$\dfrac{7.40-5.66}{0.98}$	1.63	$\dfrac{1}{\sqrt{2(3.1418)}}2.7128^{-(1.63^2/2)}$	0.150

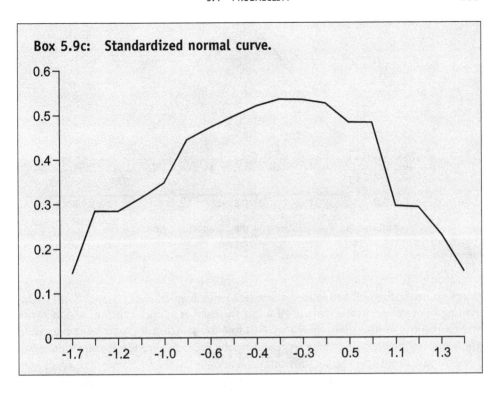

Box 5.9c: Standardized normal curve.

- What proportion or percentage of X values can be expected to occur within a certain range (e.g. between -1.75 standard deviations less than the mean and $+2.0$ standard deviations greater than the mean)?

- What is the probability that any one X value or Z score will fall within a certain range (e.g. greater than $+1.75$ standard deviations from the mean?

Such questions are phrased a little differently to those relating to integer attributes, which are concerned with the probability of a certain number of discrete outcomes. Nevertheless, they share the same basic purpose of trying to specify the probability of some outcome occurring.

The traditional method for determining the Normal Distribution probabilities associated with both types of question was to consult a table of probabilities for different values of Z. These tables, which have appeared as appendices in numerous texts on statistical techniques, can be presented in slightly different formats but their essential purpose is to link the proportions of the area under the normal curve with different Z score values. These proportions provide the probabilities because the area under the normal is treated as finite, despite the tails approaching but never

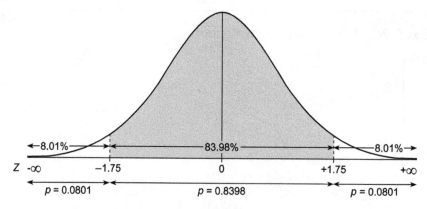

Figure 5.3   Probabilities and the Standardized Normal Curve.

reaching the horizontal axis, and this area is assigned an arbitrary value of 1.0 representing the total probability of all values. If the total area under the curve equates to the probability of all values, then a vertical line drawn from the peak of the curve to the point on the horizontal axis representing the mean, where the Z score equals 0.0, bisects the area into two equal portions each containing 0.5 or 50 per cent of all possible values. The example in Figure 5.3 will help to show how this works. The shaded area in Figure 5.3 is composed of two parts one of each side of the central vertical line: the part on the left goes from −1.75 to 0.0 and the one on the right from 0.0 to +1.75 on the Z score scale. The proportions of the area under the curve associated with these two sections from a Z table are 0.4199 and 0.4199. Summing these proportions indicates that the whole of the shaded area accounts for 0.8398, which indicates generically how to answer the first type of question: in this case 83.98 per cent of all values occur between the specified Z scores. This also implies that there is 0.8398 probability of any one value lying within this range. Thus, we have answered the second type of question. Suppose interest was focused beyond this Z value range towards the negative and positive ends of the distribution. Since the area under the curve represents 100 per cent of values, the unshaded zone contains 100–83.98 per cent of values and there is 0.1602 (16.02 per cent) probability of any one value occurring in this region.

The standardized normal curve for the worked example in Box 5.9 approximates to the ideal symmetrical bell-shaped form, although the match is far from perfect. There are some geographical and geological variables that tend to follow the Normal Distribution (e.g. administrative area population totals and river basin discharge), but there are many others that are manifestly non-normal when viewing their frequency distributions. This problem can be addressed by a process of **normalization,** which involves transforming or recalibrating the raw data values. Variables having a strong positively skewed distribution (i.e. a high proportion of values at the lower end of the range and relatively few high values) may be transformed so that they display a more normal shape by transforming into logarithms, provided there are no negative or zero values. This is known as the log-normal transformation. Conversely, variables with strong negatively skewed distributions may be normalized by squaring the raw data values. One of the problems with these transformations is that interpretation of analytical results it may be difficult because the variables are no longer recorded in a way that relates to the phenomena under investigation.

5.5 Inference and hypotheses

Previous chapters have already referred answering **research questions** as a way of describing what researchers are trying to do when carrying out an investigation. As students of Geography, Earth Science and related subjects the aim of our research is to explore phenomena from a global to a local scale and to establish whether our findings are reliable and how different phenomena relate to each other. Such questions may be theoretical or empirical in their origin. In other words, they might be based on thinking about why things occur and how they are connected in the world around us or they might be founded on seeking to discover some 'factual' information about people, places and physical environments. The difference between the two approaches is very similar to the distinction made between *a priori* and *a posteriori* probability examined earlier in this chapter. Using another example, the probability of the head side of a coin lying uppermost when it was tossed could be derived in two ways: by thinking about the properties of the coin especially that there are two flat surfaces on which it could land, or by carrying out a number of coin-tossing experiments or trials recording the number of times each outcome occurred.

Irrespective of whether research questions originate from theoretical reflection or empirical observation, and are often a combination of both, the process of undertaking an investigation in a rigorous and scientific manner requires that we specify the concepts relating to the phenomena of interest to us. Sustainability is a currently 'hot topic', but what does this phenomenon mean? Student essays on the subject almost invariably define the term by quoting from the Brundtland Commission Report (United Nations, 1993), which asserted that Sustainable development is development that meets the needs of the present without compromising the ability of future generations to meet their own needs. But how can we tell whether actions undertaken at the present time are going to have a damaging impact on future generations? And even if society avoided taking such actions, what is to stop some other unforeseen events occurring that will have a harmful effect in the future?

This example illustrates the importance of translating definitions of concepts into practical and precise terms that can be incorporated into hypotheses capable of being tested statistically. It is from careful specification of these terms that we are able to identify the attributes and variables that need to be measured or quantified. The purpose of this process is to make some inference about what might happen or why something has happened. In other words, we are trying to reach some conclusion about the phenomena of interest. It is worth recalling Chapter 2 suggested that what is recorded or measured and how this is achieved will inevitably affect the outcome of a piece of research. Vague definition and inconsistent application of concepts will often produce results that are difficult or perhaps impossible to interpret. It sometimes appears that conclusions from research are commonsense, if this is the case from 'good' research; surely they are nonsense from 'bad' research.

The use of inferential statistical tests involves the investigator specifying two diametrically opposite hypotheses. These are called the Null Hypothesis and the Alternative Hypothesis, which are represented by the symbols H_0 and H_1. Most quantitative research investigations entail the specification of several pairs of Null and Alternative Hypotheses in order to accumulate a series of answers to the research question(s). Some multistage investigations will include linked sets of hypotheses, so that the decisions taken with respect to one set will influence the direction of the research and the hypotheses tested at a later stage. An investigation into whether the present actions of a company or other organization are sustainable may look at several different types of action. For example, it could examine recycling of waste materials (paper, plastic, computer consumables, etc.), type of fuel used in company vehicles (petrol, diesel, electricity, biofuel, etc.) and use of electricity in buildings (e.g. energy-efficient lighting, equipment left on standby, etc.). The decisions taken in respect of Null and Alternative Hypotheses relating to this first stage of the research are likely to dictate which pairs of hypotheses will be appropriate at the second stage, and so on perhaps to a third stage and beyond.

The exact phrasing of any pair of Null and Alternative Hypotheses will obviously depend on the research question(s) being investigated, nevertheless, it is possible to offer some guidance on their generic format. In most cases, the focus of these hypotheses can be simplified to consideration of a quantifiable difference in the values of a statistical measure between a population and a sample, or between two or more samples to indicate whether they might have come from the same population. The statistical measure may be one of those already encountered (mode, mean, standard deviation, etc.) or those yet to be examined (e.g. correlation and regression coefficients). The phrasing of the Null Hypothesis is deliberately guarded in order to give rigour to the process and to reduce the opportunity for reaching inappropriate conclusions. Caution is exercised by always assuming at the outset that the Null Hypothesis is correct, which maximizes the opportunity for accepting it and minimizes the risk of incorrectly rejecting it. The penalty for reaching such an erroneous decision is rarely life-threatening in geographical, geological and environmental research, which may not be the case in other fields. It may help to think of the Null Hypothesis as being put on trial and the test providing evidence enabling the jury (the investigator) to reach a decision on its acceptance or rejection.

Continuing with the example of investigating the sustainability of an organization's actions, the following pair of hypotheses might apply:

H_0 The amount of material recycled by organizations that have introduced a policy to promote recycling during the last year is the same as in those that have not taken this action. Any observed difference is the result of chance and is not significant.

H_1 The observed difference in the quantity of material recycled is significantly greater in organizations that have made recycling part of company policy than in those that have not done so.

Several different variables could be identified as a means of measuring the quantity of recycling, for example tonnes of recycled materials per employee, per $100\,m^2$ of floor space or per £1000 of turnover. A sample survey of organizations stratified according to whether or not they had introduced a policy to increase recycling might be carried out from which the mean (average) tonnage of recycled materials per employee, per $100\,m^2$ of floor space or per £1000 of turnover was calculated. A suitable test could be applied to reach a decision on whether the difference between the two mean values was significant or not. If the test suggested that the Null Hypothesis should be accepted then it could be concluded that the introduction of a policy to increase recycling had been a waste of time, since it produced no significant gain in the amount of recycled materials. This hypothesis is deliberately cautious because it avoids reaching the conclusion that introducing such a policy will inevitably lead to an increase in recycling, although acceptance of the Alternative Hypothesis, which suggests that the two things are linked, does not confirm that this connection necessarily exists.

The decision about whether to accept the Null or the Alternative Hypothesis enables an investigator to *infer* or tease out a conclusion in relation to the research question(s). Hence the term inferential statistics is used to describe those techniques enabling researchers to infer conclusions about their data. All statistical tests involve reaching a decision in relation to the Null and Alternative Hypotheses: if H_0 is accepted then H_1 is necessarily rejected, and vice versa. They are mutually exclusive and there is no 'third way'. The decision on which of these hypotheses to accept is made by reference to the probability associated with a test statistic that is calculated from the empirical data. The value of the test statistic depends on the number of observations (outcomes to use previous terminology) and the data values. There are many different statistical tests available that are often named using a letter, such as the Z, t and χ^2 tests, although some are also identified by reference to the statistician(s) who first devised it, for example the Mann–Whitney U test.

The main reason for using inferential statistical tests is that some degree of uncertainty exists over the data values obtained for a sample of observations and we cannot be sure of the extent to which this **sampling error** is present. Although there are a number of reasons why sampling error is present, it can be thought of as the sum total of the error contained in a set of data that has arisen because the values have been measured for a sample of observations rather than for the entire population. Chapter 2 explored the reasons why most investigations are based on one or more samples rather than a population. If there is a large amount of sampling error present in a dataset, for example such that the mean or standard deviation is much larger or smaller than in the parent population, or the modal attribute category is different to that in the population, then we might make erroneous decisions about the Null and Alternative Hypotheses. In contrast, if there is very little sampling error, the important features of the data are less likely to be obscured and valid inferences may still be reached. Therefore, what is needed is some way of knowing how much sampling error is present, but since we cannot answer this question for certain all that can be done is to assume there is a random amount present. Our examination of the Binomial, Poisson and Normal Distributions has shown that they provide the probability of a random outcome occurring in relation to dichotomous, ordinal integer and continuous outcomes. Comparing the outcomes they and other similar distributions expect to have occurred with what our data have yielded therefore provides a means of discovering if the amount of sampling error present is random or otherwise.

Given that the probability of having obtained a particular value for the test statistic indicates whether or not to accept the Null Hypothesis, it is necessary to have a criterion that enables an investigator to reach a decision on acceptance or rejection. This helps to ensure that any investigator analysing the attributes and variables in a given dataset to answer a research question would reach the same conclusion. Thus, when students on a statistical techniques module are all given the same dataset to analyse they should, in theory at least, independently reach a common decision about the Null and Alternative Hypotheses when applying a particular statistical test to a variable. This agreement is achieved by using a standard **level of significance** when applying statistical tests so that when different people follow the same standard they all reach the same decision about the Null and Alternative Hypotheses. The most common level of significance is a probability of 0.05 or 5 per cent. This means that there is a 5 in a 100 chance of having obtained the test statistic as a result of chance or randomness. Some investigations might require a rather more stringent criterion be applied for example 0.01, 0.001 or 0.0001. Again, there is a parallel with the judge and jury system for trying criminal cases in British courts. If the case relates to a particularly serious crime for example murder or rape, the judge might direct the jury should return a verdict on which they are all agreed (i.e. unanimous); whereas in lesser cases the judge might direct that a majority verdict (e.g. 11 to 1 or 91.6%) would be acceptable.

What are the probabilities 0.01, 0.001 and 0.001 expressed as percentages? How are they written if reported as a ratio?

Thus far, statistical testing has been discussed from the perspective of investigating whether different samples might have come from the same parent population or whether one sample is significantly different from its parent population. However, these questions can be reversed or 'turned on their head' to determine whether the values of statistical quantities (e.g. measures central tendency and dispersion) calculated from sample data provide a reliable estimate of the corresponding figure in the parent population. It is often the case that little is known about the characteristics of a parent population for the reasons outlined in Chapter 3 and in these circumstances the researcher's aim is often to conclude something about this population on the basis of analysing data from a sample of observations. Since it is not possible to be absolutely certain about whether statistics calculated from a sample are the same as would have been obtained if the population of observations had been analysed, the results are expressed within **confidence limits.** These limits constitute a zone or range within which an investigator can be 95 per cent, 99 per cent, 99.9 per cent or even more confident that the true, but unknown **population parameter** corresponding to the **sample statistic** lies. There will remain a small chance, in the latter case 0.1 per cent, that the population parameter falls outside this range, but this may be regarded as too small to cause any concern. Again reflecting on the analogy with a court of law, it is not feasible to ask several juries to try the same case simultaneously and for the judge to take the average outcome. So it may be sufficient to accept a majority verdict in certain less serious types of case but not in those where a narrower margin of error is more appropriate.

5.6 Connecting summary measures, frequency distributions and probability

This chapter has explored three related aspects of statistical analysis. The presentation of attributes and variables in the form of frequency distributions provides a convenient way of visualizing the pattern formed by a given set of data values in tabular and graphical formats. However, the form and shape of these distributions can be influenced by how the raw data values are categorized. This can relate to the classification schemes used for continuous and nominal data values that may be redefined in different ways. Frequency distributions not only complement single value descriptive statistical measures as a way of summarizing attributes and variables (see Chapter 4), but these measures can also be estimated for such distributions.

The juxtaposition of discussion about frequency and probability distributions in this chapter reflects the important connection between describing the characteristics of a set of data values representing a given attribute or variable and the likelihood (probability) of having obtained those values during the course of empirical data collection. Some other probability distributions will be introduced later, however, the Binomial, Poisson and Normal examined here illustrate the role of probability in statistical tests and how to discover the probabilities associated with random outcomes from different types of event. Finally, the principles of hypothesis testing and the general procedure for carrying out statistical tests have been examined. The Null and Alternative Hypotheses provide a common framework for testing whether data collected in respect of samples help to answer research questions that are phrased in different ways. Researchers will rarely have access to data for all members of a population and inferential statistics offer a way of reaching conclusions about a parent population from a sample of observations. All statistical tests make assumptions about the characteristics of sample data, for example that observations have been selected randomly from the population and variables follow the Normal Distribution. The reason for making these assumptions is to ensure that the probability of a calculated test statistic can be determined by reference to the known probabilities of the corresponding distribution so that by reference to the chosen level of significance a standard decision on acceptance or rejection of the Null Hypothesis can be reached.

There is a wide range of statistical tests available. The choice of test to use in different circumstances depends on the question being asked, the characteristics of the variable or attribute (i.e. the measurement scale employed), the number of observations, the degree of randomness in selecting the sample, the size of the sample and whether the data can be assumed to follow one of the standard probability distributions. If the more stringent assumptions demanded by certain tests are not known to be attained, for example in relation to complete random selection of sampled observations and normality, then other less-demanding tests may be available to answer the same type of question or if the sample is sufficiently large certain assumptions may be relaxed. Switching between one test and another in this way might require an investigator to convert variables into attributes, for example by categorizing a continuous variable into discrete, unambiguous classes. Following this chapter and before examining the details of different tests, a simple guide to selecting the appropriate statistical test is provided. This takes the form of a check list that students can use to help with examining the characteristics of their data so that the correct type test can be chosen. A project dataset will often include a combination of nominal attributes, and ordinal, interval or ratio scale variables and the analysis focuses on considering one in relation to another. For example, whether men and women have different fear of crime when in a shopping centre, whether arable or pastoral farming produces a greater build up of nitrates in rivers and streams and whether organizations with or without a recycling policy are more sustainable. Therefore, it is important that the check list is treated as a starting point for planning the data collection and statistical analysis in a research project. All too often investigators fired with enthusiasm for their topic will rush into the data-collection phase without thinking ahead to how the data collected will later be analysed. You cannot know in advance what the data values will be, but you can plan how you will analyse them once they have been collected.

Box 5.10: Common sequence of stages in application of statistical tests.

State Null and Alternative Hypotheses.

Select the appropriate type of statistical test.

State level of significance to be applied.

Calculate appropriate test statistic (usually obtained from statistical or spreadsheet software).

Decide whether there is any certain, undisputed reason for believing that the difference being tested should be in one direction or the other (e.g. whether one sample should necessarily have a greater mean than another); known as a one-tailed test.

Determine the probability associated with obtaining the calculated value of the test statistic or a larger value through chance (usually provided by statistical software, but not necessarily by spreadsheet programs).

Decide whether to accept or reject the Null Hypothesis. (Note: if the Null Hypothesis is rejected the Alternative Hypothesis is necessarily accepted.)

Each statistical test should be applied to address a specific type of question, for example to determine whether two samples of the same type of observation from different locations or times (e.g. soil samples from the side slopes and floor of a valley) are significantly different from each other in respect of the variable(s) under investigation (e.g. pH, particle size, organic content, etc.) Nevertheless, there is a common sequence of stages or tasks involved in applying most tests as shown in Box 5.10. Most of the statistical tests examined in the next two chapters follow this sequence and any exceptions will be highlighted.

Section 2
Testing times

6

Parametric tests

Parametric tests comprise one of the main groups of statistical techniques available to students of Geography, Earth and Environmental Science and related disciplines. Different tests are examined according to the numbers of samples and/or variables involved and whether measures of central tendency or dispersion are the focus of attention. Equivalent spatial statistics are included that focus on investigating the randomness or otherwise of the patterns formed by the distribution of spatial features capable of being represented as points. The computation of confidence limits to estimate population parameters from sample data is also examined.

Learning outcomes

This chapter will enable readers to:

- apply parametric statistical tests to variables measured on the ratio or interval scales;

- reach decisions on Null and Alternative Hypotheses according to appropriate levels of significance;

- undertake parametric procedures with spatial and nonspatial data;

- plan statistical analyses using parametric tests relating to spatial and nonspatial data as part of an independent research investigation in Geography, Earth Science and related disciplines.

Practical Statistics for Geographers and Earth Scientists Nigel Walford
© 2011 John Wiley & Sons, Ltd

6.1   Introduction to parametric tests

There are a number of different ways of introducing students of Geography, Earth Sciences and cognate subjects to the array of statistical techniques that are available for use in their investigations. The choice involves deciding how to steer a path through the many and varied tests and other techniques in a way that enables students to make a sound selection of which to apply in their own research projects and to be aware of how other more advanced procedures should be used in anticipation of a potential requirement to undertake other research subsequently. Having determined that this chapter will concentrate on parametric tests and the next on nonparametric ones, the question remains of how to subdivide each of these major groups into 'bite-size chunks'. The approach adopted here is to group the tests according to whether one, two or three or more variables are under scrutiny and within these sections to examine procedures separately that compare means and variances or standard deviations.

A particular issue in relation to introducing quantitative statistical analysis to geographers and other Earth scientists concerns how to include the group of techniques known as spatial statistics. Should these be relegated (or elevated) to a separate chapter or included alongside those similar but, nonetheless, fundamentally different procedures dealing with nonspatial data. The latter seems preferable, since although spatial statistics have some distinctive features their essential purpose of enabling researchers to determine whether spatial patterns are likely to have resulted from random or nonrandom processes is similar to the aim of equivalent nonspatial techniques. The main difference between spatial and nonspatial statistics is that the former treat the location of the phenomena under investigation, measured into terms of their position within a regular grid of coordinates, as a variable alongside other thematic attributes and variables (see Chapter 2).

One key assumption of parametric statistical tests is that the variables being analysed follow the Normal Distribution in the parent population of phenomena. This assumption of normality in the parent population means that the values of a given variable display the same, or at least very nearly the same, frequency distribution and therefore probabilities that are associated with the normal curve (see Chapter 5). In these circumstances, the general purpose of testing a set of values for a variable in a sample is to determine whether their distribution and the descriptive statistics calculated from them are significantly different from the normally distributed parent population. Parametric tests are only applicable to continuous variables measured on the interval or ratio scales. However, any dataset of variables can of course include some that adhere to this assumption of normality, whereas others fall short. It is therefore likely that your analysis will involve a combination of parametric and nonparametric techniques. Variables that are normally distributed may need to be combined with others that fail to meet this criterion or with attributes describing characteristics measured on the nominal scale. Chapter 5 explored ways of classifying the data values of continuous variables into groups or categories so that they can be analysed alongside other types of data.

The focus in this chapter is on parametric statistical tests that progress through different situations according to the number of samples under investigation (one, two or three plus samples). For example, the next section examines the Z and t parametric tests as applied to a single variable measured in respect of one sample of observations selected from a population. Each of the procedures relates to examining a single variable, although the same techniques may be applied to more than one variable in any particular project. You might wonder why comparable techniques jointly analysing two or more variables are not also considered here. Why might it not be appropriate, for example, to test the means of two, three or more variables for a single sample with respect to their corresponding parameters in the same parent population? There is no reason at all, but this would not imply that they were tested together, rather that they were examined concurrently or sequentially. What, for example, would be the point of comparing the mean of length of commuter journey with distance between home and workplace? The data values for one variable would be in hours/minutes and the other in kilometres. Joint analysis of these variables only makes any sense if the research questions and hypotheses are about whether the variables are related to or associated with each other. In other words whether the pairs of values display a pattern in the way they vary along their respective measurement scales. A different set of statistical techniques are required to address such questions, which are explored in Chapters 8 and 9.

6.2 One variable and one sample

It would be very unusual for a project dataset to contain only one variable, although it is quite common to focus on a single sample of observations, especially in an investigation where time and other resources are limited. Our exploration of parametric tests therefore starts by looking at the relatively straightforward situation of asking whether the data values of one continuous interval or ratio scale variable measured for a sample of observations are representative of their parent population, whose values are assumed to follow the normal probability distribution. Even when data for several variables have been collected and the ultimate aim is to examine relationships between them, it is often useful as a first stage in the analysis to explore the information about one variable at a time. The differences under scrutiny in this situation are between the descriptive statistics (mean, variance and standard deviation) for the sample and the equivalent actual or hypothesized quantities for the parent population. We have already seen that the statistical characteristics of a population, its mean variance, standard deviation, etc. are often difficult or impossible to determine and so it is necessary to hypothesize what these might be by reference to theory. This is similar to determining the *a priori* probabilities associated with tossing a coin, throwing a die or some other similar event considered in Chapter 5, which indicate the expected mean, variance and standard deviation of a series of trials carrying out this event. Thus, if a die was thrown 120 times, we would expect that each face would land facing upwards 20 times (i.e. 20 ones, 20 twos, etc.) giving a mean of 3.5 and their variance and standard deviation would both be 2.94 and 1.71, respectively. These situations deal with discrete or integer outcomes and when dealing with continuous variables it is sometimes more difficult to theorize about what the population parameter might be. However, it may be possible to hypothesize about the direction of change: for example, the effect of irradiation from the sun during daytime can reasonably be expected to increase ambient temperature and result in some melting of glacial ice leading to an increase in the flow rate of a melt water stream. It would therefore be possible to hypothesize that quantity of water would be higher in the afternoon than the early morning.

Some investigations divide a single sample into two discrete parts, for example research on new treatments for medical conditions often divides sampled observations into those receiving the new drug and those receiving a placebo. This division is not the same as selecting two different samples and comparing them, since the basis of the Null and Alternative Hypotheses is that the sampled phenomena are divided in two parts or measured twice. Similar division of a sample may also be useful in geographical and Earth scientific research. A random sample survey of **all** individuals in a town might divide the respondents into males and females when examining attitudes towards a proposed new shopping centre. A random sample of **all** occurrences of a particular plant species might be divided into those above or below 250 m altitude. The point is that the whole sample was selected at random and then divided into two parts, which is different from random sampling two sets of observations. A single sample can also be measured before and after an event, for example soil samples might be weighed before and after heating to a certain temperature in order to determine if there is a significant change in the quantity of organic matter present.

6.2.1 Comparing a sample mean with a population mean

Data values for a sample of observations will almost always include an unknown quantity of sampling error, which can cause the sample mean to under- or overestimate the true population mean. When carrying out an investigation using simple random sampling with replacement (see Chapter 2), the sample of observations selected is but one of a potentially infinite number of samples that could have been obtained. If there were sufficient time and resources available, and the characteristics of interest of the observations were stable over time, then many random samples could be selected and the means of any given variable be calculated. These individual sample means could be added together and divided by the number of samples to estimate the population mean. This value is known as the **grand mean** or the 'mean of the means'. In theory, the greater the number of sample means added into this calculation, the closer the grand mean comes to the true population mean, but the individual means are still likely to be different from each other and there will be some dispersion or spread in the sample mean values: it is extremely unlikely they will all be the same. This dispersion of the means can be quantified by a statistic known as the **standard**

error of the mean, which can be thought of as the standard deviation of the sample means. If a large number of sample means were to be viewed as a frequency distribution, this **sampling distribution of the mean** would follow the normal curve if the distribution of data values in the parent population are normal and would be approximately normal if the population values are skewed. In addition, larger sample sizes result in closer approximation to the Normal Distribution irrespective of the form of the frequency distribution of population values. In practice, an infinite number of samples cannot be selected from a population in order to find the true standard error of the mean, therefore its value needs to be estimated from just one set of sample data.

These features define the Central Limit Theorem, which provides the theoretical basis of the first of our parametric tests, known as the Z test. When simple random sampling of an independent set of observations has been carried out, the Central Limit Theorem applies and the probabilities associated with the Normal Distribution can be used to answer the question: what is the probability that the difference between the sample mean and population mean has occurred through chance. The Z test statistic is calculated using a similar equation to the one given in Section 5.4.3 for converting raw data values into Z scores (see Box 6.1). Because the sample mean has been obtained from a set of observations chosen at random, if the probability of difference is less than or equal to the level of significance (e.g. less than or equal to 0.05 or 5 per cent; or 0.01 or 1 per cent; etc.) then the Alternative Hypothesis would be accepted. Conversely, if the probability is greater than the level of significance, the Null Hypothesis should be regarded as valid.

Figure 6.1 helps to clarify the situation. It shows three samples of observations selected from a normally distributed parent population with a mean of 13.4 and a standard deviation of 3.9 and the data values in these three sets, which yield different means for variable X. The number of data values in the samples has been limited to 10 in the interests of clarity. According to the Central Limit Theorem these are just three of an infinite number of samples that could have been selected. The mean for sample A is less than the population or grand mean, for sample B it is a little higher and for sample C it is a lot higher, the corresponding differences are −1.1, 0.2 and 3.6. The means have been converted into Z scores and the probabilities associated with these scores may be obtained from Z distribution tables (see Chapter 5) or from suitable statistical analysis software. In this case the Z test values of the differences between the means for samples A, B and C and the population are 1.435, 2.735 and 6.135 and the corresponding probabilities are, respectively, 0.812, 0.436 and 0.002. These probabilities indicate that with reference to the 0.05 level of significance the difference between the mean for sample C and the population or grand mean is considerably larger than might be expected to have occurred by chance. This implies that a Z test value as large as 6.135 might be expected to occur with only two samples in 1000. The usual levels of significance correspond to certain Z scores or standard deviation units, thus 0.05 is ±1.96, 0.01 is ±2.57 and 0.001 is ±3.303. So the difference between sample C's mean and the population mean is one of those relatively unusual cases in one of the tails of the normal distribution lying beyond 2.57 standard deviations from the mean. Because there was no theoretical basis for believing that a sample

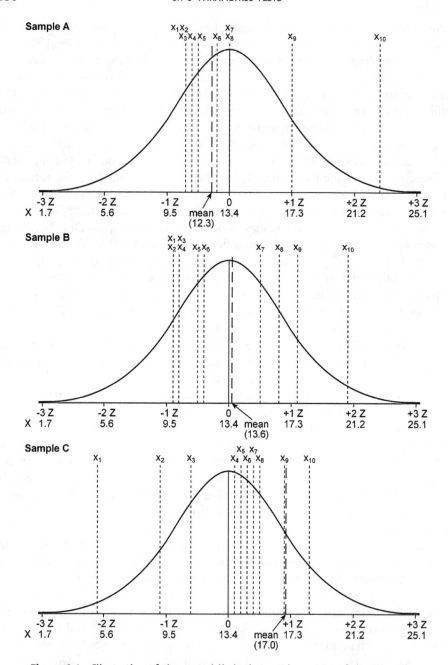

Figure 6.1 Illustration of the central limit theorem in respect of three samples.

mean should be more or less than the population mean, the level of significance probability (0.05) is split equally between the tails with 0.25 in each. The probability of the Z test statistic for sample C is a lot less than 0.05 and the difference is therefore considered statistically significant, whereas those between samples A and B, and the population mean are not, because they are so small that they could easily have occurred by chance in more than 5 samples in 100. In fact, their differences (1.1 and 0.2) may be expected to occur in 81 and 44 samples in 100 respectively.

Why is it useful to be 95 per cent sure that the difference between the mean of sample C and the population mean is significant, in contrast to the nonsignificant differences of samples A and B? This information is useful in two ways: first, it implies that the observations in sample C are to some extent different from the population in respect of their data values for the variable; and second that samples A and B are more typical of the population, at least as far as their means for this particular variable are concerned. Suppose that by chance sample C included a disproportionately large number of recent additions to the parent population, for example new households that had come to live on a housing estate during the last five years. An alternative interpretation can now be made of the statistically significant difference between the sample and the parent population in respect of variable X. Suppose that variable X measured gross annual household income, we might now conclude that over the five-year period new households had contributed to raising the economic status of the estate.

Reflect further on this example to suggest other ways of interpreting these results.

All statistical tests make certain assumptions about the characteristics of the data to which they may be applied. Some of the assumptions should be more strictly adhered to than others. The assumptions relating to the Z test given in Box 6.1c are seemingly so stringent as to make application of the test a rarity. The demands of the Z test are such that its application in the social and 'softer' natural sciences, such as Geography and Environmental Science, is less than in other areas. However, this is not to say that it lacks potential, since it might be regarded as the 'gold standard' of statistical testing and the possibility of relaxing some of the assumptions allows it to be used in certain circumstances. It also forms a useful starting point for examining other less-demanding tests.

One important drawback of the Z test is that it requires an investigator to know the population standard deviation so that the standard error of the mean can be calculated. Although the population mean may be hypothesized or determined by reference to other information sources, it is often impossible to know the value of the population standard deviation since this has to be calculated from the raw data values. Fortunately, an alternative test exists for use in these situations, known as the t test or more formally Student's t test after William Gosset the statistician who published it under the pseudonym Student in 1908. It substitutes the sample standard deviation

Box 6.1a: The Z test.

Z test statistic: $\dfrac{|\bar{X} - \mu|}{\sigma / \sqrt{n}}$

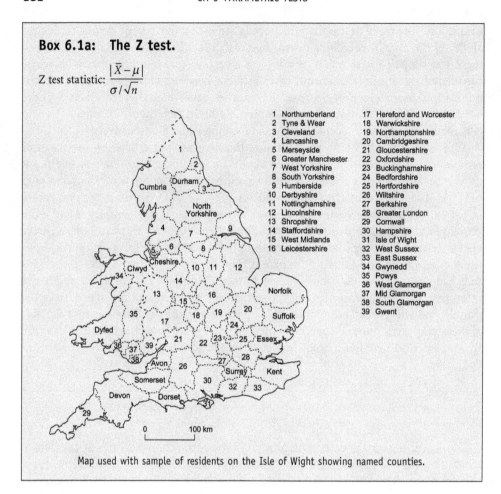

1	Northumberland	17	Hereford and Worcester
2	Tyne & Wear	18	Warwickshire
3	Cleveland	19	Northamptonshire
4	Lancashire	20	Cambridgeshire
5	Merseyside	21	Gloucestershire
6	Greater Manchester	22	Oxfordshire
7	West Yorkshire	23	Buckinghamshire
8	South Yorkshire	24	Bedfordshire
9	Humberside	25	Hertfordshire
10	Derbyshire	26	Wiltshire
11	Nottinghamshire	27	Berkshire
12	Lincolnshire	28	Greater London
13	Shropshire	29	Cornwall
14	Staffordshire	30	Hampshire
15	West Midlands	31	Isle of Wight
16	Leicestershire	32	West Sussex
		33	East Sussex
		34	Gwynedd
		35	Powys
		36	West Glamorgan
		37	Mid Glamorgan
		38	South Glamorgan
		39	Gwent

Map used with sample of residents on the Isle of Wight showing named counties.

Box 6.1b: Application of the Z test.

The Z test requires that values for the population mean (μ) and standard deviation (σ) can be specified by the investigator. The equation for the Z test statistic is a simple adaptation of the one used to convert individual data values into Z scores outlined in Chapter 5. The upper part of the Z statistic equation (the numerator) calculates the difference between the sample mean and the population mean. The lower part (the denominator) is the population standard deviation divided by the square root of the number of observations in the sample, which produces a value that is an unbiased estimate of the standard error of the mean ($s_{\bar{x}}$) when the population can be considered infinitely large. The distribution of the Z statistic follows the standardized normal curve and so the probabilities associated with different values of Z can be found in order to reach a decision on the Null and Alternative Hypotheses. The Z test is used to discover the probability of having obtained the sample mean rather than the population mean value. This probability helps an investigator decide whether the difference between the sample and population means has arisen through chance or whether there is something unusual or untypical about the particular set of sampled observations.

The worked example refers to a random sample of residents on the Isle of Wight (an island off the south coast of England), who were presented with a map showing the names of 53 counties in England and Wales. Each respondent was asked to rank the top five counties where they would like to live if they had a free choice. These ranks for the 30 counties selected by the 26 respondents (23 were not chosen) were aggregated to produce a mean score per county. For example, the sum of the ranks for Devon was 46, which produces a mean rank of 1.769 (46/30). These average scores are listed in the columns headed x_n below. The subscript n has been used to indicate that the names of the counties were shown on the map in contrast with a similar survey that was carried with maps just showing the county boundaries that will be introduced later. How do we know what the population mean and standard deviation, which are necessary to apply the Z test, might be in this case? Each survey respondent had 15 'marks' to distribute across 53 counties (5, 4, 3, 2 and 1 ranks). If these were distributed randomly, the population mean score per county for Isle of Wight residents would be 0.283 (15/53). The population standard deviation is, as usual, a little more difficult to determine and has been estimated by generating random total scores per county within the range 0 to 26, representing the extremes of nonselection (0) and any one county as fifth favourite by all respondents (26). This produced a population standard deviation of 0.306. The Z test results indicate that, on the basis of the sample, Isle of Wight residents have distinct (nonrandom) preferences for certain counties in England and Wales as places to live.

The key stages in carrying out a Z test are:

State Null Hypothesis and significance level: the difference between the sample and population mean in respect of the preferences of Isle of Wight residents for living in the counties of England and Wales has arisen through sampling error and is not significant at the 0.05 level of significance.

Calculate test statistic (Z): the calculations for the Z statistic are given below.

Select whether to apply a one- or two-tailed test: in this case there is no prior reason to believe that the sample mean would be larger or small than the population mean.

Determine the probability of the calculated Z: the probability of obtaining $Z = 2.163$ is $p = 0.015$.

Accept or reject the Null Hypothesis: the probability of Z is <0.05, therefore reject the Null Hypothesis. By implication the Alternative Hypothesis is accepted.

Box 6.1c: Assumptions of the Z test.

There are five main assumptions:
  Random sampling should be applied.
  Data values should be independent of each other in the population (i.e. the data value of the variable for one observation should not influence the value for any another).
  The data values of the variable in the population should follow the normal curve (or at least approximately). Near normality may be achieved if the sample is sufficiently large, but this raises the question of how large is sufficiently large.
  The test ignores any difference between the population and sample standard deviations, which may or may not contribute to any difference between the means that is under scrutiny.
  The need to know the population standard deviation might be difficult to achieve.

Box 6.1d: Calculation of the Z test statistic.

Counties	Rank sum	x_n	Counties ...	Rank sum ...	x_n ...
Avon	11	0.423	Kent	19	0.423
Bedfordshire			Lancashire		
Berkshire			Leicestershire		
Buckinghamshire	2	0.077	Lincolnshire		
Cambridgeshire			Merseyside		
Cheshire	4	0.154	Mid Glamorgan		
Cleveland			Norfolk		
Clwyd	5	0.192	North Yorkshire	4	0.154
Cornwall	38	1.462	Northamptonshire	1	0.038
Cumbria	8	0.308	Northumberland	1	0.038
Derbyshire			Nottinghamshire		
Devon	46	1.769	Oxfordshire	11	0.423
Dorset	33	1.269	Powys		
Durham			Shropshire	3	0.115
Dyfed	2	0.077	Somerset	10	0.385
East Sussex	3	0.115	South Glamorgan		
Essex	7	0.269	South Yorkshire		
Gloucestershire	14	0.538	Staffordshire	1	0.038
Greater London	21	0.808	Suffolk	2	0.077
Greater Manchester			Surrey	27	1.038
Gwent			Tyne & Wear		
Gwynedd	4	0.154	Warwickshire	5	0.192
Hampshire	19	0.731	West Glamorgan		
Hereford & Worcestershire	3	0.115	West Midlands		
Hertfordshire	3	0.115	West Sussex	3	0.115

Counties	Rank sum	x_n	Counties ...	Rank sum ...	x_n ...
Hertfordshire			West Yorkshire		
Humberside	11	0.731	Wiltshire	5	0.192

$$\sum x_n = 12.115$$

Mean	$\bar{x}_n = \dfrac{12.115}{30} = 0.404$
Standard error of the mean	$s_{\bar{x}} = \dfrac{\sigma}{\sqrt{n}} = \dfrac{0.306}{\sqrt{30}} = 0.056$
Z statistic	$\dfrac{\lvert \bar{x} - \mu \rvert}{\sigma / \sqrt{n}} = \dfrac{\lvert 0.404 - 0.283 \rvert}{0.056} = 2.163$
Probability	$p = 0.015$

into the standard error equation in order to estimate this statistic, which is then used to produce the t test statistic in a similar fashion to calculating Z. This change potentially introduces additional sampling error that varies according to sample size, since inequality between the sample and population standard deviations will produce different values for the standard error. This means that the shape of the t distribution curve, which although symmetrical about the mean like the (standardized) normal curve varies slightly with sample size. The difference between the standard error calculated from a small sample and that from the parent population is likely to be larger than for a sample with more observations.

The probabilities associated with different Z values are independent of sample size. Thus, the Z test statistic and the associated probability obtained for sample C above (6.135 and 0.002, respectively) would be obtained irrespective of sample size. However, from a theoretical perspective there are as many t distribution curves as there are sample sizes. In practice, the larger the sample size, the closer the t distribution curve matches with the standardized normal curve. Small sample sizes lower the peak of the t distribution curve and raise its tails in comparison with the Normal Curve (i.e. there is a change in kurtosis). Samples with larger numbers of observations are likely to produce an estimate of the standard error that is closer to the population value, because it is calculated from a higher proportion of observations in the population, and in these circumstances the t distribution provides a closer approximation to the Normal Curve. These characteristics cause the probabilities associated with different values of t to depend on the sample size, whereas with the Z statistic they are fixed. For example, the probability associated with a Z statistic of 1.96 is always 0.05 (0.025 in each tail of the distribution), whereas the probability of a t statistic of 1.96 is 0.056 when there are 50 sampled observations, 0.062 with 25 and 0.082 if there are 10. Thus, the t test statistic needs to be larger than the Z test equivalent in order to achieve the

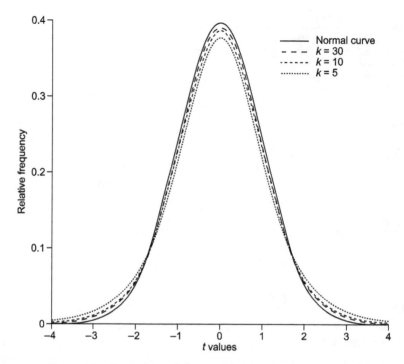

Figure 6.2 Comparison of the Normal and t distribution curves.

same level of significance. Figure 6.2 helps to clarify the differences between the Z and t distributions. There are three t distribution curves corresponding to different samples sizes (k). These can be compared with the solid line of the standardized normal curve, which does not vary in this way. A higher value of the t statistic is required for smaller sample sizes to achieve the same level of significance (0.05), 2.045 when there are 30 observations, 2.145 with 15 and 2.262 with 10. The t value associated with a probability of 0.05 approaches the equivalent Z values as sample size increases and is the same (to 3 decimal places) when sample size equals 4425 observations.

This link between sample size and t distribution probabilities introduces the concept of **degrees of freedom** (df). This refers to the number of data values in a sample that are free to vary, subject to the overall constraint that the sample mean, standard deviation or any other statistic calculated for a particular variable are determined. This concept often causes some confusion at first. After all, if you have been out in the field or laboratory and collected some data for a sample of observations, you have carefully entered the data into statistical software and produced the sample mean of your variables, surely there is no freedom for any of them to vary. The concept of degrees of freedom can be considered from a rather more theoretical perspective, which originates from the random selection of observations for entry into the sample. Suppose there was a sample containing five randomly selected observations (labelled C, H, M, P and T), which have the values 0.78, 0.26, 1.04, 0.81 and 0.76 for one of the measured variables. The mean of these values is 0.73. Now, imagine that the process of random selection had not selected observation P but one labelled J instead and its value for the variable is 1.06 rather than 0.81. Now, the five values produce a mean of 0.78. Looking at this example in another way, if the mean of the variable is truly 0.73, not just in the sample but in the population as well, then the total sum of five data values must equal 3.65. In order to achieve this total, four of the observations can take on any value, but once these are known then the fifth value is fixed. Hence there are four degrees of freedom. In this example, where there is just one variable the degrees of freedom equal the number of sampled observations minus one $(df = n - 1)$, whereas if two or more variables are being analysed they will be calculated differently. Returning to the t distribution curves in Figure 6.2 for 30, 10 and 5 sampled observations in reality these correspond to the situations where there are 29, 9 and 4 degrees of freedom respectively.

Box 6.2a: The t test.

t test statistic: $= \dfrac{|\bar{X} - \mu|}{s/\sqrt{n}}$

Box 6.2b: Application of the t test.

The t test requires knowledge of the population mean (μ) but substitutes the sample standard deviation (s) for the population equivalent used in the Z test. The numerator of the t test statistic equation is identical to the one for the Z test. The denominator is very similar with the sample standard deviation substituting for the population parameter to estimate the standard error. The t distribution varies with sample size and degrees of freedom ($df = n - 1$) and probabilities associated with different values of t and (df) enable a decision to be reached between the Null and Alternative Hypotheses. The purpose of the t test is essentially the same as the Z test, namely to decide the probability of having obtained an empirical sample mean and whether the difference between it and the known or hypothesized population mean has arisen through chance or sampling error.

The t test has been applied below to the same example used for the Z test in Box 6.1 in order to allow comparison between the results of the two tests. The population mean score per county for Isle of Wight residents is again 0.283, but the use of the sample standard deviation (0.456) to calculate the standard error of the mean produces a different value for the t test statistic. The degrees of freedom in this example are 29 (30-1): the number of counties chosen by any of the respondents minus 1. The t test results support those of the Z test, namely that the sampled respondents reveal Isle of Wight residents to have distinctly (non-random) preferences for living in certain English and Welsh counties.

The key stages in carrying out a t test are:

State Null Hypothesis and significance level: the difference between the sample and population mean in respect of the preferences of Isle of Wight residents for living the counties in England and Wales has arisen through sampling error and is not significant at the 0.05 level of significance.

Calculate test statistic (t): the calculations for the t statistic are given below.

Select whether to apply a one- or two-tailed test: in this case there is no prior reason to believe that the sample mean would be larger or small than the population mean, so a two-tailed test is required.

Determine the probability of the calculated t: the probability of obtaining t = 2.942 is $p = 0.006$ with 29 degrees of freedom.

Accept or reject the Null Hypothesis: the probability of t is <0.05, therefore reject the Null Hypothesis. By implication the Alternative Hypothesis is accepted.

Box 6.2c: Assumptions of the t test.

There are five main assumptions:

Random sampling should be applied.

Data values should be independent of each other in the population

The variable should be normally distributed, although it may be difficult to determine and low to moderate skewness is generally permissible.

A highly skewed distribution is likely to achieve a poor fit between the theoretical t distribution and the sampling distribution of the mean. A sufficiently large sample may help to overcome this difficulty.

A negatively or positively skewed distribution affects the symmetry of the t distribution and the equal division of probabilities either side of the mean.

There are occasions when a single sample of observations can be divided into two parts or be measured twice to produce a paired set of data values. In these circumstances a special version of the t test can be used, known as the paired sample t test, to investigate whether the difference between the means of the two sets of data values is significant. In the Box 6.3 the temperature measurements for the melt water stream from Les Bossons Glacier in the French Alps are treated as a single sample of observations from one water course. These measurements can be paired because they were

Box 6.2d: Calculation of the t test statistic.

Counties	x_n	$(x_n - \bar{x}_n)^2$	Counties ...	x_n ...	$(x_n - \bar{x}_n)^2$ ...
Avon	0.423	0.038	Kent	0.423	0.252
Bedfordshire			Lancashire		
Berkshire			Leicestershire		
Bucks	0.077	0.023	Lincolnshire		
Cambridgeshire			Merseyside		
Cheshire	0.154	0.006	Mid Glamorgan		
Cleveland			Norfolk		
Clwyd	0.192	0.001	North Yorkshire	0.154	0.006
Cornwall	1.462	1.520	Northamptonshire	0.038	0.036
Cumbria	0.308	0.006	Northumberland	0.038	0.036
Derbyshire			Nottinghamshire		
Devon	1.769	2.374	Oxfordshire	0.423	0.038
Dorset	1.269	1.083	Powys		
Durham			Shropshire	0.115	0.013
Dyfed	0.077	0.023	Somerset	0.385	0.024
East Sussex	0.115	0.013	South Glamorgan		
Essex	0.269	0.002	South Yorkshire		
Gloucestershire	0.538	0.096	Staffordshire	0.038	0.036
Greater London	0.808	0.335	Suffolk	0.077	0.023
Greater Manchester			Surrey	1.038	0.656
Gwent			Tyne & Wear		
Gwynedd	0.154	0.006	Warwickshire	0.192	0.001
Hampshire	0.731	0.252	West Glamorgan		
Hereford & Worcestershire	0.115	0.013	West Midlands		
Hertfordshire	0.115	0.013	West Sussex	0.115	0.013
Hertfordshire			West Yorkshire		
Humberside	0.731	0.038	Wiltshire	0.192	0.001

$$\sum x_n = 12.115 \quad \sum (x_n - \bar{x}_n)^2 = 6.017$$

Mean

$$\bar{x}_n = \frac{12.115}{30} = 0.404$$

Standard deviation

$$s = \sqrt{\frac{\sum (x_n - \bar{x}_n)^2}{(n-1)}} = \sqrt{\frac{6.017}{(30-1)}} = 0.456$$

Standard error of the mean

$$s_{\bar{x}} = \frac{s}{\sqrt{n}} = \frac{0.456}{\sqrt{29}} = 0.085$$

t statistic

$$\frac{|\bar{x}_n - \mu|}{s/\sqrt{n}} = \frac{|0.404 - 0.283|}{0.085} = 2.942$$

Probability

$$p = 0.006$$

taken at the same time of day and it is reasonable to argue that they represent two sets of data values collected simultaneously for the same stream. The Null Hypothesis in the paired t test usually maintains that the difference between the sample means in the population of observations is zero, in other words no difference exists, and any that has arisen in a particular set of sample data is the result of sampling error. Thus, the test seeks to discover whether this sampling error is so small that it might be attributed to chance or large enough to indicate a significant difference between the two occasions or two parts of the observations when they were measured.

Identify other situations in which pairs of measurements might be made in respect of geographical, environmental or geoscientific phenomena. Remember, you are looking for examples of where the sample of observations can be split into two parts or measured twice to obtain a paired set of data values.

Box 6.3a: The paired sample t test.

$$\text{t test statistic:} = \frac{|\bar{d} - 0|}{s_d / \sqrt{n}}$$

Box 6.3b: Application of the paired t test.

The paired t test normally hypothesizes that the population mean (μ) of the differences (d) between the paired data values equals zero and uses the standard deviation of this sample of differences (s_d) to estimate the population standard error. The degrees of freedom are calculated in the same manner as for the 'standard' t test ($df = n - 1$) and the fate of the Null and Alternative Hypotheses is decided by comparing the probability of the paired t test statistic with the level of significance, usually ≤ 0.05. The purpose of the paired t test is to discover the probability of having obtained the difference in means between the paired values and to reach a decision on whether this is significantly different from zero (no difference).

The application of the paired t test below relates to the paired measurements of temperature in the melt water stream flowing from an alpine glacier at the snout (x_s) and the exit of the outwash plain (x_e). The test proceeds by calculating the difference between the two sets of temperature measurements, which have been made at the same time of day and treat the stream as a single water course. Not surprisingly, the test results seem to confirm the Alternative Hypothesis that the temperature of the water near the glacier snout is lower than near the exit from the sandur plain. Since the test is examining the differences between the measurements irrespective of where they were made, the test results will be the same if the pairs of values were reversed (i.e. higher values near the snout and lower ones at the exit). It is only through interpretation of the results by the investigator that meaning can be attached to the outcome of the test.

The key stages in carrying out a t test are:

State Null Hypothesis and significance level: the mean of the differences between the pairs of water temperature measurements in Les Bossons Glacier stream and zero has arisen through sampling error and is not significant at the 0.05 level of significance.

Calculate test statistic (t): the calculations for the paired t statistic are given below.

Select whether to apply a one- or two-tailed test: in this example it is reasonable to argue that water temperature closer to the glacier snout will be lower than those recorded further away and therefore a one-tailed would be appropriate, which involves halving the probability.

Determine the probability of the calculated t: the probability of obtaining t = 7.425 is 0.000000992 with 17 degrees of freedom.

Accept or reject the Null Hypothesis: the probability of t is >0.05, therefore reject the Null Hypothesis. By implication the Alternative Hypothesis is accepted.

Box 6.3c: Assumptions of the paired t test.

There are four main assumptions are the same as for the standard t test (see Box 6.2c).

Box 6.3d: Calculation of the paired t test statistic.

	x_s	x_e	$x_e - x_e = d$	$(d - \bar{d})^2$
09.00 hrs	4.10	4.30	−0.20	10.43
09.30 hrs	4.60	4.70	−0.10	11.09
10.00 hrs	4.60	4.80	−0.20	10.43
10.30 hrs	4.70	5.20	−0.50	8.58
11.30 hrs	4.80	6.70	−1.90	2.34
12.00 hrs	5.20	10.10	−4.90	2.16
12.30 hrs	5.40	10.50	−5.10	2.79
13.00 hrs	6.30	11.20	−4.90	2.16

	x_s	x_e	$x_e - x_e = d$	$(d - \bar{d})^2$
13.30 hrs	6.90	11.40	−4.50	1.14
14.00 hrs	7.40	11.80	−4.40	0.94
14.30 hrs	7.10	12.30	−5.20	3.13
15.00 hrs	6.90	11.90	−5.00	2.46
15.30 hrs	6.30	11.40	−5.10	2.79
16.00 hrs	6.10	11.10	−5.00	2.46
16.30 hrs	5.50	10.20	−4.70	1.61
17.00 hrs	5.50	9.30	−3.80	0.14
17.30 hrs	5.30	8.80	−3.50	0.00
18.00 hrs	5.10	7.80	−2.70	0.53
			$\sum d = -61.70$	$\sum (d - \bar{d})^2 = 65.22$

Mean difference

$$\bar{d} = \frac{-61.70}{18} = -3.43$$

Standard deviation

$$s = \sqrt{\frac{\sum (d - \bar{d})^2}{(n-1)}} = \sqrt{\frac{65.22}{(18-1)}} = 1.96$$

Standard error of the mean

$$s_{\bar{d}} = \frac{s}{\sqrt{n}} = \frac{1.96}{\sqrt{18}} = 0.462$$

t statistic

$$\frac{|\bar{d} - 0|}{s_d / \sqrt{n}} = \frac{|3.43 - 0|}{0.462} = 7.425$$

Probability

$$p = 0.000\,000\,992$$

6.2.2   Analysing patterns of spatial entities

Examination of the Z and t tests has illustrated that the focus of attention in a single sample of nonspatial observations, albeit ones that exist in a certain geographical context, is on whether the descriptive statistical quantities derived from the data values for measured variables are significantly different from what might have occurred through chance. When considering the spatial distribution of observations we are also concerned with whether their pattern in location is random or otherwise. Thus, what is tested, for example in the case of a group of points, is whether their observed location within a given area of space is significantly different from what would have occurred if they were randomly distributed. An area of space may be regarded as an infinite collection of points, each of which may be occupied by a discrete occurrence of a category of spatial entity (e.g. oil wells, churches, post boxes, trees, etc.) or sets of points forming linear or areal features (e.g. roads, rivers, lakes, land covers, etc.). However, from both computational and representational perspectives a particular set of such entities cannot occur everywhere, since their location is dictated by the level of specificity or precision in the grid coordinates.

There are certain types of geographical and environmental variables, such as air temperature, elevation and depth, water content and atmospheric pressure that do exist everywhere over the Earth's surface. Even some socioeconomic variables (e.g. population density and monetary value of land (or water)) also exist over most of the Earth's terrestrial surface and over some marine areas as well. These variables create spatial entities by virtue of variations in their intensity or data values over the extent of the Earth's surface. Thus, anticyclones and cyclones are features with spatial extent within the Earth's atmosphere as a result of high- and low-pressure values existing in close proximity to each other. It will usually only be possible to measure this type of variable with any degree of certainty with respect to discrete spatial units, such as pixels from remotely sensed data, and to interpolate intermediate values by modelling.

Figure 6.3 represents a 1000 m² portion of the Earth's surface superimposed with a regular grid of 2500 squares each representing 0.4 m². Some of the individual squares in the upper version of the grid (Figure 6.3a) have been shaded to indicate the locations where a species of ground-nesting migrant bird has chosen to lay its eggs. In theory, the birds have three options when choosing a nest site: select a square at random, opt for regular equidistant spacing to maximize available living space or cluster together to provide some protection against predators. Spatial statistics usually analyse a complete set or population of spatial entities, nevertheless, the occupied squares effectively represent a sample of the potentially infinite, but often practically limited, locations where the entities could be found, in this case where the birds could have nested. Suppose there had been 2500 birds flying over this portion of the Earth's surface in search of a nesting site they could each have selected to occupy their own square. However, nesting in such close proximity to each other might increase competition for food and encourage conflict between birds, and some overflying birds might have elected to carry on to another part of the Earth's surface where there was more space. Such rationalization of the birds' selection of a nest site relates to a behavioural process operating in the species.

The lower version of the grid (Figure 6.3b) represents a different scenario. The shaded squares indicate the presence of relatively high amounts of surface water on a generally flat portion of the Earth's surface adjacent to a major river channel that experiences a diurnal change in flow leading to overtopping of the banks to produce a network of braided streams. Again, the shaded squares may be considered a sample of the population of squares that might have been occupied by sufficient surface water to form streams with a continuous flow. Just as all of the squares could have become occupied by nesting birds, it is reasonable to imagine that all of them could have been inundated with water as it flowed over the bank of the river channel. However, small-scale variations in the surface, in the flow of water and scouring of the land are likely to mean that the water bifurcates to produce a braided network. Testing a particular set of locations in both these examples, whether these are individual grid squares occupied by point entities or sets of squares forming discrete linear or areal entities, involves deciding whether the observed pattern differs significantly from what would have arisen if those entities were located randomly. Underpinning these tests is the

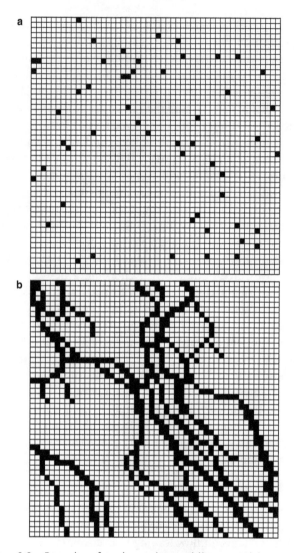

Figure 6.3 Examples of random points and linear spatial processes.

assumption that the entities are located where they are as a result of a process and the procedures seek to establish whether the process in a particular case is random or intentional. In other words, can the process be explained or understood by reference to some, perhaps unidentified, controlling factors, or is it just random.

This leads on to consideration of a group of spatial statistical techniques that are concerned with **pattern in location**. These seek to discover whether the observed pattern formed by the location of a set of spatial features is significantly different from what would have occurred if they were distributed randomly. Although not exactly the same, it is very similar to testing whether chance or sampling error accounts for a difference between a population and sample mean or variance. Many techniques coming under the broad heading spatial statistics deal with point rather than line or area patterns and these fall into two main groups: distance statistics and grid-based techniques for point sets. Both sets of procedures refer to probability distributions, some of which are used in nonspatial statistics such as the Z and t distributions, to test whether the pattern produced by a particular set of points are significantly different from a random one. Since these techniques essentially treat the counts of points in the cells of a regular grid as a frequency distribution, they are considered in Chapter 7 alongside methods dealing cross-tabulations of nonspatial data. Statistics based on distance are dealt with here.

6.2.2.1 Statistics based on distance

Techniques based on distance focus on the space between points and the extent to which they are separated from each other across space, since this potentially casts some light on the underlying process. **Neighbourhood statistics** constitute the most important group of techniques of this type and probably the most intuitive and useful of these focuses on nearest or first-order neighbours; although others such as second-order or furthest (N^{th} order) neighbour may be appropriate. The mean distance for each order in this series can be calculated up to the $(K - 1)$th order, where K represents the total number of points, which constitutes the 'global' or overall mean distance. Theoretically any distribution of points could represent the situation where they are all found at the same location (i.e. on top of each other – completely clustered), evenly or randomly spaced. These three situations translate into extreme values of the neighbour statistic R, namely: 0.0 completely clustered; 1.0 randomly distributed; and 2.1491 an evenly spaced triangular lattice. The **nearest neighbour** (NN) index (R) may be used as a descriptive measure to compare the patterns of different categories of phenomena within the same study area. For example, it may be used to investigate the patterns different species of trees in an area of woodland. Each tree of the different species is treated as being located at a specific point and the NN index provides a way of describing whether some species are more or less clustered than others. It acts as a quantitative summary of the spacing or layout of a particular set of entities, trees in this example, although comparison between two groups of either the same or different sets of entities in separate study areas is problematic. If the same set of tree species is found in another area of woodland that is larger (or smaller) than the first, then it would not be legitimate to compare their NN index values. These situations create problems because the size of the area is important when calculating the R index and the standard error, which is required for the Z statistic (see Box 6.4).

Strictly speaking a particular set of points is usually not a sample of spatial entities but the population of locations where all occurrences happen to be located (see Section 6.2.2). Nevertheless, because there are an infinite number of locations where the entities could potentially have been located, it may be necessary to discover whether the difference between the NN index value obtained from the observed set of points and that which would have been obtained if the points were randomly distributed could have arisen by chance. This difference can be tested using the Z distribution with a Null Hypothesis that R does not equal 1.0 (randomly spaced) as a result of sampling error, provided that the number of points is reasonably large (e.g. >100). The R index is the ratio between the observed and expected mean nearest-neighbour

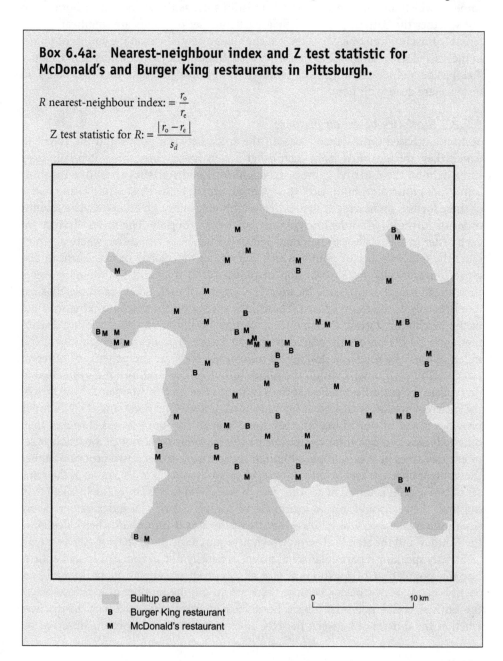

Box 6.4a: Nearest-neighbour index and Z test statistic for McDonald's and Burger King restaurants in Pittsburgh.

R nearest-neighbour index: $= \dfrac{r_o}{r_e}$

Z test statistic for R: $= \dfrac{\left| r_o - r_e \right|}{s_d}$

Builtup area
B Burger King restaurant
M McDonald's restaurant

0   10 km

Box 6.4b: Application of the nearest-neighbour index (R) and Z test statistic.

The nearest-neighbour index (R) is a summary measure describing the pattern in location of points representing a set of entities. The nearest distance (d) of each point to its closest neighbour is obtained by producing the distance matrix between all points in the set and sorting or searching the columns to determine the minimum value. The sum of these minima divided by the number of points produces the observed mean nearest-neighbour distance (r_o). The expected mean nearest-neighbour distance (r_e) is obtained from a standard equation that connects the number of points (N) and the size of the study area (A) thus, $1 / \left(2\sqrt{\dfrac{N}{A}} \right)$ – the term $\dfrac{N}{A}$ represents the density of points per unit area. The index value is the ratio between the observed and expected mean nearest distance between the points within a defined area of space.

   A given set of points under investigation is usually regarded as a population, since there is little reason for sampling spatial entities in order to analyse their pattern in location. The Z test can therefore be applied, which requires the standard error (s_d) that is obtained from the following equation $\dfrac{0.26136}{\sqrt{N(N/A)}}$. The value of Z, calculated using the definitional equation given above, and its associated probability are used to examine whether the observed mean nearest distance is different from the expected value as a result of chance or sampling error. The decision on whether to accept the Null or Alternative Hypothesis is made by reference the probability of Z in relation to the chosen level of significance.

   The mean nearest-neighbour index has been applied to the complete set of 64 McDonald's and Burger King restaurants in 816.7 km^2 of Pittsburgh with the distances and area measured in km. The distance matrix between points is tedious to calculate by hand or using standard spreadsheet or statistical software and, since there was an example in Box 4.6c relating to the points representing the location one company's restaurants, details for the full set are not provided here. The R index value indicates the points are dispersed within the continuous built-up area of the city and this value is highly significant at the 0.05 level. It would occur by chance less than 0.1 per cent of occasions. Visual scrutiny of the pattern reveals a certain amount of clustering of restaurants in parts of the urban area, but with some dispersion. The R index is reflected in an R index value of 1.69, which is somewhat lower than 1.0 (random) within the range of possible values (0.0 to 2.1491).

   The key stages in calculating the nearest-neighbour index are:

Calculate the distance matrix: this requires the distance between each point and all of the others in the set.

Calculate the observed mean nearest distance: determine the 1st order nearest distance of each point; sum these nearest neighbour distances and divide the result by the total number of points.

Calculate the expected mean nearest distance: this is obtained from the standard equation above.

Calculate the nearest-neighbour index R: this is the ratio between observed and expected mean nearest-neighbour distances.

State Null Hypothesis and significance level: the distribution of points is completely random and any appearance of clustering or regular spacing is the result chance and is not significant at the 0.05 level of significance.

Calculate test statistic (Z): the calculations for the Z statistic are given below and Z = 10.277.

Determine the probability of the calculated Z: the probability of obtaining this Z value is less than 0.001.

Accept or reject the Null Hypothesis: the probability of Z is considerably <0.05, therefore reject the Null Hypothesis and by implication accept the Alternative Hypothesis that the pattern is not likely to be the result of chance.

Box 6.4c: Tabulated nearest-neighbour distances.

Point No.	d	Point No.	d	Point No.	d	Point No.	d
1	1.3	17	25.2	33	6.1	49	1.2
2	1.3	18	1.2	34	1.7	50	1.9
3	6.2	19	8.0	35	3.9	51	0.8
4	0.9	20	1.1	36	3.2	52	1.4
5	2.9	21	9.2	37	5.8	53	1.0
6	2.2	22	1.0	38	0.9	54	2.0
7	3.0	23	2.3	39	0.9	55	3.1
8	1.3	24	1.0	40	1.2	56	4.9
9	4.4	25	3.6	41	0.8	57	4.4
10	6.6	26	2.3	42	3.0	58	3.2
11	1.1	27	5.0	43	4.1	59	5.9
12	2.5	28	1.9	44	2.1	60	3.5
13	1.0	29	2.2	45	1.0	61	3.2
14	4.6	30	1.2	46	1.0	62	2.1
15	1.1	31	2.1	47	0.7	63	1.0
16	4.0	32	2.2	48	0.6	64	1.2

$$\sum d = 191.7$$

---

distance and the Null Hypothesis therefore examines whether chance or sampling error accounts for these values being unequal.

Problems can arise with applying the NN index in two situations: if the boundary of the study area or the minimum distance is ill-suited to the particular distribution of points. The boundary of the area is likely to be problematic either when the points are within an area that includes a large amount of empty space or when the area's boundary is so tightly drawn around the entities that they are artificially inhibited

---

Box 6.4d: Calculation of the nearest-neighbour index, R, and Z test statistic.

Observed mean nearest distance	$r_o = \dfrac{\sum d}{N} = \dfrac{191.7}{64} = 3.02$
Expected mean nearest distance	$r_e = 1 \Big/ \left(2\sqrt{\dfrac{N}{A}} \right) = 1 \Big/ \left(2\sqrt{\dfrac{64}{816.7}} \right) = 1.79$
Nearest neighbour R index	$R = \dfrac{r_o}{r_e} = \dfrac{3.02}{1.79} = 1.69$
Standard error of mean nearest-neighbour distance	$s_d = \dfrac{0.26136}{\sqrt{A(A/N)}} = \dfrac{0.26136}{\sqrt{64(64/8.167)}} = 0.117$
Z statistic	$Z = \dfrac{\lvert r_o - r_e \rvert}{s_d} = \dfrac{\lvert 3.02 - 1.79 \rvert}{0.117} = 10.277$
Probability	$p < 0.001$

---

from displaying a more dispersed pattern. The sensible solution to these issues is to delimit the area with regard to the nature of the entities, for example defined with reference to continuous built-up area in the case of features distributed in 'urban space' rather than some arbitrary rectangular or circular enclosing area. Nevertheless, the boundary may be relatively arbitrary and the nearest neighbour of a feature might lie outside the defined study area. If the point locations represent dynamic features of the landscape, such as houses that are for sale or occurrences of a particular plant species, then the NN index may be subject to different values over a relatively short period of time. The minimum or nearest distance may be inappropriate if there is a series of discrete but regularly spaced groups containing a similar number of entities. In this, perhaps rather unusual, situation a higher order would be a more suitable means of analysing the overall pattern. The choice of order would be made with reference to the maximum number of points per cluster so that none of the distances used in calculating the R index came the same group of points. Another common problem associated with using a single set of distances to summarize the pattern in location of a set of points is that pairs of points are often nearest neighbours of each other, they are reflexive, which arises from the location of one being dependent upon that of another.

An alternative technique for examining patterns of points in location that addresses some of the issues associated with the NN index is **Ripley's K** function. Rather than focusing on **one** set of distances in respect of each point, for example its 1st, 2nd or $(K-1)^{th}$-order neighbour, this technique embraces all point-to-point distances. Conceptually, circles of fixed radius are drawn around each point, the number of points in each circle is counted and these counts are summed across all points. This is repeated for successive small increments of distance moving outwards from the first circle around each point in the set. The cumulative aggregate frequencies for each distance increment provides values for the function $K(d)$ where d is the radius around each point. The values of $K(d)$ can be plotted against distance to provide information on localized clustering in the set of points. Statistical testing can be applied to the $K(d)$ function, since the expected aggregate count of points for each distance zone is the overall point density multiplied by the area of circle in turn. Subtracting the expected from the observed count in each circle transforms the $K(d)$ into $L(d)$, which would equal zero for a spatially random pattern, no difference between the two counts. Thus, $L(d)$ would be approximately zero for each circle radius (d), however, the sampling distribution of $L(d)$ is unknown, in other words there is no standard probability distribution again which to test the difference between the observed and expected values.

A number of spatial statistics encounter this problem and the typical solution is to resort to **Monte Carlo simulation** to obtain a cumulative frequency distribution based on random occurrence of events. This involves simulating a set of N entirely random events (points) within a given area (A), where N and A have the same values as for the observed number of points in the region of interest and are defined by the reference coordinate space (i.e. only events possessing coordinates within the limits of region's boundary are permitted). This simulation is repeated many times (e.g. 9999) to generate probabilities by determining the maximum absolute difference between the simulated and theoretical cumulative distribution. This process may be compared with discovering the probability of events empirically (see Section 5.4). If the number of times the maximum absolute difference was greater than that between the observed and expected pattern is 499 and the number of simulation runs or trials is 9999, then:

$$p = \frac{1+499}{1+9999} = 0.05$$

This is equivalent to a 5 per cent probability of the difference having occurred through chance. Figure 6.4 shows the upper and lower confidence intervals (see Section 6.5) produced with 999 simulation trials together the observed and expected K values for the Burger King and McDonald's restaurants in Pittsburgh. The region of interest has been constrained to the continuous built-up area shown in Box 6.4a. The observed plot line lies between these limits, which indicates the overall pattern of the points' location is generally random rather than clustered or evenly spaced. This may be compared with the NN index results, which suggested the points were close to random although tending towards the clustered end of the range. Ideally, Ripley's K technique should be applied when there are at least 100 points, rather than with only 64 as in this example. Edge-correction procedures, for example weighting expected counts for the circular zones by the proportions lying inside and outside the region of interest have not been applied in this case.

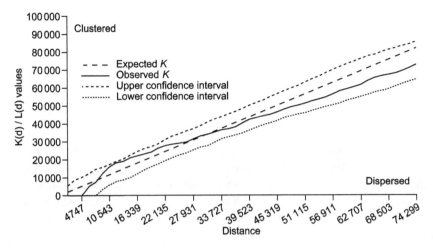

Figure 6.4 Confidence intervals resulting from Monte Carlo simulation.

6.3 Two samples and one variable

Many investigations focus on using samples of observations from two distinct populations in order to determine if they are different from or similar to each other. The next stage in our exploration of parametric statistical tests proceeds by examining such situations where an investigator has identified two populations of observations, for example outcrops of sedimentary and igneous rocks, households in a developed and developing country, or areas of predominantly coniferous and broad-leafed woodland. Again, the focus of attention is on the data values for one or more continuous interval or ratio variables capable of being measured in respect of the individual entities in these populations. These variables are assumed to adhere to the normal probability distribution and sampled observations to have been selected randomly. The differences under scrutiny by the statistical tests are between measures of central tendency and dispersion in the two samples and by implication whether they are 'genuinely' from two populations or in reality from one.

The hypothesized difference is often assumed to equal zero with the Null and Alternative Hypotheses phrased in such a way that they examine whether the two populations from which the sampled observations have been drawn are statistically different from each other. If the test results suggested that the Null Hypothesis should be accepted, the investigator should conclude that the populations are in fact the same in respect of the variable that has been examined, whereas acceptance of the Alternative Hypothesis suggests that they can be regarded as different from a statistical point of view. In other words, continuing with the example of sedimentary and igneous outcrops, if differences between mean quartz content of rock samples taken from random locations across these areas was not statistically significant, then it might be possible to conclude that there was only one population rather than two. Obviously, other variables would need to be examined to determine whether they supported this conclusion.

6.3.1　Comparing two sample means with population means

The theoretical starting point for examining the statistical tests available for answering this question is to imagine there are two populations with known means (μ_1 and μ_2) and variances (σ_1^2 and σ_2^2) in respect of a certain variable (X) where subscripts 1 and 2 denote the two populations. We have already seen that according to the Central Limit Theorem random samples with n observations selected from each of the populations would produce a series of sample means that would follow the Normal Curve. Suppose that pairs of independent random samples were drawn from the two populations, a process that could be carried on indefinitely, and the differences calculated between each pair of means in respect of the same variable. These differences could be plotted as a frequency distribution and would form the **sampling distribution of the difference in means**. Descriptive statistics (e.g. mean, variance and standard deviation) could be determined for this set of differences between the sample means for populations 1 and 2. Theoretically, the difference between the population means ($\mu_1 - \mu_2$) equals the mean of the differences obtained from the pairs of samples. However, some sampling error is likely to be present and the standard error of the differences can also be calculated.

Application of this theory to statistical testing requires knowledge of whether the sampling distribution of the difference in means follows the Normal Distribution. This is determined by two factors: the nature of the distributions of the parent populations and sample size. The sampling distribution of the difference in means will be normal in two situations: if the data values in the parent populations are normal and sample size is finite or if the parent populations are not normal but sample size is infinite. In these circumstances the Z test can be used to examine whether the observed and hypothesized difference between the means of two samples selected from populations with known variances is significantly different. However, just as in the single sample case the Z test requires knowledge of the population mean and variance, so the two sample Z test demands an investigator to know the means and variances of two populations, although the former are usually assumed to be equal with a difference of zero. It may also be difficult to achieve or determine normality in two populations and sample size will invariably be finite. An estimate of Z ($\hat{z}$) can be obtained by substituting the sample variances for the population ones if the sample sizes are sufficiently large, but this begs the question of how large is sufficiently large.

Fortunately, an alternative solution exists, namely to apply the two-sample t test, which although slightly less robust than the Z test has the advantage of using the sample variances. There are alternative versions of the two-sample t test available depending upon whether the population variances are assumed equal (homoscedastic t test) or different (heteroscedastic t test). The decision on which version of the test should be applied may be made by reference to the results of an F test of the difference between variances (see Section 6.3.2). If the population variances are known to be equal or an F test indicates that they are likely to be so, the homoscedastic version t test should be applied, which pools or combines the sample variances to estimate the standard error of the difference in means. If equality of population variances does not exist, then a heteroscedastic t test should be used. Box 6.5 illustrates the application of the two-sample t test, in this case in respect of whether residents on the Isle of Wight comprise two statistical populations in relation to their knowledge of the names of counties in England and Wales in which they might wish to live. Fortunately, statistical software will usually offer both versions of the test and the choice over which to employ simply involves carrying out a preliminary F test of difference between variances.

6.3.2 Comparing two sample variances with population variances

The tests examined so far have focused on differences relating to the mean as a measure central tendency. However, as outlined in Chapter 4, it is perfectly possible for two sets of numbers to share the same mean value but to have very different variances and standard deviations. A difference in variance between two samples from two populations with respect to a certain variable signifies some variation in the relative dispersion of their data values. But, are the samples dissimilar because of sampling error or because the populations have different characteristics? The point of departure for understanding how to test for a significant difference between two sample variances involves starting with two populations of observations that can be measured in respect of a common variable that follows the normal distribution. If their means and variances were equal to each other (i.e. $\mu_1 = \mu_2$ and $\sigma_1^2 = \sigma_2^2$), the populations would be regarded as statistically identical in respect of these parameters and the ratio between their variances is 1. Such equality of variances is unlikely to arise for populations let alone samples of 'real-world' phenomena and so one variance will nearly

---

Box 6.5a: The two-sample t test.

t test statistic: $= \dfrac{|\bar{x}_1 - \bar{x}_2|}{\sqrt{\dfrac{s_1^2}{n_1} + \dfrac{s_2^2}{n_2}}}$ unequal population variances

$= \dfrac{|\bar{x}_1 - \bar{x}_2|}{\sqrt{\dfrac{n_1 s_1^2 + n_2 s_2^2}{n_1 + n - 2}} \sqrt{\dfrac{n_1 + n_2}{n_1 n_2}}}$ equal population variances

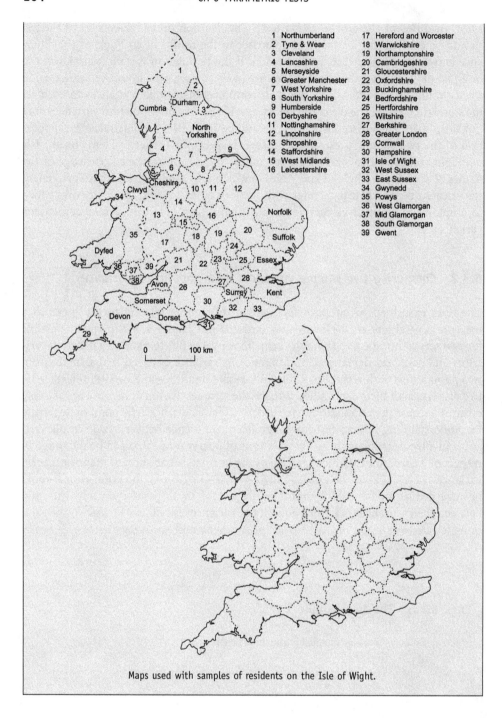

1	Northumberland	17	Hereford and Worcester
2	Tyne & Wear	18	Warwickshire
3	Cleveland	19	Northamptonshire
4	Lancashire	20	Cambridgeshire
5	Merseyside	21	Gloucestershire
6	Greater Manchester	22	Oxfordshire
7	West Yorkshire	23	Buckinghamshire
8	South Yorkshire	24	Bedfordshire
9	Humberside	25	Hertfordshire
10	Derbyshire	26	Wiltshire
11	Nottinghamshire	27	Berkshire
12	Lincolnshire	28	Greater London
13	Shropshire	29	Cornwall
14	Staffordshire	30	Hampshire
15	West Midlands	31	Isle of Wight
16	Leicestershire	32	West Sussex
		33	East Sussex
		34	Gwynedd
		35	Powys
		36	West Glamorgan
		37	Mid Glamorgan
		38	South Glamorgan
		39	Gwent

Maps used with samples of residents on the Isle of Wight.

Box 6.5b: Application of the two-sample t test.

The two-sample t test is often used to test whether the observed difference between sample means selected from two populations has arisen through sampling error and, by implication, whether the true difference is zero, since the population means are hypothesized to be equal. Alternative formulae for the standard error of the difference in means ($s_{\bar{x}_1} - s_{\bar{x}_2}$) exist for use in the two-sample t test equation depending upon whether the population variances are known or believed as a result of statistical testing to be equal. The degrees of freedom are calculated differently from the 'standard' t test ($df = (n_1 - 1) + (n_2 - 1)$), since the samples need not contain the same number of observations. A decision on the Null and Alternative Hypotheses is reached by comparing the probability of having obtained the t test statistic with the chosen level of significance, usually whether ≤0.05.

The two-sample t test has been applied to two samples of residents on the Isle of Wight who were selected at random and completed an interview survey. The respondents in one sample were shown a map of counties in England and Wales that included county names and those in the other sample were given an unlabelled map. Both sets of interviewees were asked to rank the five counties where they would like to live mark if there was no restriction. The sample seeing the labelled map named 30 counties and the other one listed 36. The samples purport to divide the Isle of Wight residents into two statistical populations on the basis of their knowledge of the administrative geography of England and Wales. The preliminary F test (see Box 6.6) indicates that the variances of the parent populations were equal and therefore the homoscedastic version of the two-sample test should be applied.

The procedure for carrying out the two-sample t test proceeds by calculating the means and variances of each sample. These calculations have been exemplified elsewhere and so Box 6.5d simply gives the raw data (mean aggregate rank scores per named (x_n) and unnamed county (x_u)) together with the values of these statistics. The standard error of the mean is calculated using a rather complicated looking equation when applying the homoscedastic version of the test. The difference between the sample means is more straightforward to determine. These values are used to produce the t test statistic. The test results indicate that the difference between the sample means is so small that it could easily have arisen by chance and that there was no significance difference in respondents' preferences for living in other counties in England and Wales in respect of whether they viewed a labelled or unlabelled map.

The key stages in carrying out a two-sample t test are:

State Null Hypothesis and significance level: the means of the sampled populations are the same and any observed difference between the samples has arisen through sampling error and is not significant at the 0.05 level of significance.

Determine whether population variances are equal or likely to be so: apply F test (see Section 6.3.2 and Box 6.6)

Calculate test statistic (t): the calculations for the two-sample t statistic with equal population variances are given below.

Select whether to apply a one- or two-tailed test: there is no prior reason for arguing that one of the samples should have a higher or lower mean than the other, therefore a two-tailed test is appropriate.

Determine the probability of the calculated t: the probability of obtaining t = 0.613 with 64 degrees of freedom is p = 0.470.

Accept or reject the Null Hypothesis: the probability of t is >0.05, therefore accept the Null Hypothesis. By implication, the Alternative Hypothesis is rejected.

Box 6.5c: Assumptions of the two-sample t test.

There are four main assumptions:
  Random sampling should be applied and observations should not form pairs.
  Data values should be independent of each other in the population
  The variable under investigation should be normally distributed in both parent populations.
  Assessment of the homogeneity or heterogeneity (similarity/dissimilarity) of population variances guides selection of the appropriate version of the two-sample test.

Box 6.5d: Calculation of the two-sample t test statistic.

Counties	x_n	x_u	Counties ...	x_n...	x_u...
Avon	0.423	0.154	Kent	0.423	0.115
Bedfordshire		0.077	Lancashire		0.154
Berkshire		0.077	Leicestershire		0.269
Bucks	0.077	0.154	Lincolnshire		0.192
Cambridgeshire			Merseyside		
Cheshire	0.154		Mid Glamorgan		
Cleveland			Norfolk		0.500
Clwyd	0.192		North Yorkshire	0.154	0.115
Cornwall	1.462	1.885	Northants	0.038	
Cumbria	0.308	0.231	Northumberland	0.038	0.077
Derbyshire		0.500	Nottinghamshire		0.038
Devon	1.769	1.115	Oxfordshire	0.423	0.115
Dorset	1.269	0.846	Powys		0.077
Durham			Shropshire	0.115	0.154
Dyfed	0.077	0.192	Somerset	0.385	0.346
East Sussex	0.115	0.385	South Glamorgan		
Essex	0.269	0.154	South Yorkshire		0.077
Gloucestershire	0.538		Staffordshire	0.038	0.077
Greater London	0.808	0.269	Suffolk	0.077	0.308
Greater Manchester		0.154	Surrey	1.038	0.192
Gwent			Tyne & Wear		
Gwynedd	0.154	0.038	Warwickshire	0.192	
Hampshire	0.731	0.692	West Glamorgan		
Hereford & Worcestershire	0.115		West Midlands		
Hertfordshire	0.115	0.192	West Sussex	0.115	0.923

Counties	x_n	x_u	Counties ...	x_n...	x_u...
Hertfordshire		0.192	West Yorkshire		
Humberside	0.731	0.154	Wiltshire	0.192	0.808
			$\sum x_n = 12.115$		$\sum x_u = 11.846$
Mean	Sample n (named counties)		$\bar{x}_n = \dfrac{12.115}{30} = 0.404$		
	Sample u (unnamed counties)		$\bar{x}_u = \dfrac{11.846}{36} = 0.329$		
Variance	Sample n (named counties)		$s_n^2 = \dfrac{\sum(x_n - \bar{x}_n)^2}{n_n - 1} = \dfrac{6.017}{29} = 0.207$		
	Sample u (unnamed counties)		$s_u^2 = \dfrac{\sum(x_u - \bar{x}_u)^2}{n_u - 1} = \dfrac{5.063}{35} = 0.145$		
Standard error of the difference in sample means			$\sqrt{\dfrac{n_n s_n^2 + n_u s_u^2}{n_n + n_u - 2}} \cdot \sqrt{\dfrac{n_n + n_u}{n_n n_u}} = \sqrt{\dfrac{30(0.207) + 36(0.145)}{30 + 36 - 2}}$ $\sqrt{\dfrac{30 + 36}{30(36)}} = 0.105$		
t statistic			$\dfrac{\lvert \bar{x}_{n1} - \bar{x}_u \rvert}{\sqrt{\dfrac{n_n s_n^2 + n_u s_u^2}{n_n + n_u - 2}}\sqrt{\dfrac{n_n + n_u}{n_n n_u}}} = \dfrac{0.075}{0.105} = 0.716$		
Degrees of freedom			$df = (n_n - 1) + (n_u - 1)$ $(30 - 1) + (36 - 1) =$ 64		
Probability			$p = 0.470$		

always be larger than the other. The ratio of the larger to the smaller variance provides a statistic, F, which quantifies the magnitude of the difference between them. A succession of samples each with a specific and possibly unequal number of observations selected from two populations may include some overlap in respect of individual entities or each pair may be entirely separate, nevertheless, they all purport to measure the variance (and mean) of their respective parent population. However, sampling error is likely to result in the sample variances being different with the larger values coming from either of the populations. The selection of pairs of such samples and calculation of their F ratios could be carried on indefinitely to produce a frequency distribution of F values. Each combination of sample sizes from two populations (e.g. 25 from population A and 50 from population B) produces a distinct asymmetrical distribution with a relatively high number of small F values and fewer large ones (see Figure 6.5) respectively indicating a high probability of similar variances and a low probability of very dissimilar ones.

The probabilities associated with these distributions form the basis for assessing a difference between sample variances using the F test. The test provides a way of determining whether the variances of the populations from which two samples have been selected may be considered equal. If the difference and therefore also the F ratio are so small that they can be regarded as having occurred through chance, then the population variances would be regarded as the same. As we have already seen, application of the F test is usually a precursor to carrying out a two-sample t test, since alternative versions of this test are used depending upon whether the population variances are equal. Calculation of the F test statistic (see Box 6.6) is relatively straightforward: it simply involves dividing the larger variance by the smaller. The probability associated with an F ratio for a particular combination of sample sizes, from which the degrees of freedom for each sample can be determined, enables an investigator to decide whether the F value is likely to have arisen through chance or is indicative of disparity in the dispersion of the data values in the two populations. If the probability associated with the calculated F statistic is less than or equal to the chosen level of

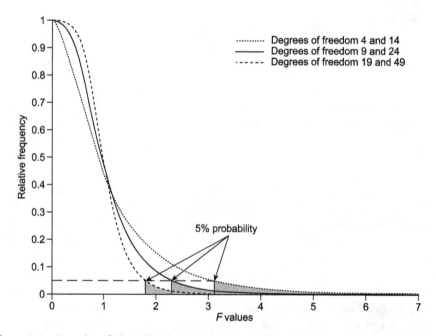

Figure 6.5 Examples of the F distribution with selected combinations of degrees of freedom.

Box 6.6a: The two-sample F test.

F test statistic: $= \dfrac{s_1^2}{s_2^2}$

Box 6.6b: Application of the F test.

The F test assesses the difference between the variances of two samples drawn from different populations in order to determine whether the population variances are equal or unequal. The Null Hypothesis argues that the population variances are the same ($\sigma_1^2 = \sigma_2^2$). The F test statistic is simply the ratio of the larger to the smaller sample variance (denoted by the subscripts 1 and s), a value that will always be greater than 1 unless the sample variances are equal, in which case an absence of sampling error is assumed. The samples do not need to contain the same number of observations and the degrees of freedom for each is $df_l = n_l - 1$ and $df_s = n_s - 1$. The fate of the Null Hypothesis is decided by reference to the probability of obtaining the F statistic or one greater in relation to the chosen level of significance, usually whether ≤ 0.05.

The F test has been applied to the samples of residents on the Isle of Wight who were shown different maps of counties in England and Wales, one with county names and one without. The samples purport to reflect a division in the Isle of Wight residents between those who have knowledge of the administrative geography of England and Wales, and can identify counties in which they would like to live without the aid of county names and those whose knowledge is less. The calculations for the sample variances were given in Box 6.5d, therefore Box 6.6d below simply illustrates the procedure for obtaining the F test statistic and the degrees of freedom. The F test results indicate that the difference in the variances is so small (insignificant) that it might arise through sampling error on 30.7 per cent of occasions.

The key stages in carrying out an F test are:

State Null Hypothesis and significance level: the variances of the populations from which samples have been selected are the same: inequality of sample variances between the two groups of Isle of Wight residents is the result of sampling error and is not significant at the 0.05 level of significance.

Calculate test statistic (F): the calculations for the F statistic appear below.

Select whether to apply a one- or two-tailed test: there is no prior reason for arguing that the variance of one population would be larger than the other therefore a two-tailed is appropriate.

Determine the probability of the calculated F: the probability of obtaining $F = 1.434$ is 0.307 with 29 degrees of freedom for the sample with the larger variance and 35 from the one with the smaller variance.

Accept or reject the Null Hypothesis: the probability of F is >0.05, therefore accept the Null Hypothesis. By implication the Alternative Hypothesis is rejected.

Box 6.6c: Assumptions of the F test.

The F test has three main assumptions:

   Sampled observations should be chosen randomly with replacement unless the populations are of infinite size.

   Data values for the variable under investigation should be independent of each other in both populations.

   The variable should be normally distributed in both parent populations since minor skewness can potentially bias the outcome of the test.

Box 6.6d: Calculation of the F test statistic.

Variance	Sample n (named counties)	From Box 6.5d	$\dfrac{6.017}{29} = 0.207$
	Sample u (unnamed counties)	From Box 6.5d	$\dfrac{5.063}{35} = 0.145$
F statistic			$\dfrac{s_n^2}{s_u^2} = \dfrac{0.207}{0.145} = 1.434$
Degrees of freedom	Sample n (named counties) larger variance		$df_n = n_n - 1 = 30 - 1 = 29$
	Sample u (unnamed counties) smaller variance		$df_u = n_u - 1 = 36 - 1 = 35$
Probability			$p = 0.307$

---

significance (e.g. 0.05 or 0.01), the Null Hypothesis would be rejected in favour of the Alternative Hypothesis stating that the population variances are significantly different.

6.4   Three or more samples and one variable

The next logical extension to our examination of parametric statistical tests continuing from the previous sections is to explore techniques involving one variable in respect of more than two populations and samples. For example, an investigator might be interested in predominantly arable, dairy, beef and horticultural farms, in people travelling to work by car, train, bus/tram, bicycle/motorcycle and by walking, or outcrops of sedimentary, igneous and metamorphic rocks. The intention is to compare measures of central tendency and dispersion calculated from the data values of the same variables measured on the interval or ratio scales for observations from each of the populations rather than to examine how two or more variables relate to each other. The observations in the populations are assumed to follow the normal distribution in respect of each variable under investigation. The statistical tests applied to such data in part are concerned with helping an investigator decide on whether the populations are really (statistically) different from each other with regard to certain variables according to a specified level of significance. The key question posed by the Null and Alternative Hypotheses associated with this type of test is whether the populations can be regarded as statistically distinct. This assessment is usually made on the basis of a sample of observations selected at random from the populations.

6.4.1 Comparing three or more sample means with population means

An extension of the two-sample F test provides one of the most important statistical tests available allowing researchers to discover whether the means of samples selected from three or more different populations are statistically different from each other. The technique seeks to determine whether the observed differences between the sample means are of such magnitude as to indicate that they could have come from the same or different populations. Rather perversely, the technique is known as analysis of variance (ANOVA), or more accurately one-way analysis of variance since only one variable is being considered, rather than by the seemingly more logical name of 'analysis of means'. Recall that the variance measures the spread of set of data values about their mean and represents the sum of the squared differences between these numbers and their mean divided by $n - 1$, where n equals the number of values. Now suppose there was a set of samples purporting to come from separate populations that were measured in respect of a single identically defined variable. The variance for each sample and the overall pooled variance of all data values could be calculated and, irrespective of whether the samples were in fact from different populations, there is no reason to suppose that dispersion of the variable would vary. In other words, if the means of the samples are different or the same, there is no reason to suppose that their variances would not be equal to each other and to the overall variance.

This suggests that the overall variance can be separated into two parts: one relates to the dispersion of the data values within each sample and the other between each one. Division of the within-groups by the between-groups variance yields the F ratio. A large between-groups variance in comparison with the within-groups variance will produce a relatively high F ratio and signify that intergroup differences account for a major part of the overall variance. In contrast a small F ratio denotes the reverse, namely that there is substantial overlap of the data values associated with the different groups. This separation of the total variance into two parts introduces the idea that some statistical explanation might account for the differences. The within-groups variance encapsulates some unexplained variation between the data values in each sample, whereas the between-groups variance relates to genuine differences between the samples. Figure 6.6 helps to explain these principles. The series of

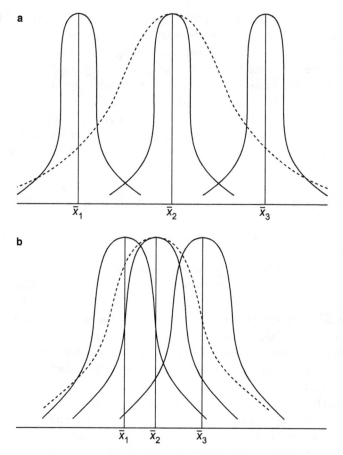

Figure 6.6   Comparison of samples with different means and variances.

identical curves in Figure 6.6a represent samples with identical variances but different means: hence they are separated from each other along the horizontal axis. The curve representing the relative frequency distribution of the data values pooled from all of the samples is wider and has a larger variance, since it has been calculated from values throughout the range. In contrast, the identical curves in Figure 6.6b are for samples with means that are relatively similar and closer together along the horizontal axis. The curve representing the complete set of data values covers a smaller range and its shape is closer to that if the individual samples. Furthermore, the sum of the squares of the differences between the sample means and the overall mean is smaller than the equivalent figure for the samples in Figure 6.6a. Is it possible that the samples might be from one single population rather than three separate ones?

The probabilities of the F distribution used in respect of the F test are also employed in ANOVA and are used to help with deciding whether the samples come from one population with reference to the chosen level of significance. Box 6.7 illustrates one-way ANOVA with respect to three samples. The test is used to examine whether the difference between the within- and between-groups variances is so small that it might well have arisen through chance or sampling error. If this is the case, then it is not only reasonable to argue that the variances of the population and sample variances

---

Box 6.7a: One-way Analysis of Variance (ANOVA).

Within-groups variance $s_w^2 = \sum \dfrac{\sum (x_{ij} - \bar{x}_j)^2}{n - k}$

Between-groups variance $s_b^2 = \dfrac{\sum n_j (\bar{x}_j - \bar{x}_t)^2}{k - 1}$

ANOVA F statistic: $= \dfrac{s_w^2}{s_b^2}$

Box 6.7b: Application of one-way ANOVA.

One-way ANOVA assesses the relative contribution of the within-groups and the between-groups variances to the overall or total variance of a variable measured for samples selected from three or more populations. The Null Hypothesis states that population means are the same and only differ in the samples as a result of sampling error. The procedure involves calculation of the within-groups and between-groups variances, respectively identified as s_w^2 and s_b^2. In the definitional formulae for these quantities above the subscripts refer to individual data values in each sample (i), the identity of the separate samples or groups (j) and the total the data values (t). Adding together the numerators of the equations for the within- and between-groups variances produces a quantity known as the total sum of squares. The ratio of the within-groups to between-groups variances produces the F statistic, which will be greater than 1 unless they happen to be equal. The degrees of freedom for the within-groups variance is defined as $df_w = n - k$ and the between-groups as ($df_b = k - 1$), where n represent the total number of data values and k the number of groups or samples. The probability of the having obtained the F statistic or a larger value helps an investigator decide between the Null and Alternative Hypotheses at the chosen level of significance, usually whether it is ≤ 0.05.

One-way ANOVA has been applied to unequal-sized samples of households in four villages (Kerry (13); Llanenddwyn (19); St Harmon (16); and Talgarth (16)) in mid-Wales. The face-to-face questionnaire survey asked respondents to say where they went to do their main household shopping for groceries. Using this information together with grid references for the households' addresses and the shopping towns, 'straight line' distances were calculated to indicate how far people travel for this purpose. These distances are tabulated in Box 6.7d with the settlements identified by the subscripts K, L, SH and T together with the sample means and variances. There is some variability in the sample means and variances within the survey suggesting that households in Talgarth travel further to shop. In contrast households in Llanenddwyn show least variability by having the smallest variance with most people travelling to one town, Barmouth, approximately 8 km away.

The detailed calculations of the within- and between-groups variances would be tedious to reproduce. These have been summarized along with calculation of the F statistic. The results suggest there is a significant difference between the mean distances travelled by the populations of residents in the four settlements to do their shopping, although the test examines the interplay between means and variances, it does not specify which sample or samples is (are) different.

The key stages in carrying out one-way ANOVA are:

State Null Hypothesis and significance level: the mean distances travelled to shop by the populations of the four villages are the same and they are only different in sampled households as a result of sampling error. This variation in sample means is not significant at the 0.05 level of significance.

Estimate the within- and between-groups variances: the results from these calculations are given below.

Calculate ANOVA statistic (F): the calculations for the F statistic appear below ($F = 3.36$)

Determine the probability of the calculated F: the probability (p) of obtaining this F value is 0.025 with 60 degrees of freedom for the within-groups variance and 3 for the between-groups variance.

Accept or reject the Null Hypothesis: the probability of F is <0.05, therefore reject the Null Hypothesis. By implication the Alternative Hypothesis is accepted.

Box 6.7c: Assumptions of one-way ANOVA.

One-way ANOVA makes three main assumptions:

Variables should be defined and measured on the interval or ratio scales identically for each of the samples.

Observations should be chosen randomly with replacement unless the populations are of infinite size.

Variables should be normally distributed in each of the populations.

---

are very nearly the same but that their means are also very similar. The calculations for the within- and between-groups variances are laborious if attempted by hand even with the aid of a calculator and Box 6.7 summarizes how these are carried out. Degrees of freedom were previously defined as the number of values that were 'free to vary' without affecting the outcome of calculating the mean, standard deviation, etc. In the case of one-way ANOVA there are two areas where such freedom needs to be considered. First, in relation to the overall mean of the entire set of data values, how many

Box 6.7d: Calculation of one-way ANOVA.

No.	Kerry (x_K)	Llanenddwyn (x_L)	St Harmon (x_{SH})	Talgarth (x_T)
1	12.41	14.89	55.05	13.21
2	11.96	8.07	1.30	13.63
3	11.96	8.07	0.45	13.96
4	11.96	0.58	4.62	13.96
5	11.55	8.07	3.67	13.96
6	0.36	7.68	1.92	19.68
7	0.36	7.74	1.92	19.68
8	0.36	7.74	3.67	19.68
9	0.36	7.75	9.11	21.61
10	0.42	15.33	4.62	19.68
11	0.42	7.74	3.67	19.68
12	0.42	0.99	3.67	21.80
13	0.36	7.75	4.43	8.42
14		7.75	4.43	5.41
15		8.45	41.10	8.42
16		7.41	4.43	5.36
17		8.80		
18		8.80		
19		8.60		
Sums	$\sum x_K = 69.20$	$\sum x_L = 152.19$	$\sum x_{SH} = 148.06$	$\sum x_T = 238.16$
Means	$\bar{x}_K = 4.84$	$\bar{x}_L = 8.01$	$\bar{x}_{SH} = 9.25$	$\bar{x}_T = 14.89$
Variances	$s_K^2 = 32.43$	$s_L^2 = 11.55$	$s_{SH}^2 = 239.87$	$s_T^2 = 31.96$

Variance	Within-groups	$s_w^2 = \sum \dfrac{\sum (x_{ij} - \bar{x}_j)^2}{n-k} = \dfrac{4697.809}{64-4} = 262.88$
	Between-groups	$s_b^2 = \dfrac{\sum n_j (\bar{x}_j - \bar{x}_t)^2}{k-1} = \dfrac{788.639}{4-1} = 78.29$
F statistic		$\dfrac{s_w^2}{s_b^2} = \dfrac{262.88}{78.29} = 3.36$
Degrees of freedom	Within-groups	$df_w = n - k = 64 - 4 = 60$
	Between-groups	$df_b = k - 1 = 4 - 1 = 3$
Probability		$p = 0.025$

of these can alter without changing the overall grand mean (mean of the means)? The answer is the number of values minus the number of samples. Secondly, in respect of the means of the individual samples, how many of these can vary without affecting the overall grand mean (mean of the means)? The answer is the number of samples minus 1. These, respectively, are known as the within- and between-groups degrees of freedom. The within-groups variance will always be larger than or equal to the between-groups variance, so the F statistic is obtained by dividing the former by the latter. The probability associated with an F statistic calculated for one-way ANOVA is the same as the F ratio for an F test for a particular combination of degrees of freedom. If the outcome from one-way ANOVA indicates that the Null Hypothesis should be rejected (i.e. the means are significantly different), this does not allow an investigator to conclude that one of the samples is 'more different' from the others, even if its mean is separated from the others.

Why is within-groups variance always larger than or equal to the between-groups variance? Hint: have another look at Figure 6.6.

The worked example in Box 6.7 uses the straight-line distance between address and shopping town. Why might this give a misleading indication of the amount of travelling people undertake to do their shopping?

6.5 Confidence intervals

A sample of observations is often our only route into discovering or estimating anything about the characteristics of its parent population. Chapter 2 explored the reasons why investigating all members of a population is rarely a feasible option for researchers, irrespective of the scale of resources available to the study. If a sample is the only source of information about the population, then it would be useful to have some way of knowing how reliable are the statistics produced from its data values. How confident are we that the sample data are providing a truthful message, or are they misleading us about the population's characteristics? Thinking about the sampled measurements of glacial melt water stream temperatures taken at the entry and exit points of the outwash plain, how confident are we that the means produced from these data values (5.7 °C and 9.1 °C, respectively) are a true reflection of the population parameters?

The answer to this question lies in calculating **confidence limits**, which represent a range within which we can be confident to a certain degree that the population parameter falls. We cannot be absolutely certain that it lies within this range, since sampling error may be present: therefore, we express our degree of confidence in percentage or proportionate probability terms. Such confidence limits are typically

95 or 99 per cent confidence, which, respectively, indicate that we are prepared to be wrong about the population parameter lying within the limits 5 or 1 per cent of times. In the cases examined in this chapter, these limits are calculated from the probabilities associated with the Normal, t or F distributions. In the case of confidence limits in respect of a statistical measure such as the mean, they comprise two numbers, one larger and one smaller than the mean that provide a range within which we can be 95 or 99 per cent confident that the true population parameter falls. The wider this range, the less precise is our figure for the population mean, whereas if the difference between the upper and lower limits is narrow, the sample provides a more constrained estimate of the parameter. Thus, there is an interaction between the size of the difference and the probability or confidence level: it is possible to be 95 or 99 per cent confident in respect of a wide or narrow confidence limit range.

Calculation of the confidence limits for the Z and t distributions (see Box 6.8) are similar, although in the latter case it is necessary to take into account that the critical

---

Box 6.8a: Confidence limits and interval.

Z distribution confidence limits: $\bar{x} - Z_p \sigma_e$ to $\bar{x} + Z_p \sigma_e$

t distribution confidence limits: $\bar{x} - t_{p,df} s_{\bar{x}}$ to $\bar{x} + t_{p,df} s_{\bar{x}}$

Box 6.8b: Application of confidence limits.

The value of Z in the confidence limits equation depends simply on the chosen level of confidence required ±1.96 for 95 per cent and ±2.57 for 99 per cent. The value of *t* differs according to the level of confidence and the degrees of freedom: thus the *t* values corresponding to the 95 and 99 per cent confidence levels with 10 and 25 degrees of freedom are respectively ±2.23/±2.06; and ±3.17/±2.79. Both types of confidence interval entail multiplying these Z or *t* values by the standard error of the mean (the population standard error in the case of Z confidence limits) or equivalently by the standard deviation divided by the square root of the number of observations. These results are subtracted from and added to the sample mean ($\bar{x}$) to produce a value lower than and another higher than the population mean. The sample mean thus estimates that the true population mean lies within this range with the specified level of confidence.

In practice, it is unlikely that the population standard error σ_e will be known without also knowing the population mean μ, thus confidence limits based on the Z distribution are rarely needed. The t distribution limits are much more useful, since the population mean is frequently missing. However, both sets of calculations are given in Box 6.8d with respect to the sample of Isle of Wight residents who were given the maps showing county names. The mean score per county was 0.404 from the sample of respondents. The confidence limits based on the Z distribution are narrower than those from the t distribution. The 'known' population mean (0.283) lies outside of the interval produced from the Z confidence limits, but within the t distribution limits. This result indicates the relative strength of these distributions.

The key stages in calculating confidence limits (intervals) are:

Decide on degree of confidence: the usual levels are 95 and 99 per cent corresponding to 0.05 and 0.01 probabilities.

Determine suitable probability distribution: Z or t distributions are the main ones used with normally distributed variables measured on the interval or ratio scales.

Identify critical value of distribution: this value will vary according to the degrees of freedom with certain distributions.

Insert critical value into confidence limits equation: calculate limits using equation.

Box 6.8c: Assumptions of Z and t confidence limits.

The assumptions pertaining to the Z and t tests also apply to their confidence limits:
Variables should be normally distributed in the population or the sample so large that the sampling distribution of the mean can be assumed normal (Z distribution).
Observations should be independent of each other.
Variables should be defined and measured on the interval or ratio scales.
Observations should be chosen randomly with replacement unless the populations are of infinite size.

Box 6.8d: Calculation of confidence limits.

Z distribution confidence limits	$\bar{x} - Z_p\sigma_e$ to $\bar{x} + Z_p\sigma_e = (0.404) - 1.96(0.559)$ to $0.404 + 1.96(0.559) = -0.294$ to $+0.513$
t distribution confidence limits	$\bar{x} - t_{p,df}s_{\bar{x}}$ to $\bar{x} + t_{p,df}s_{\bar{x}} = 0.404 - 2.045(0.846)$ to $0.404 + 2.045(0.846) = -0.231$ to $+0.576$

---

value of *t* associated with a particular level of probability varies according to the degrees of freedom. The values of *Z*, *t* or *F* inserted into the equations for calculating confidence limits represent the probability of the population parameter occurring in the tail of the distribution. Thus, the 95 per cent confidence limits signify that there is a 5 per cent chance of the population parameter lying beyond this range, which is normally split equally between the tails of the distribution, 2.5 per cent in each. In the case of the Z distribution this point always occurs at a Z value of ±1.96, whereas the t statistic corresponding to the same probability or confidence limit varies according to the degrees of freedom. Calculation of confidence limits based on the Z distribution requires the population mean and standard deviation to be known, but if this is the case there seems little point. Confidence limits based on the t distribution are more useful because they can be calculated from sample data. Confidence limits calculated with reference to the Z distribution will be narrower than those based on the t distribution for the same set of data values. Smaller sample sizes will usually produce wider confidence limits than larger ones and thus be less clear about the true value of the population parameter.

6.6 Closing comments

This chapter has indicated that as investigators we should raise questions about the data values obtained and the descriptive statistics that can be produced from sampled data. In particular, it has focused on variables measured on the interval or ratio scales and the parametric statistical tests that can be applied to help with reaching decisions about how much confidence can be placed on the information and whether results derived from a sample are significantly different from what might have occurred had random forces been 'at work'. When applying statistical testing techniques, it is rarely if ever possible to be certain that we have taken the right decision with the respect to the Null and Alternative Hypotheses, even if the rules of the particular test have been followed rigorously and we have not carried out hand calculations but used computers to undertake the computations. There is always a risk of committing a type-I or a type-II error. This arises because there remains a chance the results obtained from the sample data upon which the calculations have been based is one of those rare instances when:

- the true hypothesis is rejected (type I); or

- the false hypothesis is accepted (type II).

The risk of making a type-I error is given by the level of significance (e.g. 0.05 or 0.01). It might therefore seem prudent to choose a very low significance level in order to minimize the chance of making such errors. However, a low significance level increases the probability is committing a type-II error. We can never be certain which hypothesis is true and which is false, so it is not possible to know whether one of these two types of erroneous decision has been taken.

When carrying out analyses on features distributed across the Earth's surface, apart from examining the values of their variables and attributes, we might be interested in the patterns formed by the locations of these phenomena. Analysis of patterns in location and their spatial characteristics will normally be applied to all instances of such features rather than a sample. In general terms there is little point and even less justification for randomly or arbitrarily selecting a subset. One of the main reasons for analysing spatial patterns is to determine whether their distribution is random or tending towards dispersal or clustering. If the location of all features is known, why discard some instances, especially when the analytical procedures can be carried out by computer software rather than hand calculation. The spatial analysis techniques examined in this chapter have been concerned with the location of and distances between points in coordinate space. There are techniques available for analysing the patterns formed by lines and areas, although these are perhaps less well developed than those for points. Linear features are of two main types, simple paths across space or networks, with the latter further divided into branching and circular varieties. The shape of areas or polygons is perhaps their most obvious characteristic to merit analysis, for example measuring complexity by vmeans of perimeter per unit area (see de Smith, Goodchild and Longley, 2007).

7

Nonparametric tests

Chapter 7 has a similar overall structure to the previous chapter, but in this case the focus is on nonparametric tests. These are examined in relation to the numbers of samples and/or attributes (variables) and the types of summary measure under scrutiny. It mainly deals with frequency counts of nominal and ordinal attributes as opposed to scalar (ratio/interval) variables. Grid-based spatial statistics, which present an alternative to distance measures for examining the patterns in location of spatial entities, are also examined. The statistical procedures covered in this chapter are some of the most widely used by students and researchers in Geography, Earth and Environmental Science and related disciplines.

Learning outcomes

This chapter will enable readers to:

- apply nonparametric tests to nominal and ordinal attributes;

- decide whether to accept or reject the Null and Alternative Hypotheses in the case of nonparametric tests;

- explore spatial patterns using grid-based techniques;

- continue planning an independent research investigation in Geography, Earth Science and related disciplines making use on nonparametric procedures.

Practical Statistics for Geographers and Earth Scientists   Nigel Walford
© 2011 John Wiley & Sons, Ltd

7.1   Introduction to nonparametric tests

The assumptions and requirements of many of the parametric techniques examined in the previous chapter often means that they are difficult to apply in Geography, Earth and Environmental Sciences. This is especially the case where the intention is to investigate the connections between people and institutions, and the physical, environmental context in which they act. Interest in these connections has become an increasingly important focus of research in recent years. Sustainability is currently a major area for research and has become central to policy at international, national and local scales. It provides a 'classic' example of an issue relating to the interactions between human and physical environments to which geographers, Earth and environmental scientists are making important contributions. The difficulties of applying parametric procedures are particularly common where the focus of attention is on the actions and behaviour of humans in different geographical contexts. It is rare for some let alone all of the attributes and variables relevant in such investigations to conform to the assumptions of normality associated with parametric statistical tests.

Nonparametric tests do not impose the requirement that the variables under scrutiny in a sample's parent population should conform to the Normal Distribution, and are sometimes referred to as 'distribution-free'. This assumption is required in parametric tests so that a valid comparison may be made between the test statistic calculated from the sample data and the corresponding value in the Normal Distribution or at least its 'close relative' the t distribution, which have known probabilities associated with them. We have seen that parametric procedures are often applied to samples of data values in order to estimate a population parameter (e.g. the mean), hence the name. However, most nonparametric procedures examine the frequency counts of data values produced when samples of observations are distributed between categories or classes. When the data values of scalar (interval or ratio) variables are divided between classes, it is possible to estimate the mean value even when the individual values are unknown as explained in Chapter 5. Thus, although nonparametric techniques do not directly estimate the value of a population parameter with a known probability expressed as a confidence interval, they do help to decide whether the distribution of observed values is significantly different from a random one and whether the approximated mean or an other measure of central tendency is in the correct range. In general terms, the Null Hypothesis in most nonparametric tests may be stated as examining whether the observed distribution of frequency counts is significantly different from what would have occurred if the entities had 'fallen' randomly into the finite number of available categories.

This chapter parallels the structure of Chapter 6 by grouping nonparametric tests according the number of samples and attributes (variables) under examination. Most of the procedures are used with either nominal or ordinal frequency distributions and the following sections progress through those dealing with one, two or three plus samples. The Pearson's Chi Square test features in several sections since it may be employed in different circumstances. In contrast, some tests, such as the enticingly named Kolmogorov–Smirnov test that deals with a single sample of observations from which a frequency distribution of a ranked or sequenced attribute can be produced, only feature once. Wilcoxon's signed ranks test for a single sample of observations measured twice on a variable, which is the nonparametric equivalent of the paired-sample t test with less stringent assumptions, is also included in this chapter.

Some spatial statistics are also examined, in particular those focusing on the distribution of spatial phenomena within a regular square grid, which provide an alternative way of exploring whether they possess a clustered, random or dispersed pattern in location. There are close links between these spatial statistics and those used for nonspatial distributions, although in the former case the categories or classes are defined within the framework of grid coordinates and in the latter with respect to the values of the attributes or variables. The location of each grid square within this framework and its count of observations are jointly of crucial importance in spatial statistics, since these define the spatial pattern and density of the entities. In some nonspatial nonparametric tests the order of the categories is also important when these represent a numerical sequence or ranking, such as occurs when they are based on measurements of distance, time, elevation, currency, etc.: in these situations where each class occurs in the sequence (its relative position) matters when investigating the frequency distribution.

7.2 One variable and one sample

Most investigations carried out as a student project or full-scale research enquiry will involve the analysis of more than one variable, although attention may quite legitimately focus on a single sample of observations. In parallel with our examination of parametric tests we begin looking at nonparametric procedures by considering the relatively simple question of whether the frequency distribution produced by placing the data values of a single attribute for a sample of observations into a set of categories is significantly different from what would have occurred if they had been spread randomly across the classes. The difference to be examined in this situation, as in most nonparametric tests, does not relate to a single value but to the contrasts between the observed and expected frequencies. Just as with parametric tests it is necessary to discover the probability of a difference having arisen by chance. If the test results indicate that the difference is too large to have occurred through chance or sampling error, then the researcher needs to explain it in terms of some human or physical process.

This raises the now familiar question of how to determine the probabilities associated with what would occur if a random process was operating, although in this case such a process would produce a distribution of counts within a series of categories. The framework of rows and/or columns into which a particular set of entities are distributed provides three types of constraint – the numbers of categories, attributes (variables) and observations – that control the probabilities. The choice between

Table 7.1   Alternative ways of obtaining expected frequency counts.

Area size class (ha)	Count from previous census	Proportion	Expected count for sample of 100 farms based on:		
			Empirical proportions	Equal proportions	Randomness
0.00–99.9	20 923	0.867	87	20	16
100.0–299.9	2389	0.099	10	20	24
300.0–499.9	523	0.022	2	20	25
500.0–699.9	155	0.006	1	20	17
700.0 & over	118	0.005	0	20	18
	24 108	1.000	100	100	100

Note: Expected counts rounded to integers.
Source: Department of the Environment, Food and Rural Affairs (2003).

a posteriori and *a priori* probabilities exists, although the frequency distribution derived from population data with which to compare a sample is often difficult to determine. In some circumstances there may be frequency distributions of the same type of observations categorized with respect to identical attributes that were produced from data collected previously or in respect of a different locality. For example, a sample-based frequency distribution may be compared with tables of statistics from an earlier national census or survey. These statistics can be used to give the empirical probabilities or proportions of the total number of observations that previously occurred. An alternative approach is to argue that the probability of an observation occurring in a particular category is in proportion to the number of categories or cells available. Table 7.1 illustrates three ways of obtaining the expected counts for a sample of 100 farms in five size categories. The first is based on empirical probabilities, where the proportions of farms in the size groups that occurred in a previous census dictate the expected number in the sample. The second approach simply assigns an equal probability (0.2) to each of the cells. Finally, a random number generator was used to obtain 100 integers between 1 and 5 corresponding to the five categories, which has resulted in some clustering in the second and third categories and rather less in the first.

7.2.1   Comparing a sample mean with a population mean

The basis of the distinction between parametric and nonparametric statistical tests is not simply that the former are concerned with summary quantities (e.g. mean and variance) and the latter with frequency counts. It is more a question that parametric tests assume variables follow the Normal Distribution, whereas nonparametric ones do not. Thus there are some nonparametric tests that make less severe assumptions than their parametric counterparts when focusing on similar statistical measures. The

Wilcoxon Signed Ranks test illustrates this point since it is used to investigate whether the difference in mean for a set of paired measurements with respect to a scalar variable for one sample of observations are significantly different from each other. It is the nonparametric equivalent of the paired-sample t test.

Box 7.1a: The Wilcoxon Signed Ranks W test.

W test statistic: $= \dfrac{|\bar{d} - 0|}{s_d / \sqrt{n}}$

Box 7.1b: Application of the Wilcoxon Signed Ranks W test.

The Wilcoxon Signed Ranks test hypotheses that if there is no difference between the means of the paired sets of data values for a single variable in a sample of observations (i.e. subtracting one mean from the other equals zero), the sum of the ranks for the paired measurements for each observation also equals zero. Calculations for the test statistic W exclude pairs of measurements where the difference is zero. The probabilities associated with possible values of W vary according to number of nonzero differences. The purpose of the W test is to discover the probability of obtaining the observed difference in the ranks as a result of chance and to decide whether the difference is significantly different from zero.

The calculations for the test statistic are relatively simple from a computational point of view, although they do involve a number of steps. First, calculate the differences between the paired measurements and then, excluding any zero values, sort these from smallest to largest ignoring the positive and negative signs (i.e. +0.04 counts as smaller than −1.25). Now assign the rank score of 1 to the smallest difference, 2 to the next and so on, applying average rank scores to equal differences. Now assign the + or − signs associated with the difference values to the rank scores and separately sum the positive and negative ranks: the smaller of these totals is the required test statistic, W. Depending on the nature of differences under scrutiny a one- or two-tailed W test is needed, with the former involving the probability associated with the test statistic being halved. A decision on the fate of the Null and Alternative Hypotheses may be made according to the chosen level of significance either by reference to the probability of obtaining the W test statistic itself or by transforming W into Z and then determining its probability. Transformation of W to Z is preferable when there are more than 20 nonzero differences, since the W distribution approximates the normal distribution in these circumstances.

Information obtained from the samples of residents on the Isle of Wight who were asked about their residential preferences using maps of counties in England and Wales with and without names is used to illustrate the application of the Wilcoxon Signed Ranks Test. The first two data columns in Box 7.1d below represent the mean score of each county (excluding those not favoured by any survey respondent) using the named (x_n) and unnamed maps (x_u). Having calculated the differences between these mean scores, the remaining columns represent the different steps involved in calculating W. The sums of the positive and negative signed ranks are respectively 438.5 and −464.5 leading to W equalling 438.5. Since there are more than 20 differences this has been transformed in Z in order to obtain the test statistic's probability. The probability of obtaining a Z (or W) test value of this size is extremely low and thus at the 0.05 level of significance the Null Hypothesis can be rejected. There does not appear to be a genuine difference between people's residential preferences when looking at maps with and without names.

The key stages in carrying out a Wilcoxon Signed Ranks test are:

State Null Hypothesis and significance level: the difference between the mean rank scores of the two sets of Isle of Wight residents and zero is the result of sampling error and is not significant at the 0.05 level of significance.

Calculate test statistic (W): the results of the calculations are given below.

Select whether to apply a one- or two-tailed test: in this case there is no sound reason to argue that the difference should be in one direction from zero or the other since people's residential preferences should not change simply because they have looked at a map with county names as opposed to one without.

Determine the probability of the calculated W (Z): the probability of obtaining $Z = 0.149$ is 0.881.

Accept or reject the Null Hypothesis: the probability of Z is >0.05, therefore accept the Null Hypothesis and reject the Alternative Hypothesis.

Box 7.1c: Assumptions of the Wilcoxon Signed Ranks W test.

There are three main assumptions of the W test:

The differences are independent of each other.

Zero-value differences arise because rounding values to a certain number of decimal places creates equal values, thus their exclusion is arbitrary since they would disappear if more decimal places were used.

The test is less powerful than the equivalent parametric paired t-test, although the decision in respect of the Null and Alternative Hypotheses may be the same.

7.2.2 Comparing a sample's nominal counts with a population

It is now time to introduce one of the most useful and versatile nonparametric tests generally known as the Chi-square test, or more correctly as **Pearson's Chi-square test**. It is used in different guises with nominal (categorical) counts, although the chi-square (χ^2) probability distribution that provides the means for assessing prob-

Box 7.1d: Calculation of the Wilcoxon Signed Rank W test statistic.

County	x_n	x_u	$x_n - x_u = d$	Sorted counties	Sorted absoluted	Rank	Signed rank
Avon	0.154	0.423	0.269	Hants.	0.038	4.5	4.5
Beds	0.077	0.000	−0.077	North Yorks.	0.038	4.5	4.5
Berks.	0.077	0.000	−0.077	N'hants.	0.038	4.5	4.5
Bucks.	0.154	0.077	−0.077	Northum.	0.038	4.5	−4.5
Ches.	0.000	0.154	0.154	Notts.	0.038	4.5	−4.5
Clwyd	0.000	0.192	0.192	Shrops.	0.038	4.5	−4.5
Cornwall	1.885	1.462	−0.423	Staffs.	0.038	4.5	−4.5
Cumbria	0.231	0.308	0.077	Somerset	0.038	4.5	4.5
Derbs.	0.500	0.000	−0.500	Beds.	0.077	12	−12
Devon	1.115	1.769	0.654	Berks.	0.077	12	−12
Dorset	0.846	1.269	0.423	Bucks.	0.077	12	−12
Dyfed	0.192	0.077	−0.115	Cumbria	0.077	12	12
East Sussex	0.385	0.115	−0.269	Herts.	0.077	12	−12
Essex	0.154	0.269	0.115	Powys	0.077	12	−12
Glos.	0.000	0.538	0.538	South Yorks.	0.077	12	−12
Greater London	0.269	0.808	0.538	Essex	0.115	17.5	17.5
Greater Manch.	0.154	0.000	−0.154	Dyfed	0.115	17.5	−17.5
Gwynedd	0.154	0.154	0.115	Gwynedd	0.115	17.5	17.5
Hants.	0.692	0.731	0.038	Heref & Worcs.	0.115	17.5	17.5
Heref. & Worcs.	0.000	0.115	0.115	Ches.	0.154	21	21
Herts.	0.192	0.115	−0.077	Greater Manch.	0.154	21	−21
Hum'side	0.192	0.000	−0.192	Lancs.	0.154	21	−21
Kent	0.115	0.731	0.615	Clwyd	0.192	24.5	24.5

County	x_n	x_u	$x_n - x_u = d$	Sorted counties	Sorted absolute	Rank	Signed rank
Lancs.	0.154	0.000	-0.154	Hum'side	0.192	24.5	-24.5
Leics.	0.269	0.000	-0.269	Lincs.	0.192	24.5	-24.5
Lincs.	0.192	0.000	-0.192	Warks.	0.192	24.5	24.5
Norfolk	0.500	0.000	-0.500	Suffolk	0.231	27	-27
North Yorks.	0.115	0.154	0.038	Avon	0.269	29	29
N'hants.	0.000	0.038	0.038	Leics.	0.269	29	-29
Northum.	0.077	0.038	-0.038	East Sussex	0.269	29	-29
Notts.	0.038	0.000	-0.038	Oxfords.	0.308	31	31
Oxfords.	0.115	0.423	0.308	Dorset	0.423	32.5	32.5
Powys	0.077	0.000	-0.077	Cornwall	0.423	32.5	-32.5
Shrops.	0.154	0.115	-0.038	Derbs.	0.5	34.5	-34.5
Somerset	0.346	0.385	0.038	Norfolk	0.5	34.5	-34.5
South Yorks.	0.077	0.000	-0.077	Glos.	0.538	36.5	36.5
Staffs.	0.077	0.038	-0.038	Greater London	0.538	36.5	36.5
Suffolk	0.308	0.077	-0.231	Kent	0.615	38.5	38.5
Surrey	0.192	1.038	0.846	Wilts.	0.615	38.5	-38.5
Warks.	0.000	0.192	0.192	Devon	0.654	40	40
West Sussex	0.923	0.115	-0.808	West Sussex	0.808	41	-41
Wilts.	0.808	0.192	-0.615	Surrey	0.846	42	42
Sum of the negative ranks				-464.5			
Sum of the negative ranks				438.5			
W statistic				438.5			

$$z = \frac{\left|W - \mu_W\right|}{\sigma_W} = \frac{\left|438.5 - \dfrac{42(42+1)}{4}\right| - 0.5}{\sqrt{\dfrac{42(42+1)((2\times 42)+1)}{24}}}$$

$$z = 0.147$$
$$p = 0.881$$

abilities relates to measurements rather than frequencies. We have seen that the data values of a normally distributed variable can be transformed or standardized into Z scores (see Chapter 5). A series of n such Z scores (Z_1, Z_2, Z_3, ... Z_n) can be squared and summed to produce a value known as chi-square (χ^2), thus:

$$\chi^2 = (Z_1)^2 + (Z_2)^2 + (Z_3)^2 + \ldots + (Z_n)^2$$

The value of chi-square depends on the amount of variation in the series of Z scores, if they are all identical the statistic would equal zero whereas an infinite amount of variation would produce a maximum value of plus infinity. The number of data values (the sample size) provides the degrees of freedom, which influences the form and shape of the χ^2 distribution. Small sample sizes (degrees of freedom) produce a distribution curve that is highly skewed to the left (negative), whereas once the number of cases approaches 40 or more the shape and form is similar to that produced by the Normal Distribution (Figure 7.1). Given that the form and shape of the curve varies with sample size, but that individual values of χ^2 can be obtained from different sets of data values, it follows that the probabilities associated with the test statistic are related to the degrees of freedom. The probability of obtaining a χ^2 value of 11.34 is 0.50 when there are 12 degrees of freedom (nonsignificant at 0.05) and 0.01 if $df = 3$ (significant at 0.05). The implication is that it is more difficult to obtain a larger chi-square with a relatively small number of observations. Since the minimum value χ^2 is always zero and the curve extends towards plus infinity at the extreme right of the chart, the probabilities relate to the chance of obtaining the particular χ^2 value or a larger one. Thus, if 100 samples with 3 observations were selected randomly from the same population, five samples should produce χ^2 with a value of at least 11.34; similarly 100 samples with 12 observations in each should result in 50 with a χ^2 of at least 11.34.

The chi-square statistic can be used to test for a difference between variances, but much more often the closely related Pearson's chi-squared distribution is used with counts of (n) observations distributed across (j) groups or categories. The purpose of the Pearson's Chi-square test is to examine whether the difference between the observed and expected frequency distributions of counts is more or less than would be expected to occur by chance. Pearson's Chi-square test statistic, commonly but incorrectly referred to as χ^2, is also calculated from the summation of a series of discrete elements:

$$\text{Pearson's chi-square } \chi^2 = \frac{(O_1 - E_1)^2}{E_1} + \frac{(O_2 - E_2)^2}{E_2} + \frac{(O_3 - E_3)^2}{E_3} + \ldots + \frac{(O_j - E_j)^2}{E_j}$$

where O and E, respectively, refer to the observed and expected frequency counts in each of j groups or categories. The result of summing the elements in this series is to produce a statistic that very closely approximates the χ^2 distribution and its associated probabilities with ($j - 1$) degrees of freedom.

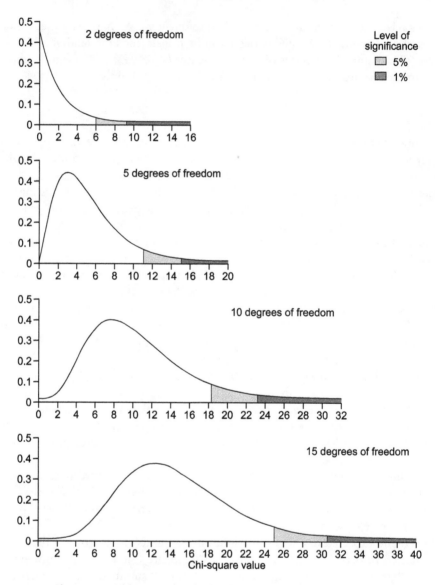

Figure 7.1 Chi-square distribution for selected degrees of freedom.

7.2.3 Comparing a sample's ordinal counts with a population

The requirement that the Pearson's Chi-square test should only be used with attributes or variables whose categories are simple nominal groups and not ordered or ranked in any way might initially seem unproblematic. After all, there must be lots of attributes and variables that simply differentiate between observations and do not

Box 7.2a: The Pearson's Chi-square (χ^2) test.

Pearson's Chi-square (χ^2) test statistic: $= \sum\limits_{j}^{1} \dfrac{(O_j - E_j)^2}{E_j}$

Box 7.2b: Application of the Pearson's Chi-square (χ^2) test.

The Pearson's Chi-square test adopts the now familiar cautious stance (hypothesis) that any difference between the observed (O) and expected (E) frequency counts has arisen as a result of sampling error or chance. It works on the difference between the observed and expected counts for the entire set of observations in aggregate across the groups. Calculations for the test statistic do not involve complex mathematics, but, as with most tests, it can readily be produced using statistical software. The degrees of freedom are defined as the number of groups or cells in the frequency distribution minus 1 ($j - 1$). The probabilities of the test statistic depend on the degrees of freedom. The calculations involve dividing the squared difference between the observed and expected frequency count in each cell by the expected frequency and then summing these values to produce the test statistic. Comparison of the probability of having obtained this test statistic value in relation to the chosen level of significance enables a decision to be made regarding whether to accept or reject the Null Hypothesis, and by implication the Alternative Hypothesis.

The Pearson's Chi-square test has been applied to households in four villages in mid-Wales that have been pooled into a single sample in respect of the frequency distribution of the mode of transport used when doing their main household shopping for groceries. This comprises the univariate frequency distribution reproduced in Box 7.2d below. The expected frequencies have been calculated in two ways: on the basis that households would be distributed in equal proportions between the four modes of transport; and alternatively that they would occur in the unequal proportions obtained from the results of a national survey of households' shopping behaviour. The test statistics for these two applications of the procedure are, respectively, 9.12 and 14.80, whose corresponding probabilities are 0.023 and 0.002. These are both lower than the 0.05 significance level and so the Alternative Hypothesis is accepted.

The key stages in carrying out a Pearson's Chi-square test are:

State Null Hypothesis and significance level: the difference between the observed and expected frequencies has arisen as a result of sampling error and is not significant at the 0.05 level of significance. The sample frequency distribution is not significantly different from either an equal allocation between the categories or what would be expected according to a previous national survey.

Calculate test statistic (χ^2): the results of the calculations are given below.

Determine the degrees of freedom: in this example there are $4 - 1 = 3$ degrees of freedom.

Determine the probability of the calculated χ^2: the probability of obtaining $\chi^2 = 9.12$ and $\chi^2 = 14.80$ each with 3 degrees of freedom are, respectively, 0.023 and 0.002.

Accept or reject the Null Hypothesis: the probability of both χ^2 values is <0.05, therefore reject the Null Hypotheses and accept the Alternative Hypothesis.

Box 7.2c: Assumptions of the Pearson's Chi-square (χ^2) test.

There are five main assumptions of the Pearson's Chi-square test:

Simple or stratified random sampling should be used to select observations that should be independent of each other.

The number of categories or classes has a major effect on the size of the test statistic. In the case of nominal categories (as in this transport to shop example) these may be predetermined, although they can be collapsed but not expanded (e.g. family car and other car could be merged). However, there are many possibilities for classifying continuous interval or ratio scale variables (see Chapter 5), although this should not produce ordered classes.

Frequencies expressed as percentages (relative frequencies) will produce an inaccurate probability for Pearson's Chi-square statistic: if the sample size is <100 it will be too low and if >100 to high. If, as in the worked example, the sample size happens to equal 100 then absolute and relative frequencies are the same and the probability is accurate.

The minimum expected frequency count per cell is 5 and overall no more than 20% of the cells should be at or below this level. The number of cells with low expected frequencies may be adjusted by altering classes or categories.

The categories or classes should not be ordered or ranked (e.g. a sequence of distance groups or ordered social classes) because the same χ^2 will be obtained irrespective of the order in which categories are tabulated.

Box 7.2d: Calculation of the Pearson's Chi-square (χ^2) test.

Transport mode	Observed sample frequencies	Exp. freq. (equal props)	$\dfrac{(O-E)^2}{E}$	Exp. freq. (props from national survey)	$\dfrac{(O-E)^2}{E}$
Family car	36	25	4.84	40	0.40
Other car	18	25	1.96	10	6.40
Public transport	18	25	1.96	30	4.80
Walk or cycle	28	25	0.36	20	3.20
Total	100	100	9.12	100	14.80
Pearson's chi-square statistic	Equal proportions		$\chi^2 = 9.12$	Proportions from national survey	$\chi^2 = 14.80$
Probability	Equal proportions		$p = 0.023$	Proportions from national survey	$p = 0.002$

put them in order. However, the more you think about it the more classified variables measuring such things as distance, area, temperature, acidity/alkalinity, volume and population as well as various financial quantities (land value, income, expenditure, rental value, etc.) do in fact summarize a sequence of data values. For example, parcels of land classified according to different types of use and assigned to zones from a city centre constitute an ordered sequence of distance zones. Perhaps there are only relatively few truly nominal categories such as gender, household tenure, mode of transport, ethnicity, aspect, etc. Nevertheless, some form of statistical test is required that enables investigators to get around the assumption that the Pearson's Chi-square test should not be used with ordered categories. One such test is the **Kolmogorov–Smirnov D test**.

One of the most typical applications of this test involves comparing an observed set of frequency counts for a ranked attribute or variable with the frequencies expected to arise by applying a probability distribution, such as the Poisson distribution. This test focuses on the cumulative observed and expected probabilities of the number of observations in each of the series of ordered classes. The D statistic is simply the maximum absolute difference between the cumulative observed and expected probabilities. Because the observations are distributed between an **ordered** series of classes, the cumulative probabilities for each successive class represent the proportion of observations that have occurred or would be expected to occur according to the Poisson distribution up to that point. Figure 7.2 illustrates how the two cumulative

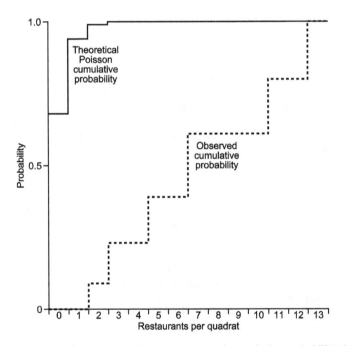

Figure 7.2 Comparison of observed and Poisson expected cumulative probability distributions.

probabilities increment in a stepwise fashion and that the maximum vertical difference between them indicates the goodness of fit between the two distributions. Application of the Kolmogorov–Smirnov is these circumstances is not universally accepted as legitimate by statisticians, since calculation of the Poisson expected frequencies relies on using the mean or average density of the sampled observations, which contravenes the expectation that they are obtained independently of the sample data. Despite these reservations, provided that a sufficiently rigorous level of significance is applied and caution is exercised in marginal cases, then the D test can prove to be a useful way of testing a ranked frequency distribution.

Box 7.3 illustrates use of the test to examine whether the distance travelled by households in four villages in mid-Wales to the settlement where they do their main shopping is similar to what might occur if the process of choosing where to go was random or otherwise. The distances have been classified into seven groups with the

---

Box 7.3a: The Kolmogorov–Smirnov D test.

Kolmogorov–Smirnov D test statistic: $= \text{MAX Abs}(\text{Cum } P_{\text{Obs}} - \text{Cum } P_{\text{Exp}})$

Box 7.3b: Application of the Kolmogorov–Smirnov D test.

The starting point for the Kolmogorov–Smirnov test is that any difference between the cumulative observed and expected frequency distributions of an ordered (ranked) attribute or classified variable for a single sample of observations is a consequence of sampling error. Application of the test proceeds by calculating the difference between the cumulative observed (Cum P_{Obs}) and expected (Cum P_{Exp}) probabilities across the ordered sequence of classes and the largest difference provides the D test statistic. The probability of having obtained the particular value of D, which varies according to sample size, n, helps with deciding on whether to accept or reject the Null Hypothesis at a chosen level of significance.

The Kolmogorov–Smirnov test has been applied to the distance travelled to the main settlement for shopping by households in the group of mid-Wales villages, with the original continuous variable classified into bands (groups) 10 km wide. Having tabulated the survey data into the observed frequency counts (f_{Obs}), the observed and expected probabilities are required. Box 5.7 showed how the Poisson probability distribution could be applied to produce such probabilities and has been used in this case. The observed probabilities are obtained by dividing the count of households in each class by the total (e.g. $64/35 = 0.55$). There are seven classes and 64 households producing an overall mean density of households per ranked class (λ) of 0.11 which has been inserted into the Poisson distribution equation to produce the expected probabilities. Each set of probabilities is accumulated across the classes and taking the absolute difference between these cumulative probabilities yields a column of differences. The D statistic can be identified as the maximum difference, which is 0.77 in this case. Since the probability of having obtained a maximum difference of this size is 0.17, which is larger than the standard significance level (0.05), the Null Hypothesis is accepted.

The key stages in carrying out a Kolmogorov–Smirnov test are:

State Null Hypothesis and significance level: sampling error accounts for the difference between the cumulative observed and expected frequencies and a difference this size is likely to have occurred more often than 95 times in 100. The observed ordered frequency distribution from the sample of observations is not significantly different from what would be expected according to the Poisson probability distribution

Calculate test statistic (D): the tabulated ordered frequency distributions and calculations are given below showing the maximum difference is 0.77.

Determine the probability of the calculated D: the probability of the test statistic varies according to sample size and in this case with a sample of 64 observations (households) it is 0.17.

Accept or reject the Null Hypothesis: the probability of D is >0.05, so the Null Hypothesis should be accepted with the consequence that the Alternative Hypothesis is rejected.

Box 7.3c: Assumptions of the Kolmogorov–Smirnov (D) test.

There is one main assumption of the Kolmogorov–Smirnov test:

Observations should be selected by means of simple random sampling and be independent of each other.

Box 7.3d: Calculation of Kolmogorov–Smirnov D statistic.

Distance (km)	f_{Obs}	P_{Obs}	P_{Exp}	Cum P_{Obs}	Cum P_{Exp}	Abs (Cum P_{Obs} – Cum P_{Exp})
0.0–0.5	8	0.13	0.90	0.13	0.90	**0.77**
0.5–10.4	35	0.56	0.10	0.67	0.99	0.32
10.5–20.4	17	0.27	0.01	0.94	1.00	0.06
20.5–30.4	2	0.03	0.00	0.97	1.00	0.03
30.5–40.4	0	0.00	0.00	0.97	1.00	0.03
40.5–50.4	1	0.02	0.00	0.98	1.00	0.02
50.5–60.4	1	0.02	0.00	1.00	1.00	0.00
	64	1.00	1.00			

Kolmogorov–
Smirnov statistic $D = \text{MAX Abs(Cum } P_{Obs} - \text{Cum } P_{Exp}) = 0.77$

Probability $p = 0.17$

first containing the count of households travelling less than 0.5 km, regarded for this purpose as zero up to the seventh with just one household that journeyed in the range 50.5–60.4 km. The test results in this case indicate there is a moderate probability of obtaining D equal to 0.77 with 64 observations and so, despite some statisticians' reservations about the procedure, it may be considered reasonably reliable in this case.

The Kolmogorov–Smirnov test is used with an ordered series of classes in a frequency distribution, in other words the observations have been allocated to a set of integer classes but the test pays no attention to the order in which their outcomes occurred. This characteristic can be examined by the **Runs (or Wald–Walfowitz) test,** which focuses on the sequence of individual dichotomous outcomes through time and across space and, as the name suggests, whether 'runs' of the same outcome occur or if they are jumbled up. The test can be used to examine whether the outcomes for a sample of observations can be regarded as independent of each other. The test is founded on the notion that the outcomes of a certain set of events can only be ordered in a finite series of combinations, since the number of events is not open-ended. If the number of runs observed in a particular sample exceeds or falls short of what might be expected then the independence of the observations is brought into question. A run is defined as either one or an uninterrupted series of identical outcomes. Suppose an event involves tossing a coin four times in succession to find out how many combinations or sequences of outcomes there are that include a total of two heads and two tails: the answer is that there are six:

HHTT; HTHT; HTTH; TTHH; THTH; and THHT

How many runs of the same outcome are there in each of these sequences?

Box 7.4 applies the Runs test to a rather more geographical problem. Box 7.4b shows a 250 m stretch of road in a residential area along which there a series of 50 segments. If there was at least one car or other vehicle parked at the kerb in a 5 m segment it was counted as occupied, otherwise it was unoccupied. There were respectively 19 and 31 occupied and unoccupied segments on the day when the data were recorded, indicating that 38% had vehicles parked in 25 runs. Driving along the road from one segment to the next would mean that a new run started when going from a segment with parked vehicles to one without or vice versa, whereas adjacent segments with the same outcome would be part of the same run. The Runs test is applied in this case to determine whether there is sufficient evidence to conclude that the number of runs along the stretch of road on this occasion is random. The test seeks to establish whether the observed number of runs is more or less than the expected mean given that the overall sequence of dichotomous outcomes is finite. The comparison is made using the Z statistic and its associated probability distribution. Figure 7.3 relates to the example in Box 7.4 where there are a total 50 outcomes in the sequence and illustrates the distinctive shape of probability distribution. There are two areas where the combinations of dichotomous outcomes (e.g. total segments with and without

---

Box 7.4a: The Runs (Wald–Walfowitz) Test.

R = number of changes of dichotomous outcomes in complete sequence

Conversion of R to Z test statistic: $= \dfrac{R - \mu_R}{\sqrt{\sigma_R^2}}$

Box 7.4b: Application of the Runs (Wald–Walfowitz) Test.

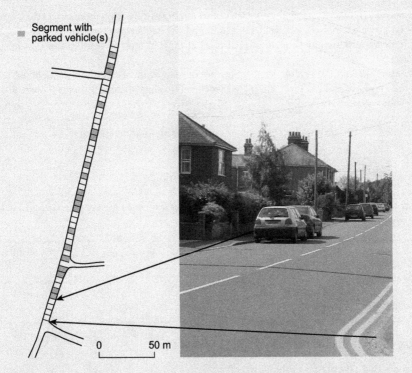

The Runs test focuses on the sequence of individual dichotomous data values and compares the observed number of runs (R) with the average (μ_R) that would be expected if changes in the sequence of outcomes had occurred randomly. The expected mean (μ_R) and variance (σ_R^2) are calculated empirically from the sample data and are used to convert the observed number of runs into a Z statistic. The Null Hypothesis is evaluated by comparing the probability of Z in relation to the chosen level of significance, typically 0.05.

Cars and other vehicles parked at kerbs along roads going through residential areas are commonplace: from the motorists' perspective they represent an obstacle impeding and slowing progress, for residents they constitute a convenient parking place and for pedestrians they present a potential hazard when attempting to cross a road. If we consider any given stretch of road capable of being divided into segments of equal length (say 5 m), the Runs test can be used to examine the hypothesis that the presence or absence of one or more vehicles in a sequence of segments are mutually independent of each other. The Runs test has been applied to test such a hypothesis in respect of the stretch of road shown above together with a photo to illustrate kerbside parking. The sequence of segments occupied and unoccupied by cars or other vehicles is tabulated in Box 7.4d. The 25 runs were of various lengths ranging from single segments up to a run of five adjacent segments without any parked vehicles. The expected number of runs was 24.56 and the variance 10.85, which produce a Z statistic with high probability and so the Null Hypothesis of a random number of runs is upheld. So, it would seem that the outcome in each segment of the road is independent of the others.

The key stages in performing a runs test and calculating the associated Z statistic are:

State Null Hypothesis and significance level: the number of runs is no more or less than would be expected to have occurred randomly and the sequence of outcomes indicates the observations are independent of each other at the 0.05 level of significance.

Determine the number of runs, R: the sequence of road segments with and with parked vehicles is given below and the alternate shaded and unshaded sections indicate there are 25 runs.

Calculate Z statistic: the Z statistic requires that the mean and variance of the sequence be obtained before calculating the test statistic itself, which equals 0.134 in this case.

Determine the probability of the Z statistic): this probability is 0.897.

Accept or reject the Null Hypothesis: the probability of Z is much greater than 0.05 and so the Null Hypothesis should be accepted.

Box 7.4c: Assumptions of the Runs test.

There is one assumption of the Runs test:

Observations are assumed independent of each other (the test examines whether this is likely to be the case).

Box 7.4d: Calculation of the Runs test.

Segment	1	2	3	4	5	6	7	8	9	10	11	12
Car	NC	NC	NC	NC	C	C	NC	NC	C	C	NC	C
Runs	1	1	1	1	2	2	3	3	4	4	5	6
Seg. (cont.)	13	14	15	16	17	18	19	20	21	22	23	24
Car (cont.)	NC	C	NC	NC	C	NC	NC	NC	C	C	C	C
Runs (cont.)	7	8	9	9	10	11	11	11	12	12	12	12
Seg. (cont.)	25	26	27	28	29	30	31	32	33	34	35	36
Car (cont.)	NC	NC	C	NC	C	C	NC	NC	NC	C	C	NC
Runs (cont.)	13	13	14	15	16	17	17	17	17	18	18	19
Seg. (cont.)	37	38	39	40	41	42	43	44	45	46	47	48
Car (cont.)	NC	NC	NC	C	NC	NC	NC	NC	C	C	C	NC
Runs (cont.)	19	19	19	20	21	21	21	21	21	22	22	23
Seg. (cont.)	49	50										
Car (cont.)	C	NC										
Runs (cont.)	24	25										

Segs. with cars $N_C = 19$

Total runs $R = 25$

Segs. without cars $N_{NC} = 31$

Mean

$$\mu_R = \left(\frac{(2(N_C \times N_{NC})}{N}\right)+1 = \left(\frac{2(19 \times 31)}{50}\right)+1 = 24.56$$

Variance

$$\sigma_R^2 = \frac{(2(N_C N_{NC})(2(N_C N_{NC} - N_C - N_{NC}))}{(N)^2(N-1)} = \frac{2(1178(1178-19-31))}{250(49)} = 10.85$$

Z

$$Z = \frac{R-\mu_R}{\sqrt{\sigma_R^2}} = \frac{25-24.56}{\sqrt{10.85}} = 0.134$$

Probability

$$p = 0.897$$

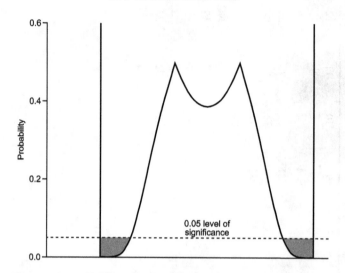

Figure 7.3   Probability distribution associated with the Runs test.

parked vehicles) fall below the 0.05 significance level (9 and 41; 10 and 40; 11 and 39; 12 and 38; and 13 and 37), where the Null Hypothesis would be rejected. The example of 19 and 31 segments in the two types of outcome is not one of these and so the Null Hypothesis is accepted. It is worth noting before moving on from the Runs test that the sequence of occupied and unoccupied road segments in this example may display **spatial autocorrelation**, a tendency for similar data values to occur close together in space, a topic that we will look at in Chapter 10. However, for the time being it is useful to consider the idea that in a sequence of 50 dichotomous outcomes a combination of 19 and 31 runs of the two types could produce a number of different spatial patterns in addition to the one shown in Box 7.4b.

7.2.4   Analysing patterns of spatial entities

Chapter 6 included an introduction to spatial pattern analysis and explored neighbourhood techniques that use measurements of the distances between observations. Quadrat or grid-based techniques offer another approach to analysing such patterns, particularly in respect of phenomena represented as points. These involve overlaying the entire set of points with a regular, but not necessarily rectangular, grid and counting the number of entities falling within each of its cells. In certain respects, this is analogous to crosstabulating a set of observations in respect of two attributes or categorized variables, and assigning them to a set of cells. The main difference is that the grid lines represent spatial boundaries and thus summarize the overall pattern of the points' location. The frequency distribution of the number of point entities per grid cell can be assessed in relation to what would be expected according to the Poisson probability distribution (see Chapter 5). In general, if the observed and expected frequency distributions are sufficiently similar, then it might be concluded that the entities are randomly distributed.

A useful feature of the Poisson distribution is that its mean and variance are equal, which enables us to formulate a Null Hypothesis stating that if the observed pattern of points is random then according to the Poisson distribution the **Variance Mean Ratio (VMR)** will equal 1.0. In contrast, a statistically significant difference from 1.0 suggests some nonrandom process at least partially accounts for the observed pattern, which must necessarily tend towards either clustering or dispersion. Chapter 6 introduced the idea that any particular set of points, even if it includes all known instances of the entity, should be considered as but one of a potentially infinite number of such sets of the phenomena in question: it may therefore be regarded as a sample. A finite number of points distributed across a fixed number of grid cells will have a mean that does not vary (λ) irrespective of whether they are clustered, random or dispersed. The three grids in Figure 7.4 represent situations where a set of 18 points distributed across nine cells is either clustered, random or dispersed and there is an overall average of two points per cell in each case. The nature of the pattern is therefore indicated by differences in the variances of the point patterns. A variance to mean ratio significantly greater than 1.0 suggests clustering and conversely less than 1.0 uniformity or dispersion. Since any one set of points constitutes a sample, the VMR should be tested for its significance, which can be achieved by conversion into a t or χ^2 test statistic.

We have already come across the illustration in Box 7.5b, which shows a nine-cell square grid overlain on the pattern of Burger King's restaurants in Pittsburgh. The numbers in the quadrats represent the observed frequency count of these restaurants, which seem to display some evidence of clustering in the pattern of their locations. The question is whether such visual evidence is supported by spatial statistics. The outcome of applying the Variance Mean Ratio procedure in Box 7.5 suggests that the pattern is not significantly different from what might be expected to occur if a random process was operating. However, there remains an issue over whether this result is an artefact of the number of quadrats and where the grid is located. The size of the

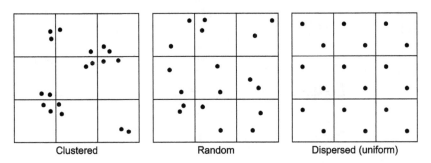

Figure 7.4 Clustered, random and dispersed spatial distributions of point features.

Box 7.5a: The Variance to Mean Ratio.

$$\text{VMR} = \frac{s^2}{\lambda} = s^2 = \frac{1}{n-1}\left(\sum f_i x_i^{\ 2} - \frac{\left(\sum f_i x_i\right)^2}{n} \right) : \lambda = \frac{n}{n_q}$$

Conversion of VMR to t test statistic: $= \dfrac{\text{VMR}-1}{s_{\text{VMR}}}$

Box 7.5b: Application and Testing of the Variance to Mean Ratio.

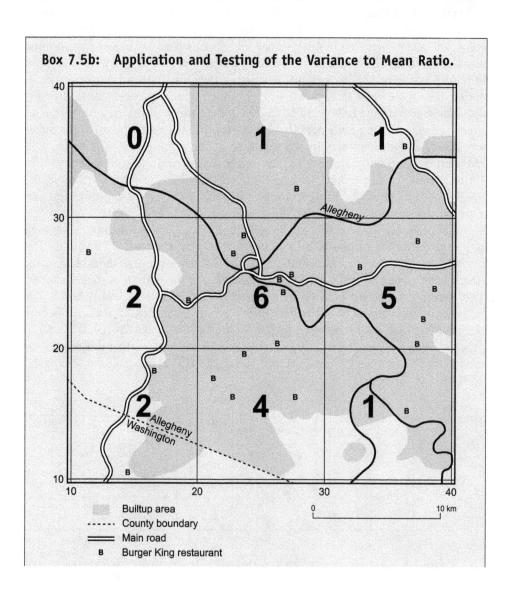

A regular grid of quadrats is overlain on a map showing the location of the point features and a count of points per quadrat is made. However, before the Variance to Mean Ratio (VMR) can be computed, it is necessary to calculate the variance and mean of the points in the quadrats. Calculating the mean or average (λ) number of points per quadrat is relatively easy, it is the number of points (n) divided by the total quadrats (k). The equation for calculating the variance is not unlike the one used to estimate the variance for a nonspatial frequency distribution in Chapter 5. The number of points per quadrat in the variance formula is denoted by x in each of i frequency classes going from 0, 1, 2, 3 ... i and the frequency count of quadrats in each class is f. There are two alternative test statistics that can be calculated in order to examine the significance of the VMR – the t statistic and the continuous, χ^2, (not Pearson's Chi-square) in both cases with $k - 1$ degrees of freedom. The probabilities associated with these can then be used in the usual way to decide whether to accept or reject the Null Hypothesis.

A regular grid containing nine cells or quadrats has been superimposed over the map of Pittsburgh showing the location of the Burger King restaurants and a count of restaurants per quadrat has been made. The observed counts of restaurants in the centre and centre-right quadrats are relatively high (6 and 5, respectively) whereas the remainder are less well served. But is this any different from what would be expected if the restaurants were randomly distributed? Or are they, as it might appear, mostly located in the main built-up area of the city? The mean number of restaurants per quadrat is 2.444 (22/9) and the variance 1.630 resulting in a VMR equal to 0.667, which suggests the Burger King restaurants tend towards being dispersed in a fairly uniform pattern. Calculating both types of test statistic with respect to this VMR value and their associated probabilities and with 8 degrees of freedom indicates that we really cannot conclude that this marginally dispersed pattern is any more likely than would be expected if a random locational process was operating.

The key stages in applying the variance to mean ratio and associated t and χ^2 tests are:

State Null Hypothesis and significance level: the restaurants are distributed at random in the city and any suggestion from the VMR ratio that they are either clustered or dispersed (uniform) is the result of sampling error and its value is not significant at the 0.05 level.

Calculate VMR ratio: the tabulated ordered frequency distribution of Burger King's restaurants per quadrat and calculations to obtain the variance and mean, prerequisites of the VMR are given below (VMR = 0.667).

Calculate t and/or χ^2 test statistic(s): the t statistic requires that the standard error of the mean be obtained before calculating the test statistic itself, which equals -1.333 in this case. The χ^2 statistic is simply VMR multiplied by the degrees of freedom and here equals 5.336.

Determine the probability of t and/or χ^2 test statistic(s): these probabilities are, respectively, 0.220 and 0.721 with 8 degrees of freedom.

Accept or reject the Null Hypothesis: both probabilities are such as to clearly indicate that the Null Hypothesis should be accepted at the 0.05 level of significance.

Box 7.5c: Calculation of Variance to Mean Ratio and associated t statistic.

Number of Burger King restaurants (x)	Frequency count (f)	fx	fx^2
0	1	0	0
1	3	3	3
2	2	4	8
3	0	0	0
4	1	4	16
5	1	5	25
6	1	6	36
	9	22	88

Sample variance

$$s^2 = \frac{1}{n-1}\left(\sum f_i x_i^2 - \frac{\left(\sum f_i x_i\right)^2}{k}\right) = \frac{1}{22-1}\left(88 - \frac{(22)^2}{9}\right) = 1.630$$

Sample mean

$$\lambda = \frac{n}{k} = \frac{22}{9} = 2.444$$

Variance to Mean Ratio

$$\text{VMR} = \frac{s^2}{\lambda} = \frac{1.630}{2.444} = 0.667$$

Standard error of the mean

$$\text{VMR}_s = \frac{2^2}{(k-1)} = \frac{2^2}{(9-1)} = 0.500$$

t statistic

$$t = \frac{\text{VMR}-1}{s_{\text{VMR}}} = \frac{0.667-1}{0.5} = -1.333$$

Chi-square statistic

$$\chi^2 = (k-1)\text{VMR} \qquad\qquad (9-1)0.667 = 5.336$$

Probability   t statistic $p = 0.220$ Chi-square statistic $p = 0.721$

quadrats, 10 km, has been determined in an arbitrary fashion, as have the orientation and origin of the grid. A grid with smaller or larger quadrats and/or rotated clockwise 45° is likely to have produced entirely different observed and therefore expected frequency distributions, although the underlying location of the points (restaurants) is unaffected. Help with deciding quadrat size is provided by the rule-of-thumb that these should be in the range D to $2D$, where D is the study area divided by the number of points. In this example there are 22 points in a 900 km² study area $D = 40.9$), which suggests the quadrats should have been in the range 40.9 to 8.18 km²: the cells in the grid in Box 7.4b are slightly larger than this upper guideline at 100 km².

An alternative grid-based way of examining the spatial pattern of a set of point observations to see if they are randomly distributed across a set of quadrats is by carrying out a Pearson's Chi-square test using the Poisson distribution to generate the

Box 7.6:   Calculation of Pearson's Chi-square statistic.

Number of Burger King restaurants per grid square (x)	Frequency count O	Expected frequency count E	$(O - E)$	$\dfrac{(O - E)^2}{E}$
0	1	0.78	−0.78	0.78
1	3	1.91	1.09	0.62
2	2	2.33	1.67	1.19
3	0	1.90	−1.90	1.90
4	1	1.16	2.84	6.93
5	1	0.57	4.43	34.58
6	1	0.23	5.77	143.80
	9	8.89		189.81

Pearson's chi-square statistic	$\chi^2 = 189.81$
Probability	$p < 0.000$

expected frequencies. Thus, the expected frequencies reflect what would have occurred if a random Poisson process was at work. The Pearson's Chi-square test for a univariate frequency distribution has already been examined and Box 7.6 simply illustrates the calculations involved with applying the procedure to the counts of Burger King's restaurants in Pittsburgh. The observed and expected frequency counts relate to the quadrats with different numbers of restaurants (i.e. 0, 1, 2, 3, 4, 5 and 6) not to the numbers of restaurants themselves. So, with a total of 22 restaurants and nine quadrats the average number per cell is 2.444. Although this example deals with fewer points than would normally be the case, it is realistic to assume that the probability of an observation occurring at a specific location is small. There are two factors reducing the degrees of freedom in this type of application of the test, not only is the total count known but also calculating the expected frequencies using the Poisson distribution employs the sample mean. Thus, the degrees of freedom are the number of classes minus 2 and equal 5 in this case. With a Null Hypothesis that the observed distribution of points is random and referring to the results shown in Box 7.6, we can see that there is an extremely low probability (<0.000) of having obtained a χ^2 test statistic as high as 189.81 with 5 degrees of freedom by chance and so the Alternative Hypothesis can be accepted that restaurants are not randomly distributed.

7.3   Two samples and one (or more) variable(s)

Many investigations are concerned with examining the differences and similarities between two populations through separate random or stratified random samples rather than the more limited case of focusing on a single set of observations. Our exploration of nonparametric tests now moves on to such situations as samples drawn from populations of arable and dairy farms, of acidic and alkaline soils, of sedimentary and igneous rocks, and of coastal and inland cities. The majority of questions to be examined in these situations involve nominal or ordinal frequency counts rather than measurements for continuous variables and are essentially concerned with deciding the fate of Null Hypotheses that claim the observations from the two samples are randomly distributed amongst the categories of an attribute or classified variable. Such tests seek to establish if the samples are genuinely from two significantly different populations or whether they should really be viewed as just one. The tests work by comparing the observed frequency counts with those that would be expected if they were distributed randomly or proportionately between the cells of a crosstabulation.

Some of the procedures examined in this section can also be applied in situations where there is only one sample. In these circumstances, rather than concentrating on the differences between two samples, the objective is to examine the distribution of observations between two attributes or categorized variables. For example, the data values in a set of water samples collected at different locations along a river might be crosstabulated in respect of an attribute recording the land use at the sample point and the acidity/alkalinity of the water with the aim of investigating any association between these characteristics. Most investigations using the nonparametric procedures examined here will carry out a series of such analyses in order to explore the data values in the sample(s) so as to reach conclusions about the research questions.

7.3.1 Comparing two attributes for one sample (or one attribute for two or more samples)

Section 7.2.2 outlined the theoretical background to Pearson's Chi-square test and illustrated how it could be applied to a univariate frequency distribution. The same test can be used in two further ways:

- to examine whether a sample of observations crosstabulated on two (or more) attributes or classified variables are independent or associated with each other;

- to discover whether the frequency distributions of observations from two (or more) samples in respect of the same attribute or categorized variable are significantly different from each other.

When used in the second situation Pearson's Chi-square test is the nonparametric equivalent of one-way analysis of variance (see Chapter 6). Although conceptually different the observations in each situation are normally displayed for the purpose of statistical analysis in the form of a crosstabulation comprising rows (r) and columns (c), which define the size of the table in terms of total cells. The simplest such table is a two-by-two crosstabulation where the total observations have been distributed across four cells. There is no reason why the two dimensions of a table must be equal and it is entirely feasible for the observations in a sample to be tabulated in respect of one attribute with three categories and another with four to produce a three-by-four table with 12 cells.

How many cells would there be in a crosstabulation of five samples by an attribute with three categories?

The total count of observations in the cells of a table is finite in any particular dataset and so the more cells there are the greater the chance that some will have few or no observations. In many cases the number of cells is also finite, being determined by the number of attribute/variable categories and/or samples. However, there are occasions when it is justifiable to introduce some flexibility by combining or collapsing the categories of an attribute or variable in order to increase counts in at least some of the cells. Farms in the parishes stretching across the South Downs in East and West Sussex have been allocated to two sets of size classes in Table 7.2 according to their land areas as recorded in the 1941 Agricultural Census. The first classification has 13 groups with eight of these referring to farms under 100 ha, whereas the second is more general with only two classes below this size threshold. The total number of farms in each classification is the same and so the

Table 7.2   The effect of adjusting attribute or variable categories on frequency distribution counts.

Detailed classification	East Sussex	West Sussex	Total	General classification	East Sussex	West Sussex	Total
0.0–1.9	3	1	4	0.00–49.9	174	177	351
2.0–4.9	54	54	108				
5.0–9.9	39	38	77				
10.0–19.9	30	27	57				
20.0–29.9	23	23	46				
30.0–39.9	17	18	35				
40.0–49.9	8	16	24				
50.0–99.9	40	42	82	50–99.9	40	42	82
100.0–199.9	48	49	97	100.0–299.9	61	68	129
200.0–299.9	13	19	32				
300.0–499.9	4	17	21	300.0 and over	8	20	28
500.0–699.9	3	1	4				
700.0 and over	1	2	3				
Total	283	307	590	Total	283	307	590

Source: MAFF Agricultural Statistics.

low counts in some cells have been overcome by collapsing some of them together. The interaction between the number of rows and columns and the total count of observations not only defines the number of cells in a table, but also constrains how the bivariate distribution is free to vary, in other words they control the degrees of freedom. Referring to Table 7.2, only one cell is free to vary within each column, because total farms in the South Downs parishes in the two counties are fixed. However, when distributing the observations in a single sample across the intersecting rows and columns defined by two categorized attributes/variables, the degrees of freedom are defined as $(r - 1)(c - 1)$.

How many degrees of freedom are there in		A	B	C	D	E
this crosstabulation?	1					
	2					
	3					

The data collected for investigations in geography, Earth and environmental sciences together with many of the social sciences will often includes variables that are unsuited to analysis by parametric procedures. Pearson's Chi-square test can offer a flexible and conceptually straightforward alternative means of addressing research questions in these subject areas. Pearson's Chi-square test is broadly concerned with testing a Null Hypothesis that the distribution of the observations in respect of two attributes or two or more samples and one attribute is random and they are unconnected with each other. In other words, observations are independently allocated to the cells in the crosstabulation. Thus, in the case of a two-by-two table, the fact that an observation is assigned to the top row has no influence on whether it is assigned to left or right column. Pearson's Chi-square test is reasonably robust, although since the probabilities associated with the test statistic are influenced by sample size its use is generally not recommended if there are fewer than 30 observations. However, even this number is problematic if the dimensions of the table result in more than 20 per cent of cells having an expected count of less than 5. Box 7.7 illustrates how to apply Pearson's Chi-square test when there are two or more samples and one attribute using data from the survey of households in four settlements in mid-Wales. The application examines whether people in the four settlements use different modes of transport to carry out their major shopping trip, or equivalently whether different types of transport are favoured by residents in the sampled settlements.

Box 7.7a: The Bivariate Pearson's Chi-square (χ^2) test.

Pearson's Chi-square (χ^2) test statistic: $= \sum\limits_{j}^{1} \dfrac{(O_j - E_j)^2}{E_j}$

Box 7.7b: Application of the Bivariate Pearson's Chi-square (χ^2) test.

The Pearson's Chi-square tests focuses on the overall difference between the observed (O) and expected (E) frequencies. The test statistic formula calculates the sum of the squared differences between the observed and expected counts divided by the expected for each cell. The first step in carrying out the test is to distribute the observations into a crosstabulation and then to calculate the expected frequencies. Calculating the expected frequencies for any individual cell involves multiplying the corresponding marginal row and column totals by the total number of observations (see the upper part of Box 7.7d) for each cell in the cross-tabulation. There is an observed count of 8 households in the case of the top left cell (travel by family car and Barmouth sample) and the expected value is $7.2 \left(\dfrac{(36 \times 20)}{100} \right)$: the difference is very small (0.8) in this cell, which contrasts with the lower-right cell where the difference is 4.8. The second step involves squaring the difference in each cell and dividing by the expected count (lower part of Box 7.7d) and then summing these to obtain χ^2. The probability of obtaining the χ^2 value varies according to the degrees of freedom, which equals 9 in this example. A decision on the Null Hypothesis is reached by comparing this probability with the chosen level of significance. If the probability of having obtained the particular test value is equal to or less than 0.05 (or another selected significance level), the Null Hypothesis should be rejected.

The samples of households in the four settlements in mid-Wales have been tested with Pearson's Chi-square to see if there is a significant difference in the mode of transport used by households when doing their main shopping trip. The χ^2 statistic is 10.440, which has a probability of 0.316 with 9 degrees of freedom. This implies that the Null Hypothesis should be accepted since for a four-by-four crosstabulation (i.e. 9 degrees of freedom) the test statistic value would have had to reach at least 16.619 for a significant difference at the 0.05 level. The test points to the conclusion that households in the four settlements tend to be divided between the different modes of transport in a random fashion.

Application of Pearson's Chi-square test has six main stages:

Declare the Null Hypothesis and significance level: any differences between the observed and expected frequencies are a consequence of sampling error and they are not significant at the 0.05 level.

Compute the expected frequencies (E): these usually come from dividing the product of the marginal totals corresponding to each cell by the overall total number of observations.

Calculate test statistic (χ^2): this is the sum of the squared differences between O and E divided by E.

Determine the degrees of freedom: in this example $df = (4 - 1)(4 - 1)$ and so there are 9 degrees of freedom.

Determine the probability of the calculated χ^2: the test statistic equals 10.440, which has a probability of 0.316 with 9 degrees of freedom.

Accept or reject the Null Hypothesis: the Null Hypothesis should be accepted since 0.316 is >0.05 and would be obtained in approximately 32 samples out of a 100, if households in the four settlements were sampled on that number of occasions.

Box 7.7c: Assumptions of the Pearson's Chi-square (χ^2) test.

There are five main assumptions of the Pearson's Chi-square test:

Observations selected for inclusion in the sample(s) should be selected by simple or stratified random sample and be independent of each other.

Adjustment of attribute/variable classes may be carried out, although care should be taken to ensure the categories are still meaningful.

The total count of observations in the four samples from the mid-Wales settlements happen to equal 100 and the cell counts are percentages of the overall total. However, the test should not be applied to data recorded as relative frequencies (percentages or proportions) and these should be converted back to absolute counts before carrying out a Pearson's Chi-square test.

Low expected frequencies, usually taken to mean <5 counts in no more than 20% of cells, are likely to mean that the χ^2 test statistic is unreliable. The problem may be overcome by redefining attribute/variable classes.

The test should not be applied if any of the attributes or grouped variables are ordered or ranked at all, since the calculated test statistic will be the same regardless of the ranking.

7.3.2 Comparing the medians of an ordinal variable for one or two samples

The basis of the **Mann–Whitney U test** is that a set of ordered data values for a variable can be partitioned into two groups. This can arise in two ways:

- there are two samples that have been measured in respect of an ordinal variable;

- there is one sample containing a dichotomous attribute, which may be spatial or temporal, enabling the observations to be split into two groups across an ordinal variable.

The test works by pooling the ordinal measurements across the complete set of observations and then compares the medians for each group or sample. The test establishes whether the quantitative difference between the medians is small enough simply to be the result of sampling error. If the test is applied to variables that were originally collected on the interval or ratio scales of measurement, then some loss of information is implied by sorting into ascending order then assigning rank scores. However, where the assumption of normality of the variables cannot realistically be believed, the Mann–Whitney test provides a robust alternative to the (unpaired) two sample t test. Box 7.8 applies the test to the sample of road segments with or without vehicles analysed previously using the Runs test.

Box 7.7d: Calculation of the Pearson's Chi-square (χ^2) test.

Transport mode		Sample settlement Barmouth	Brecon	Newtown	Talgarth	Row totals
Family car	O	8	6	16	6	36
	E	$\frac{(36\times20)}{100}=7.2$	$\frac{(36\times26)}{100}=9.4$	$\frac{(36\times35)}{100}=12.6$	$\frac{(36\times19)}{100}=6.8$	36
Other car	O	4	3	6	5	18
	E	$\frac{(18\times20)}{100}=3.6$	$\frac{(18\times26)}{100}=4.7$	$\frac{(18\times35)}{100}=6.3$	$\frac{(18\times19)}{100}=3.4$	18
Public transport	O	3	5	8	2	18
	E	$\frac{(18\times20)}{100}=3.6$	$\frac{(18\times26)}{100}=4.7$	$\frac{(18\times35)}{100}=6.3$	$\frac{(18\times19)}{100}=3.4$	18
Walk or cycle	O	5	12	5	6	28
	E	$\frac{(28\times20)}{100}=5.6$	$\frac{(28\times26)}{100}=7.3$	$\frac{(28\times35)}{100}=9.8$	$\frac{(28\times19)}{100}=5.3$	28
Column totals		20	26	35	19	100
Family car		$\frac{(8-7.2)^2}{7.2}=0.09$	$\frac{(6-9.4)^2}{9.4}=1.21$	$\frac{(16-12.6)^2}{12.6}=0.92$	$\frac{(6-6.8)^2}{6.8}=0.10$	
Other car		$\frac{(4-3.6)^2}{3.6}=0.04$	$\frac{(3-4.7)^2}{4.7}=0.60$	$\frac{(6-6.3)^2}{6.3}=0.01$	$\frac{(5-3.4)^2}{3.4}=0.73$	
Public transport		$\frac{(3-3.6)^2}{3.6}=0.10$	$\frac{(5-4.7)^2}{4.7}=0.02$	$\frac{(8-6.3)^2}{6.3}=0.46$	$\frac{(2-3.4)^2}{3.4}=0.59$	
Walk or cycle		$\frac{(5-5.6)^2}{5.6}=0.06$	$\frac{(8-7.2)^2}{7.2}=0.09$	$\frac{(5-9.8)^2}{9.8}=02.35$	$\frac{(6-5.3)^2}{5.3}=0.09$	

Pearson's chi-square statistic $\chi^2 = 10.440$

Degrees of freedom $df = (r-1)(c-1) = (4-1)(4-1) = 9$

Probability $p = 0.316$

Box 7.8a: Mann–Whitney U test.

Mann–Whitney U test statistic: $= \sum r - \dfrac{n_r(n_r+1)}{2}$

Conversion of U to Z test statistic: $= \dfrac{|U - \mu_U|}{\sigma_U}$

Box 7.8b: Application of Mann–Whitney U test.

The Mann–Whitney U test is commonly used to examine the difference between the medians of an ordinal variable for two samples or two groups of observations from one sample. The objective in both instances is to investigate whether the difference is of such a size that it is likely to have arisen from sampling error, in other words that the true difference is zero. In general terms, the Null Hypothesis is that the rank scores and therefore the medians of observations in the two samples or groups from a sample are such that they balance each other and so they really originate from one undifferentiated population in respect of the ordinal variable being examined. The calculations for the test statistic are relatively straight-forward. First, the observations are sorted in ascending order in respect of the variable and their ranks (R) are assigned (smallest value assigned rank 1, second rank 2 and so on). The second step involves sorting the observations' rank scores according to the sample or group to which they belong. The rank scores are summed for these two groups of observations and the smaller total is designated $\sum r$. The value of this quantity is inserted into the equation for the test statistic together with n_r – the number of observations in the group that provides $\sum r$.

The test statistic's probability is determined in different ways depending on the number of observations in each group, although using statistical software to apply the test usually removes the need to choose between the methods. The probability distribution of U becomes approximately Normal when there are more than 20 observations in each sample or group. The probability of the test statistic is obtained by transforming U into Z having calculated a population mean and standard deviation of U given the known, fixed number of observations in each sample or group (see calculations below). The fate of the Null and Alternative Hypotheses is decided by reference to the specified significance level, typically whether ≤0.05.

The Mann–Whitney test has been applied to the 250 m stretch of road divided into 5 m segments used in Box 7.4. The distance from the start of the sequence of segments increments by 5 m when progressing from one segment to the next and they are categorized into two groups by means of the dichotomous attribute denoting whether or not there are parked vehicles on at least one side of the road: these are denoted by the subscripts C and NC (car and no car), respectively. The raw data (distance measurements and rank scores) are given below. The Null Hypothesis is that the median distance of segments in the two groups from the start of the stretch of road is the same and that sampling error accounts for the observed difference. Although strictly speaking there are just too few observations in the group with parked vehicles ($n_C = 19$), the U test statistic has been converted to Z to illustrate the procedure. The Z value 0.290 has a probability of 0.772, which clearly results in the Null Hypothesis being accepted at the 0.05 significance level.

Application of the Mann–Whitney test has seven main stages:

State the Null Hypothesis and significance level: any difference two groups of road segments is the result of sampling error and it is not significant at the 0.05 level.

Sort and assign ranks (R): since the segments are numbered from the start of the length of road and the distances increment in 5 m units from this point, the sorted rank scores are the same as the segment identification numbers. Sort the segments again according to dichotomous grouping attribute or sample membership.

Sum the rank scores of each group to identify $\sum r$: the results of these additions are shown below.

Calculate test statistic (U): insert $\sum r$ into the equation and calculate the test statistic, which equals 470.

Determine the probability of the calculated U: in this example, since the number of observations in each group is very nearly >20, U as been converted into Z (0.290), which has a probability of 0.772.

Select one- or two-tailed probability: if there is any prior, legitimate reason for arguing that one group of segments be nearer to the start of the road than the other, apply a one-tailed test by halving the probability associated with Z before comparing with the significance level.

Accept or reject the Null Hypothesis: the Null Hypothesis should be accepted since there is an extremely high probability of having obtained the test statistic by chance; division of the road segments according to the presence or absence of parked vehicles seems to be random along the entire length of the road (0.772 > 0.050).

Box 7.8c:　Assumptions of the Mann–Whitney U test.

There are three main assumptions of the Mann–Whitney U test:

Observations should be selected by means of simple random sampling.

The observations should be independent of each other in respect of the ordinal, ranked variable. (Note: it is arguable whether this is the case in this example.)

The test is robust with small and large sample sizes, and is almost as powerful as the two sample t test with large numbers of observations.

Box 7.8d: Calculation of the Mann–Whitney test.

Segment	Vehicle parked	Distance (x) in metres	Overall rank (R)	Segment ID sorted	Sorted rank
1	N	5	1	1	1
2	N	10	2	2	2
3	N	15	3	3	3
4	N	20	4	4	4
5	Y	25	5	7	7
6	Y	30	6	8	8
7	N	35	7	11	11
8	N	40	8	13	13
9	Y	45	9	15	15
10	Y	50	10	16	16
11	N	55	11	18	18
12	Y	60	12	19	19
13	N	65	13	20	20
14	Y	70	14	25	25
15	N	75	15	26	26
16	N	80	16	28	28
17	Y	85	17	30	30
18	N	90	18	31	31
19	N	95	19	32	32
20	N	100	20	33	33
21	Y	105	21	36	36
22	Y	110	22	37	37
23	Y	115	23	38	38
24	Y	120	24	39	39
25	N	125	25	41	41
26	N	130	26	42	42
27	Y	135	27	43	43
28	N	140	28	44	44
29	Y	145	29	45	45
30	N	150	30	48	48
31	N	155	31	50	50
32	N	160	32	5	5
33	N	165	33	6	6
34	Y	170	34	9	9
35	Y	175	35	10	10
36	N	180	36	12	12
37	N	185	37	14	14
38	N	190	38	17	17
39	N	195	39	21	21
40	Y	200	40	22	22
41	N	205	41	23	23
42	N	210	42	24	24

Segment	Vehicle parked	Distance (x) in metres	Overall rank (R)	Segment ID sorted	Sorted rank
43	N	215	43	27	27
44	N	220	44	29	29
45	N	225	45	34	34
46	Y	230	46	35	35
47	Y	235	47	40	40
48	N	240	48	46	46
49	Y	245	49	47	47
50	N	250	50	49	49

Sum of the ranks of segments with parked vehicles	$\sum R_C = 470$
Sum of the ranks of segments without parked vehicles	$\sum R_{NC} = 805$
Mann–Whitney U statistic	$U = \sum r - \dfrac{n_r(n_r+1)}{2} = 470 - \dfrac{19(19+1)}{2} = 280$
Population mean U	$\mu_U = \dfrac{n_C n_{NC}}{2} = \dfrac{19(31)}{2} = 294.50$
Standard deviation U	$\sigma_U = \sqrt{\dfrac{n_C n_{NC}(n_C + n_{NC}+1)}{12}} = \sqrt{\dfrac{19 \times 31(19+31+1)}{12}} = 50.03$
Z test statistic	$Z = \dfrac{\lvert U - \mu_U \rvert}{\sigma_U} = \dfrac{\lvert 280 - 294.50 \rvert}{50.03} = 0.290$
Probability	$p = 0.772$

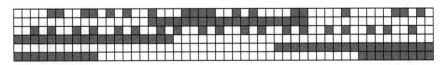

Figure 7.5 Combinations of road segments in groups with 19 and 31 observations.

The results of the Mann–Whitney test in Box 7.8 support the previous conclusion that the occurrence of the two types of segments is not significantly different from what might be expected if they were to have occurred randomly along the length of the road. However, there are many different ways in which 50 observations, in this case road segments, could be allocated to two groups containing 19 and 31 cases. The top row of Figure 7.5 reproduces the situation shown in Box 7.4, while the others present five rather more distinctive patterns. The clearest difference in medians occurs in the fourth row where all the segments without parked vehicles are at the start and the probability of such extreme differentiation occurring by chance is very small. Even the arrangement in rows 2, 5 and 6 with all the occupied segments clumped in one or two groups have relatively high probabilities of occurring randomly (>0.750). The regular, almost systematic spacing of vehicles along the stretch of road represented in row 3 is the one most likely to occur through a random process ($p = 0.928$). These examples help to reveal that the results of the test are dependent not only on the degree to which the ordinal variable separates observations between the two groups or samples, but also on the size of the groups. Rows 4 and 5 as just as differentiated as each other, but in one case the occupied segments are at the start and in the other at the end.

7.4 Multiple samples and/or multiple variables

The procedures examined in this section progress our exploration of statistical testing to those situations where there are two or more samples and two or more attributes, or one sample and three or more attribute or classified variables. These nonparametric techniques are less demanding in terms of the assumptions made about the underlying distribution of the data values than their parametric counter-parts, but, nevertheless, they can provide important insights into whether the observed data are significantly different from what might have occurred if random processes were operating. The data values are either measurements recorded on the ordinal scale (i.e. by placing the observations in rank order, but not quantifying the size of the differences between them) or categorical attributes or classified variables that allow the observations to be tabulated as counts. The crosstabulations of data considered here have at least three dimensions. For example, a single sample of households might be categorized in respect of their housing tenure, the economic status of the oldest member of the household and whether or not anyone in the household had an overseas holiday during the previous 12 months. Supposing there were four tenure categories, six types of economic status and by definition the holiday attribute is dichotomous, then there would be 48 cells in the crosstabulation ($4 \times 6 \times 2$). Such a table is difficult to visualize and would become even more so with additional dimensions. Nevertheless, some statistical tests exist to cope with these situations. The Pearson's Chi-square test examined in Section 7.3.1 can be extended in this way and the Kruskal–Wallis H test is an extension of the Mann–Whitney test.

7.4.1 Comparing three attributes for one sample (or two attributes for two or more samples)

Extension of the Pearson's Chi-square test to those situations where the crosstabulation of counts has three or more dimensions can occur in two situations where there is (are):

- one sample and three (or possibly more) categorized attributes;

- three or more samples and two (or more) categorical variables.

Both situations are represented in the three-dimensional crosstabulation shown in Table 7.3. Attributes 1 and 2 are categorized into five (I, II, III, IV and V) and three (1, 2 and 3) classes, respectively. The third dimension is shown as jointly representing either three samples or three classes a third attribute (A, B, and C). Overall, Table 7.3 represents a five-by-three-by-three crosstabulation. The counts in the crosstabulation are for illustrative purposes only and each sample (or group for attribute 3) conveniently contains 100 observations, thus the individual cells and the marginal totals in each part of the table sum to 100. The calculation of Pearson's Chi-square test statistic proceeds in the same way in this situation as with a two-dimensional crosstabulation. Once the expected frequencies have been computed, the squared difference between the observed and expected count divided by the expected $\left(\dfrac{(O-E)^2}{E}\right)$ is calculated for each cell. The sum of these gives the test statistic. The degrees of freedom are defined as $(X_1-1)(X_2-1) \ldots (X_j-1)$, where X represents the number of categories in attributes 1 to j. In the case of the crosstabulation in Table 7.3 there are 16 degrees of freedom $(5-1)(3-1)(3-1)$. The probability associated with the test statistic with these degrees of freedom enables a decision to be reached regarding the Null Hypothesis. Since most multidimensional crosstabulations will have more cells than the simpler two dimensional ones, there is a greater likelihood that low expected frequencies might occur. It is therefore important to ensure that sample sizes are sufficiently large to reduce the chance of this problem occurring during the planning stage of an investigation.

Table 7.3 Three-dimensional crosstabulation with three categorized attributed or three samples with two categorized attributes.

			Attribute 1					
			I	II	III	IV	V	Total
Attribute 3, group A	Attribute 2							
OR		1	15	8	4	3	5	35
Sample A		2	7	6	2	5	0	18
		3	6	10	4	11	13	44
		Total	28	24	10	19	18	100

			Attribute 1					
			I	II	III	IV	V	Total
Attribute 3, group B	Attribute 2							
OR		1	6	5	3	10	4	28
Sample B		2	5	7	11	9	4	36
		3	9	2	1	14	10	36
		Total	20	14	15	35	18	100

			Attribute 1					
			I	II	III	IV	V	Total
Attribute 3, group C	Attribute 2							
OR		1	11	6	10	15	0	42
Sample C		2	7	7	3	5	6	28
		3	7	3	0	4	16	30
		Total	25	16	13	24	22	100

7.4.2   Comparing the medians of an ordinal variable for three or more samples (or one sample separated into three or more groups)

The **Kruskal–Wallis H** test represents an extension or generalization of the Mann–Whitney test that enables rather more complex types of question to be investigated. The test can be used in two situations where the focus of attention is on differences in the ranks of an ordinal variable in respect of:

- one sample and three (or possibly more) ordered or ranked variables;

- three or more samples and one ordered variable.

The H test is essentially the nonparametric equivalent of the one-way ANOVA procedure examined in Section 6.3.1. One-way ANOVA explores whether the differences in the means of a continuous variable for three or more samples are statistically significant, the Kruskal–Wallis test performs an equivalent task in respect of differences in the medians. The ordinal (or sorted continuous) variable provides a unifying thread to each of these situations. The data values for this variable are pooled, just as with the Mann–Whitney test, across the three or more samples or groups from a single sample and rank scores are assigned with average ranks used for any tied observations. The general form of the Alternative and Null Hypotheses for the Kruskal–Wallis test, respectively, assert that the difference in medians between the three or more samples or groups from a sample in respect of the ordinal variable under investigation either are or are not significant. Just as with the Mann–Whitney test, the Kruskal–Wallis test explores the extent to which the complete set of observations is interleaved or interspersed with regard to their data values on the ordinal variable(s).

The Kruskal–Wallis test has been applied to the samples of households in mid-Wales and the length of their major shopping journeys to investigate whether there are differences between the populations of households in the four sampled settlements. Box 7.9 examines this application of the test procedure and reaches the conclusion that there is a significant difference between the median length of the journeys (in km) made by households in the four settlements where the sample surveys were carried out. Examination of the sample data reveals that the median distance for Talgarth households is greater than the other three settlements, however, from the perspective of the test results it cannot be concluded that this necessarily indicates that households in this settlement stand out from the rest. Just as the one-way ANOVA technique focused on the overall difference in the means of three or more samples in respect of a ratio or interval scale variable, so the Kruskal–Wallis test examines the overall, not specific, differences in medians between the samples.

Box 7.9a: Kruskal–Wallis (H) test.

Kruskal–Wallis H test statistic: $= \dfrac{12 \sum N(\bar{R}_i - \bar{R})^2}{N(N+1)}$

Box 7.9b: Application of Kruskal–Wallis (H) test.

The Kruskal–Wallis test is used in two ways, either where one sample has been subdivided into three or more groups or where there are three or more samples. In both cases there is an ordinal variable that has been measured across the groups or samples. The calculations arising from the equation for Kruskal–Wallis test statistic (H) are relatively simple summation and division, although performing these by hand can be tedious, especially if the numbers of observations in the samples or groups is fairly large, because of the amount of data sorting required. The complete set of observations are pooled and sorted in respect of the ordinal variable, and then ranks are determined. Ties are dealt with by inserting the average rank, thus four observations with the same value for the ordinal variable in the 4th, 5th, 6th and 7th positions would each be assigned a rank score of 5.5. The sums of the rank scores for each sample or group are computed and designated scores R_1, R_2, R_3 ... R_i where i denotes the number of samples. The means of the rank sum for each sample or group are calculated as

$\bar{R}_1, \bar{R}_2, \bar{R}_3, \ldots \bar{R}_i$ with the number of observations in each represented as $n_1, n_2, n_3, \ldots n_i$: both of these sets of values are inserted in the equation used to calculate H.

Examination of the probability associated with the calculated H test statistic in conjunction with the chosen level of significance indicates whether the Null Hypothesis should be accepted or rejected. Provided that there are more than 5 observations in each of the samples or groups, the probability distribution of the test statistic approximates Chi-square with $k - 1$, where k corresponds to the number of samples or groups. The Kruskal–Wallis test has been applied to the four samples of households in mid-Wales settlements to examine whether any differences in the median distance travelled to carry out their main shopping trip are significantly greater than might have occurred randomly or by chance.

Application of the Kruskal–Wallis test can be separated into six main steps:

State the Null Hypothesis and significance level: any differences between the observations in the samples (or groups from one sample) have arisen as a result of sampling error and are no greater than might be expected to occur randomly at least 6 times in 100 (i.e. using a 0.05 significance level).

Sort and assign ranks (R): pool all the observations, sort in respect of the ordinal variable (distance to 'shopping town'), assign rank scores to each, averaging any tied values.

Sum the rank scores of each sample (group): the sum of the ranks for each sample or group are calculated and identified as R_1, R_2, R_3 ... R_i, where the subscript distinguishes the groups.

Calculate test statistic (H) and apply correction factor if >25 per cent of observations are tied: calculations using the equation for the test statistic are given below, which equals 21.90 in this example. Corrected H equals 31.43.

Determine the probability of the calculated H: the number of observations in each sample is >5 and so the probability of having obtained the test statistic by chance can be determined from the Chi-square distribution with 3 degrees of freedom and is <0.001.

Accept or reject the Null Hypothesis: the test statistic's probability is lower than the chosen significance level (0.05) and therefore the Null Hypothesis should be rejected. It is very unlikely that the difference in medians between the samples (groups) has arisen by chance.

Box 7.9c: Assumptions of the Kruskal–Wallis (H) test.

There are three main assumptions of the Kruskal–Wallis test:

Observations should be selected by means of simple random sampling.

Sampled entities should be independent of each other and selected from populations of observations that have approximately the same shape in respect of the ordinal variable.

If the number of ties is large (more than 25 per cent of the ordinal values), then the calculated H statistic should be divided by a correction factor $C = 1 - \dfrac{(T^3 - T)}{n_s^3 - n_s}$, where T and n_s are, respectively, the number of tied and total number of observations.

Box 7.9d: Calculation of the Kruskal–Wallis test.

Settlement	Distance (km)	Rank (R)	Settlement (sorted)	Rank (R) (sorted)
Kerry	0.36	3	Kerry	3
Kerry	0.36	3	Kerry	3
Kerry	0.36	3	Kerry	3
Kerry	0.36	3	Kerry	3
Kerry	0.36	3	Kerry	3
Kerry	0.42	7	Kerry	7
Kerry	0.42	7	Kerry	7
Kerry	0.42	7	Kerry	7
St Harmon	0.45	9	Kerry	44
Llanenddwyn	0.58	10	Kerry	46
Llanenddwyn	0.99	11	Kerry	46
St Harmon	1.30	12	Kerry	46
St Harmon	1.92	13.5	Kerry	48
St Harmon	1.92	13.5	Llanenddwyn	10
St Harmon	3.67	16.5	Llanenddwyn	11
St Harmon	3.67	16.5	Llanenddwyn	26
St Harmon	3.67	16.5	Llanenddwyn	27
St Harmon	3.67	16.5	Llanenddwyn	29
St Harmon	4.43	20	Llanenddwyn	29
St Harmon	4.43	20	Llanenddwyn	29
St Harmon	4.43	20	Llanenddwyn	32
St Harmon	4.62	22.5	Llanenddwyn	32
St Harmon	4.62	22.5	Llanenddwyn	32
Talgarth	5.36	24	Llanenddwyn	35
Talgarth	5.41	25	Llanenddwyn	35
Llanenddwyn	7.41	26	Llanenddwyn	35
Llanenddwyn	7.68	27	Llanenddwyn	39
Llanenddwyn	7.74	29	Llanenddwyn	40
Llanenddwyn	7.74	29	Llanenddwyn	41.5

Settlement	Distance (km)	Rank (R)	Settlement (sorted)	Rank (R) (sorted)
Llanenddwyn	7.74	29	Llanenddwyn	41.5
Llanenddwyn	7.75	32	Llanenddwyn	54
Llanenddwyn	7.75	32	Llanenddwyn	55
Llanenddwyn	7.75	32	St Harmon	9
Llanenddwyn	8.07	35	St Harmon	12
Llanenddwyn	8.07	35	St Harmon	13.5
Llanenddwyn	8.07	35	St Harmon	13.5
Talgarth	8.42	37.5	St Harmon	16.5
Talgarth	8.42	37.5	St Harmon	16.5
Llanenddwyn	8.45	39	St Harmon	16.5
Llanenddwyn	8.60	40	St Harmon	16.5
Llanenddwyn	8.80	41.5	St Harmon	20
Llanenddwyn	8.80	41.5	St Harmon	20
St Harmon	9.11	43	St Harmon	20
Kerry	11.55	44	St Harmon	22.5
Kerry	11.96	46	St Harmon	22.5
Kerry	11.96	46	St Harmon	43
Kerry	11.96	46	St Harmon	63
Kerry	12.41	48	St Harmon	64
Talgarth	13.21	49	Talgarth	24
Talgarth	13.63	50	Talgarth	25
Talgarth	13.96	52	Talgarth	37.5
Talgarth	13.96	52	Talgarth	37.5
Talgarth	13.96	52	Talgarth	49
Llanenddwyn	14.89	54	Talgarth	50
Llanenddwyn	15.33	55	Talgarth	52
Talgarth	19.68	58	Talgarth	52
Talgarth	19.68	58	Talgarth	52

Talgarth	19.68	58	Talgarth
Talgarth	19.68	58	Talgarth
Talgarth	19.68	58	Talgarth
Talgarth	21.61	61	Talgarth
Talgarth	21.80	62	Talgarth
St Harmon	41.10	63	Talgarth
St Harmon	55.05	64	Talgarth

	Kerry	Llanenddwyn	St Harmon	Talgarth
Sum of the ranks $\sum R_K, \sum R_L, \sum R_{SH}, \sum R_T$	266	633	389	792
Number of households $n_K\ n_L\ n_{SH}\ n_T$	13	19	16	16
Mean of the ranks $\bar{R}_K, \bar{R}_L, \bar{R}_{SH}, \bar{R}_T$	20.5	33.3	24.3	49.5

Kruskal–Wallis H statistic

$$H = \frac{12\sum n_i(\bar{R}_i - \bar{R})^2}{N(N+1)} = \frac{12(13(20.5-32.5)^2 + 13(33.3-32.5)^2 + 13(24.3-32.5)^2 + 13(49.5-32.5)^2}{64(64+1)} = 21.90$$

Correction factor

$$C = 1 - \frac{(T^3 - T)}{n_s^3 - n_s} = 1 - \frac{(75\,907 - 43)}{262\,144 - 64} = 0.697$$

Corrected H

$$H_{Cor} = \frac{H}{C} = \frac{21.90}{0.697} = 31.43$$

Degrees of freedom

$$df = k - 1 = 4 - 1 = 3$$

Probability

$$p < 0.001$$

7.5 Closing comments

This chapter has examined a set of nonparametric statistical tests that parallel the parametric ones covered in Chapter 6. The main differences between the two groups of techniques relate to whether the data values for the variable(s) support the assumption of normality and whether they are continuous measurements on the ratio or interval scales in the case of parametric procedures. These techniques are concerned with testing numerical quantities, such as the mean, variance and standard deviation with respect to a population or between samples. The equivalent nonparametric tests either focus on ordinal measurements, in which case the main concern is with the median, or on frequency counts. Techniques concerned with frequencies generally look at whether the difference between observed and expected counts is more than might have occurred by chance through sampling error. Sometimes, these tests are used to compare samples and populations, although the role of the population data is generally to calculate the expected frequencies. For example, if an earlier population census had recorded certain proportions of households in five tenure categories, the expected frequencies for the sample of observations could be calculated using these.

Grid-based techniques provide another way of examining the pattern in location of spatial phenomena, especially those that can be represented as points. The process of counting points in a set of quadrats is similar to that of distributing nonspatial observations to set of cells in a crosstabulation. However, the key difference is that the distribution of points captures their pattern in two-dimensional geographical space, whereas the categories or classes used with a nonspatial crosstabulation by definition do not represent the pattern in location of the observations. The next chapter moves our exploration of statistical analysis in respect of spatial and nonspatial data into the realms of examining relationships between variables or attributes. However, this does not mean that we have finished with statistical testing, since the techniques used to calculate quantities to summarize the relationships between variables or attributes may well need to be tested to discover whether the value might have arisen by chance. Given that such quantities are often only produced using sampled data, it may be necessary to generate confidence limits within which the population parameter lies.

A new dataset was introduced in this chapter depicting a sequence of 5 m road segments with or without parked vehicles along a 250 m stretch of road. By way of an introduction to the topic correlation and its application in respect of spatial data examined in Chapter 8, it is worth thinking about whether the values held by each segment (i.e. vehicle or no vehicle) in this sequence are really independent of each other. Is it not possible that vehicles may well be parked outside residential properties and these are distributed with a certain regularity along the road? Are there less likely to be vehicles parked where there are parking restrictions, such as near a school or pedestrian crossing? What do these factors says about the possible lack of independence in the data?

Section 3
Forming relationships

8

Correlation

Correlation is an overarching term used to describe those statistical techniques that explore the strength and direction of relationships between attributes or variables in quantitative terms. There are various types of correlation analysis that can be applied to variables and attributes measured on the ratio/interval, ordinal and nominal scales. The chapter also covers tests that can be used to determine the significance of correlation statistics. Correlation is often used by students and researchers in Geography, Earth and Environmental Science and related disciplines to help with understanding how variables are connected with each other.

Learning outcomes

This chapter will enable readers to:

- carry out correlation analysis techniques with different types of variables and attributes;

- apply statistical tests to find out the significance of correlation measures calculated for sampled data;

- consider the application of correlation techniques when planning the analyses to be carried out in an independent research investigation in Geography, Earth Science and related disciplines.

Practical Statistics for Geographers and Earth Scientists   Nigel Walford
© 2011 John Wiley & Sons, Ltd

8.1   Nature of relationships between variables

Several of the statistical techniques that we have explored in previous chapters have concentrated on one variable at a time and treated this in isolation from others that may have been collected for the same or different sets of sampled phenomena. However, generally the more interesting research questions are those demanding that we explore how different attributes and variables are connected with or related to each other. In statistics such connections are called relationships. Most of us are probably familiar, at least in simple terms, with the idea that relationships between people can be detected my means of analysing DNA. Thus, if the DNA of each person in a randomly selected group was obtained and analysed, it would be possible to discover if any of them were related to each other. If such a relationship were to be found, it would imply that the two people had something in common, perhaps sharing a common ancestor. In everyday terms, we might expect to see some similarity between those pairs of people having a relationship, in terms of such things as eye, skin or hair colour, facial features and other physiological characteristics. If these characteristics were to be quantified as attributes or variables, then we would expect people who were related to each other to possess similar values and those who are unrelated to have contrasting values.

It is always a mistake to push an analogy too far, nevertheless, this illustration of what it means for people to be related and thereby possess similar values for certain physiological attributes does start to help us in trying to understand what is meant by a relationship in statistics. Techniques for statistically analysing relationships are broadly divided into **bivariate** and **multivariate** procedures. The former deal with situations where two attributes or variables (or one attribute and one variable) are the focus of attention: in contrast, multivariate refers to investigations where many attributes/variables are under scrutiny at the same time. Complementary aspects of analysing such relationships, namely correlation and regression, are explored in this and the following chapter.

Correlation is often defined as a way of measuring the direction and strength of the relationship between two attributes or variables. But what is meant by the phrase 'direction and strength of the relationship'? The simplest way of explaining these terms is by means of a hypothetical example. Suppose there is a single sample of observations that have been measured in respect of two variables, here designated as X and Y. The data values for these have been recorded as numbers according to either the interval or ratio measurement scales, which imply a difference of magnitude. In theory, these data values could range from minus to plus infinity, although the measurements for most of the variables familiar to geographers, Earth and environmental scientists will occur within a more limited section of the full numerical range. Arbitrarily we will assume the sample of observations has produced values for X in the range 15.4 to 63.2 and for Y between 23.3 and 47.2. The direction of the relationship between X and Y refers to the pattern between one set of data values and the

Table 8.1 Paired data values for observations in respect of variables X and Y.

Observation	Set A		Set B		Set C	
	X	Y	X	Y	X	Y
A	21.8	36.7	21.8	44.1	15.4	27.5
B	32.2	47.2	15.4	47.2	21.8	23.3
C	47.3	44.1	31.0	43.5	32.2	31.2
D	15.4	34.6	36.2	36.7	31.0	32.2
E	63.2	38.0	32.2	38.0	36.2	26.7
F	52.0	31.2	44.1	32.2	40.5	34.6
G	31.0	43.5	40.5	34.6	47.3	38.0
H	36.2	23.3	52.0	27.5	44.1	44.1
I	44.1	32.3	47.3	31.2	52.0	43.5
J	40.5	27.5	63.2	23.3	63.2	47.2
	Answer 'Yes' to question 1, 2 or 3?		Answer 'Yes' to question 1, 2 or 3?		Answer 'Yes' to question 1, 2 or 3?	

other. Examine the three sets (A, B and C) of paired X and Y data values in Table 8.1 and decide for which set you would answer 'Yes' with respect to each of the following questions:

1. Do the values of X and Y for the observations differ in parallel with each other, so that values towards the lower end of the X range are matched with Y values at the lower end of its range, and vice versa?

2. Do the values of X and Y for the observations differ in an irregular way, so that values towards the upper end of the X range stand just as much chance of being matched with Y values at the upper, middle or lower part of its range, and vice versa?

3. Do the values of X and Y for the observations differ in opposite directions to each other, so that values towards the lower end of the X range are matched with Y values at the upper end of its range, and vice versa?

If the answer to question 1 is 'Yes', this indicates a positive relationship, whereas 'Yes' in answer to question 3 implies a negative relationship is present. An affirmative answer to question 2 denotes the absence of direction in the relationship. In Table 8.1 it is fairly clear how 'Yes' answers to the three questions match with the three sets of data values A, B and C. However, in many cases it might not be so easy. Most of the low-X values are matched with low-Y ones, but some are not; or perhaps the majority of low-X values are paired with high-Y ones, but a few are not. These rather less-definite situations appear to indicate positive and negative relationships, respectively, but there remains some doubt. This conundrum may be resolved by recognizing that positive and negative relationships can be either strong or weak: if they are very weak then this suggests the data values for X and Y differ in an irregular way (see question 2).

The various methods of correlation share one important characteristic, they all produce a correlation coefficient that acts as a statistical measure of the direction and strength of the relationship between variables or attributes. Rather like the mean and variance are descriptive measures of central tendency and dispersion, a correlation coefficient describes the relationship between two attributes or variables as revealed by their data values for a set of observations. A correlation coefficient is a standardized measure whose value will always fall within the range -1.0 to $+1.0$. These, respectively, denote perfect negative and positive relationships. Standardization of the coefficient enables different sets of paired data values to be compared with each other irrespective of differences in the numerical range or the units of measurement.

Questions 1, 2 and 3 all included the phrase 'and vice versa': so question 1 could have been stated as:

1.　Do the values of Y and X for the observations differ in parallel with each other, so that values towards the lower end of the Y range are matched with X values at the lower end of its range, and vice versa?

Referring again to Table 8.1 to answer this new version of the question should result an affirmative 'Yes' answer being given for the same set of data values as previously. What does this tell us about the nature of the relationship between X and Y? The answer is that it is reciprocal. It does not matter which set of values is labelled X and which Y, the direction and strength of the relationship is the same: since the answer to both versions of question 1 is 'Yes', they are connected in a positive way. The correlation coefficient calculated for the paired data values would be the same irrespective of which were labelled X and Y. Chapter 9 will reveal that such flexibility in labelling variables does not exist with regression analysis, since one is assumed to exert a causal or controlling influence over the other. No such distinction exists in correlation analysis between which variable is the cause and which the effect, whereas it is part of the basic rationale for carrying out regression.

The application of correlation analysis to pairs of X and Y variables in research investigations will rarely, if ever, result in a correlation coefficient exactly equal to -1.0 or $+1.0$, although figures close to 0.0 are perhaps less remote. The values of -1.0 and $+1.0$ represent the ideal or perfect form of relationship in statistical terms. Far more common are correlation coefficient values somewhere within the range between these extremes. The sign denotes the direction of the relationship and its size or magnitude the strength of the **covariation** between the data values. Interpretation of the strength of a correlation coefficient is a matter of relativities. It is fairly clear that a coefficient of +0.51 is stronger than +0.34 and weaker than +0.62, but what about +0.66, +0.78 and +0.84? Should all three of these be treated as indicating the presence of strong relationships, although +0.84 is stronger than +0.66? Suppose the correlation coefficients for three pairs of variables calculated as −0.22, −0.36 and −0.46. In comparison with the previous set of values, these all seem to indicate weak negative relationships, but −0.46 rounded up to −0.50 seems a little stronger. These examples are intended to demonstrate that the interpretation of correlation coefficients is not a simple matter, especially when trying to decide whether one is stronger than another. Generally speaking, a certain amount of judgement is required when interpreting correlation coefficients, although we shall see later that statistical testing can help with deciding whether they are significant or not.

A correlation coefficient constitutes a statistical quantity measuring the direction and strength of a relationship: however, a helpful starting point before undertaking this analysis is a visual examination of the scatter of observations across the two-dimensional space defined by the X and Y attributes or variables. The 'unlikely to occur' instances of a perfect negative or positive relationship, which would, respectively, have the observations aligned along a diagonal line from top and bottom left of the chart, are not shown in Figure 8.1. The more typical outcome is that the observations form a scatter of points distributed across the chart in such a way that they indicate the general presence of a negative or positive relationship, albeit a relatively weak or strong one. Figures 8.1a and b illustrate these situations using hypothetical data values with respect to two variables X and Y. The general direction or trend of the points in Figure 8.1a is from upper left to lower right, although they are clearly not perfectly aligned along the diagonal. Conversely, the scatter of points in Figure 8.1b generally follows the lower left to upper right direction, but deviates from a perfect positive relationship. Figures 8.1c and 8.1d show a similar pair of relatively strong negative and positive relationships, but in this case for two attributes measured on the nominal scale labelled M and N. The obvious difference between the two sets of figures is that the categories of the attributes are discrete rather than continuous measurements. Thus, the observations are depicted in columns and rows with some data points being replaced by numbers to signify how many observations possessed that particular combination of categorical nominal values (codes). For example, there are three observations classed as C for attribute M and 2 for attribute N in Figure 8.1c.

There are a number of different types of correlation analysis and the main factors when deciding which to use are the manner of sampling employed and the scale of measurement of the attributes and variables being investigated. Simple random sampling with interval or ratio scale variables allows more robust techniques to be applied, whereas stratified random sampling and nominal scale attributes would really only be legitimate with weaker forms of correlation. Progression through the different measurement scales (interval/ratio (scalar), ordinal and nominal) represents a reduction of information from a situation where differences of magnitude between individual data values are known, through one in which their sequence or order is known

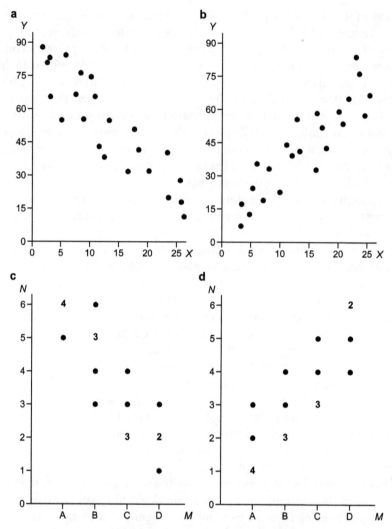

Figure 8.1 Scatter graphs showing different forms of relationship between two scalar variables (a and b) and between two categorical attributes (c and d) (X and Y).

to one where they are simply allocated to labelled categories. Figure 8.2 illustrates the effect of reducing the level of information recorded for two variables. The scatter of points in Figure 8.2a indicates the presence of a relatively strong positive correlation between the data values for variables X and Y. Now, let us reveal that one of the variables (Y) has not been measured on the ratio/interval scale, but is in fact an attribute with a large number of categories, for example representing nominal codes for 12 industrial sectors. The correlation analysis technique to be used in these circumstances should reflect the fact that there are different scales of measurement present

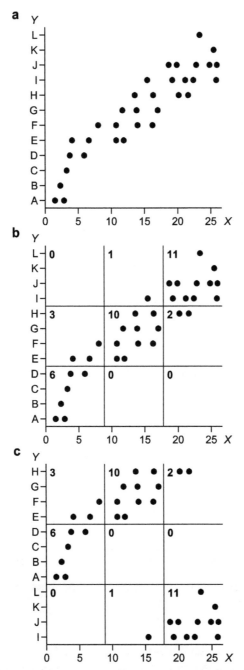

Figure 8.2   Connections between scatter graphs and crosstabulation of data values and form of relationship between two variables (*X* and *Y*).

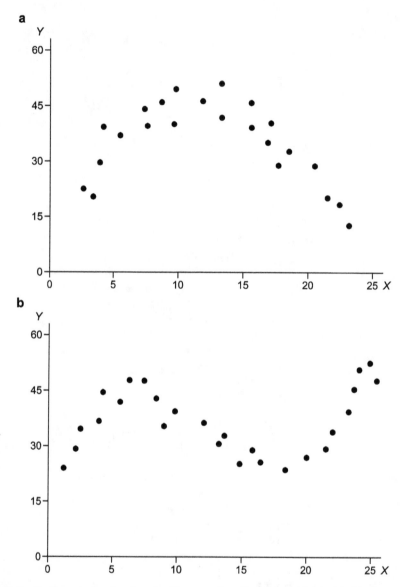

Figure 8.3 Scatter graphs showing curvilinear and cyclical forms of relationship between two scalar variables (*X* and *Y*).

in the two variables. Such techniques usually work on data expressed as a crosstabulation and there are many different ways in which the raw data values could be allocated to classes, one of which is shown in Figure 8.2b by means of a grid overlain over the scatter of points and annotated with their frequency count. Testing for any association (as opposed to correlation) between the classified variables by means of the Pearson's

Chi-square test produces a χ^2 statistic equalling 18.846 with 4 degrees of freedom and a probably of 0.000 84, which indicates the Null Hypothesis should be rejected at the 0.05 level of significance. Given that Y represents a nominal attribute there is no reason why the labels for each industrial sector should be shown in any particular order. Figure 8.2c shows the scatter of points with a similar overlain grid, but this time the sequence of industrial sector categories has been altered. The consequence of this change is that the scatter of points does not indicate the presence of such a strong correlation but the outcome of applying Pearson's Chi-square test again is to reject the Null Hypothesis ($\chi^2 = 18.846, p = 0.0084$). The change has altered the relative location of the cells in each row of the crosstabulation, for example those originally in the bottom row have moved to the centre. However, this does not alter the observed and expected frequencies of each cell since the row and column totals stay the same.

What does Figure 8.2 reveal about the connection between analyses of association as examined in a crosstabulation and of correlation as displayed in a scatter graph?

Another reason for examining the scatter of X and Y data values for a set of observations before carrying out a particular type of correlation analysis is that the scatter of points might indicate that the variables are related in a nonlinear fashion. Figures 8.3a and b show two scatters of data values for points for different sets of observations measured in respect of X and Y variables. Both examples clearly suggest the relationship between the variables is not a simple linear one and is more complex than implied by questions 1 and 3 presented earlier. The scatter of points in Figure 8.3a displays a curved form with middle range values of X generally related with high values of Y, and high and low values of X typically connected with low Y values. The general trend of points in Figure 8.3b appears wavy or cyclical, in other words there are peaks and troughs moving left to right across the scatter graph. The situations represented in Figure 8.3 require the application of rather more advanced correlation techniques than can reasonably be covered in an introductory statistics text, although Chapter 9 will examine regression techniques for these situations.

8.2   Correlation techniques

Application of correlation analysis and the calculation of a correlation coefficient between a pair of variables, indeed between several pairs, are not restricted to situations where the ratio or interval scales of measurement have been employed. However, we will start by exploring a procedure designed for such variables just as in Chapter 6 the Z test provided a starting point for examining univariate statistical tests. The Z test was billed as a robust and powerful statistical test to discover the significance of any difference between sample and population means, although its usefulness is limited in certain subject areas because of the stringent assumptions made about the method of sampling and the normality of the data. The Pearson's Product Moment Correlation Coefficient sits on a similar pinnacle with respect to the different correlation techniques used to examine bivariate linear relationships. Next in line are Spearman's Rank Correlation and Kendall's Tau Rank Correlation Coefficients, which, given their name, unsurprisingly deal with pairs of ranked or ordinal data values. Lastly, we come to three correlation techniques, the Phi Coefficient, Cramer's V and the Kappa Index of Agreement that can be used when nominal attributes are to be correlated. All of these techniques have a common purpose, namely to quantify the direction and strength of bivariate relationships.

8.2.1 Pearson's product moment correlation coefficient

Chapter 4 discussed use of the variance to examine the dispersion or spread of the data values for individual ratio or interval scale variables. The variance quantifies the extent to which the data values of a set of observations are spread out or concentrated together in relation to the mean. A key concept in Pearson's Product Moment Correlation is **covariance**, which may be thought of as a 'two-dimensional' variance measuring the degree to which the values of two variables for a group of observations are jointly dispersed or spread out. More formally, the means of the two variables locate a point known as the **centroid** that lies within the two-dimensional statistical space of the variables: covariance represents their mean (average) deviation from this point. There are slightly different equations for calculating covariance depending on whether the observations are a population or a sample, the main difference being the denominator is N (the number of observations) in the case of a population and $n - 1$ for a sample.

$$\text{Population} = \frac{\sum (X - \mu_X)(Y - \mu_Y)}{N} \qquad \text{sample} = \frac{\sum (x - \bar{x})(y - \bar{y})}{n - 1}$$

Note that the individual deviations of each X and Y value from their respective means are not squared before summation since the multiplication of one difference by the other means that the numerator does not sum to zero (compare with variance equation in Chapter 4). Figure 8.4 shows a number of combinations of dispersion for a sample of 10 observations measured in respect of variables X and Y. These combinations depict the situations where data values for:

* X and Y are widely dispersed along their observed numerical ranges and the points are distant from the centroid (Figure 8.4a);

* X are widely dispersed and for Y are close together (Figure 8.4b);

* X are close together and for Y are spread out (Figure 8.4c);

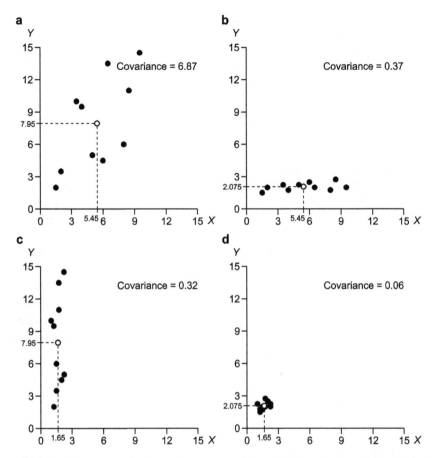

Figure 8.4 Scatter graphs showing effect on covariance of different dispersion of X and/or Y variable data values.

- X and Y are close together along their numerical ranges and the points are clustered around the centroid (Figure 8.4d).

The direction of the relationship is broadly positive (i.e. trending lower left to upper right), although clearly its strength is different, in each case. The covariance of each scatter of points has been included and reveals the most dispersed pattern to have the highest value and the least dispersed the lowest one.

A correlation coefficient calculated in respect of a sample of observations, just like any other statistic may have been affected by sampling error. In other words, the value obtained is not identical with or even close to the equivalent population parameter and so we need to apply a suitable statistical test to discover whether the difference is more or less than might be expected to have arisen through random perturbations. Turning this problem on its head, if the true value of the population parameter is unknown, it is useful to know how much confidence can be placed in the sample statistic. This requires the calculation of confidence limits for the sample correlation coefficient. These issues inevitably raise questions of probability. How probable is it that the difference between the correlation coefficient derived from a sample and its parent population are purely random? What is the probability that the population correlation coefficient lies within 5 per cent of the sample statistic? The Normal Distribution discussed previously is a univariate probability distribution dealing with one variable. The Pearson's Correlation Coefficient is a bivariate technique and so a probability distribution encompassing two variables simultaneously is needed. This is known as the **bivariate Normal Distribution**.

The Normal Distribution is often referred as being bell-shaped, although its graphical representation is really a cross-section or slice through the 'bell' (see Chapter 5). The bivariate Normal Distribution may be thought of as representing a three-dimensional bell shape, although not necessarily with a perfectly circular form around the central point when viewed from above. This central point corresponds to the centroid of the distribution, at the intersection of mean X and Y data values. This three-dimensional shape is achieved by the **conditional** and **marginal frequency** distributions of the X and Y variables in the population following the Normal Distribution. This somewhat confusing terminology needs some explanation. The conditional distributions of X and Y refer to the distributions of their values for each and every cross section drawn through the 'bell shape'. These cross sections are drawn parallel to the axes for the X and Y variables for each value of Y and of X respectively, and in all cases they will be normally distributed and symmetrical about their own means. The marginal distributions of X and Y are, respectively, the frequency distributions for **all** values of X and Y in the population. Horizontal slices through the three-dimensional 'bell shape' representing the joint frequency distribution of the population of X and Y data values will not necessarily be circular, but may be ellipsoidal, since the 'bell shape' may be elongated in one direction. This occurs because both marginal distributions are required to be normal but not necessarily with an identical variance and standard deviation (e.g. one may be more dispersed than the other). The variance and standard deviation of X may be larger than Y because its data values in relative terms are more spread out around their mean. This difference in dispersion results in the shape of the frequency distribution being stretched along the numerical scale of the X-axis in comparison with the Y-axis.

In Chapter 5 we saw that according to the Central Limit Theorem the sampling distribution of a normally distributed ratio or interval scale variable will also be normal irrespective of sample size (i.e. the means of many samples from the population will themselves follow the Normal Distribution). The sampling distribution of the bivariate Normal Distribution is affected by sample size and the population correlation coefficient. Provided that the population correlation coefficient does not equal zero, then the sampling distribution becomes closer to being normal and symmetrical with increasing sample size. Using this information, there are two ways of testing whether a sampled product moment correlation coefficient is statistically significant. The first hypothesizes that the population correlation coefficient equals zero and therefore that the variables are unrelated. It tests whether the difference between zero and the sample coefficient is the result of sampling error by calculating a t statistic. There are some occasions when it is reasonable to argue that the population coefficient is not zero but some particular value, which might have been obtained from some previous analysis of the same population of observations. The alternative test can be used in this situation to investigate whether the difference between some known population coefficient and the sample coefficient has arisen through sampling error. The sample product moment coefficient is transformed into a quantity known as **Fisher's Z**, which is not itself the same as Z discussed earlier, although it needs to be converted into a Z score so that its probability can be determined from the Z distribution.

Box 8.1 illustrates the application of Pearson's Product Moment correlation technique to two variables that have been measured in respect of the Burger King restaurants in Pittsburgh. The grid coordinates of the city centre have been determined (24.5, 26.0) and the distance of each of each restaurant has been calculated to this point using the calculations associated with Pythagoras' theorem outlined elsewhere (see Chapter 4 Box 4.6). The second variable, although recording hypothetical values, represents the daily footfall of customers in the same 22 restaurants. The example explores the idea that outlets closer to the city centre might have a higher number of customers, possibly on account of the population being inflated by workers in the downtown area during the day. This idea connects with the broader notion of retail and catering outlets competing for space and prime locations within urban centres, and of consumers' behaviour in journeying to fast-food outlets. Customers' visits to a

---

Box 8.1a: Pearson's product moment correlation coefficient.

Pearson's Correlation Coefficient

$$\text{Population: } \rho \text{ or } R = \frac{\dfrac{\sum(X-\mu_X)(Y-\mu_Y)}{N}}{\sqrt{\dfrac{\sum(X-\mu_X)^2}{N}}\sqrt{\dfrac{\sum(Y-\mu_Y)^2}{N}}} \qquad \text{Sample: } r = \frac{\dfrac{\sum(x-\bar{x})(y-\bar{y})}{(n-1)}}{\sqrt{\dfrac{\sum(x-\bar{x})^2}{n-1}}\sqrt{\dfrac{\sum(y-\bar{y})^2}{n-1}}}$$

Alternative, simpler equations:

$$\text{Population: } \rho \text{ or } R = \frac{N\sum XY-(\sum X)(\sum Y)}{\sqrt{(N\sum X^2)-(\sum X)^2}\sqrt{(N\sum Y^2)-(\sum Y)^2}}$$

$$\text{Sample: } r = \frac{n\sum xy-(\sum x)(\sum y)}{\sqrt{(n\sum x^2)-(\sum x)^2}\sqrt{(n\sum y^2)-(\sum y)^2}}$$

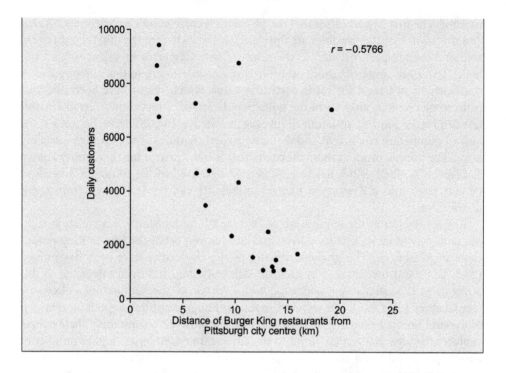

Box 8.1b: Application of Pearson's product moment correlation coefficient.

The symbols for a population and sample Pearson's Product Moment Correlation Coefficient are respectively ρ (Rho or R) and r. Two versions of the definitional equations for these quantities are given above: the upper part (nominator) of the first pair of equations represents the covariance and the lower (denominator) the product of the standard deviations. Manual calculation of the coefficient is of course unnecessary when using statistical software for your analysis, nevertheless, in the application of the technique shown below only the simpler version involving the data values for X and Y variables is used. There are two ways of testing the significance of the Pearson's Product Moment Correlation Coefficient obtained from a sample of data depending on whether the population parameter is hypothesized as zero or some other value. The Null Hypothesis states that the difference between the sample and population coefficient has arisen through sampling error. In the first case the difference is converted into a t statistic, which is assessed with $n - 2$ degrees of freedom, in the latter r is transformed into Fisher's Z, which is itself turned into the more familiar Z statistic. The probability of having obtained this difference is assessed in relation to the chosen level of significance in the normal way in order to reach a decision on the fate of the Null Hypothesis.

The Pearson's Product Moment coefficient has been calculated for the Burger King restaurants in Pittsburgh. The two variables are the distance of each restaurant from the city centre and the number of daily customers (hypothetical data values). The scatter graph suggests that the number of customers declines as distance from the city centre increases. This seems to indicate the presence of a distance decay function. The result of calculating Pearson's correlation coefficient seems to confirm this impression with a moderately strong negative relationship between the two variables. Testing this result by converting the coefficient into a t statistic reveals that there is an extremely low probability that the coefficient value has arisen through chance, so although the relationship is not very strong it is possible to be reasonably confident that it is reliable. However, the value of r^2 (the coefficient of determination) is only 33.2%, which suggests that other factors also contribute to the variation in customer numbers.

The key stages in applying and testing the significance of Pearson's Product Moment Correlation Coefficient are:

Calculate correlation coefficient, r: use this quantity to indicate the direction and strength of the relationship between the variables;

State Null Hypothesis and significance level: the difference between the sample r (-0.5766) and population (0.0) correlation coefficients in respect of distance of Burger King restaurants from the centre of Pittsburgh and the daily number of customers has arisen through sampling error and is not significant at the 0.05 level of significance.

Calculate either the t or Fisher's Z test statistic: the calculation of the t test statistic is given below.

Select whether to apply a one- or two-tailed test: in this case there is no prior reason to believe that the sample correlation coefficient would be larger or small than the population one.

Determine the probability of the calculated t: the probability is 0.000 00537

Accept or reject the Null Hypothesis: the probability of t is <0.05, therefore reject the Null Hypothesis.

Box 8.1c: Assumptions of the Pearson's product moment correlation coefficient.

There are three main assumptions:

Random sampling with replacement should be employed or without replacement from an infinite population provided that the sample is estimated at less than 10 per cent of the population.

In principle the data values for the variables X and Y should conform to the bivariate Normal Distribution: however, such an assumption may be difficult, if not impossible, to state in some circumstances. It is generally accepted that provided the sample size is reasonably large, which is itself often difficult to judge, the technique is sufficiently robust as to allow for some unknown deviation from this assumption.

The observations in the population and the sample should be independent of each other in respect of variables X and Y, which signifies an absence of autocorrelation. This assumption may not be entirely valid in this example.

Box 8.1d: Calculation of the Pearson's product moment correlation coefficient.

No.	x	y	xy	x^2	y^2
1	13.54	1200.00	16 244.38	183.25	1 440 000.00
2	19.14	7040.00	134 729.12	366.25	49 561 600.00
3	11.67	1550.00	18 092.56	136.25	2 402 500.00
4	6.26	4670.00	29 257.47	39.25	21 808 900.00
5	9.66	2340.00	22 596.45	93.25	5 475 600.00
6	10.31	8755.00	90 244.44	106.25	76 650 025.00
7	2.75	9430.00	25 932.50	7.56	88 924 900.00
8	7.16	3465.00	24 805.62	51.25	12 006 225.00
9	2.50	8660.00	21 650.00	6.25	74 995 600.00
10	6.18	7255.00	44 869.70	38.25	52 635 025.00
11	2.40	7430.00	17 832.00	5.76	55 204 900.00
12	1.80	5555.00	10 014.42	3.25	30 858 025.00
13	10.00	4330.00	43 300.00	100.00	18 748 900.00
14	2.69	6755.00	18 188.39	7.25	45 630 025.00
15	6.50	1005.00	6532.50	42.25	1 010 025.00
16	7.50	4760.00	35 700.00	56.25	22 657 600.00

17	15.91	1675.00	26 655.67	253.25	2 805 625.00
18	14.60	1090.00	15 917.36	213.25	1 188 100.00
19	13.87	1450.00	20 104.87	192.25	2 102 500.00
20	13.12	2500.00	32 811.01	172.25	6 250 000.00
21	12.66	1070.00	13 545.12	160.25	1 144 900.00
22	13.65	1040.00	14 193.24	186.25	1 081 600.00

$$\sum x = 203.88 \qquad \sum y = 93\,025.00 \qquad \sum xy = 683\,216.81 \qquad \sum x^2 = 2420.07 \qquad \sum y^2 = 574\,582\,575.0$$

r (product moment correlation coefficient)

$$r = \frac{n\sum xy - \left(\sum x\right)\left(\sum y\right)}{\sqrt{\left(n\sum x^2\right) - \left(\sum x^2\right)}\sqrt{\left(n\sum y^2\right) - \left(\sum y^2\right)}}$$

$$= \frac{22(683\,216.81) - (203.88)(93\,025.00)}{\sqrt{(22(2420.07) - 203.88^2}\sqrt{(22(57\,458\,255.00) - 93\,025.00^2}}$$

$$= \frac{-3\,933\,880.3}{(108.04)(63\,144.01)} = -0.5766$$

t statistic

$$t = \frac{(|r|\sqrt{(n-2)})}{\sqrt{(1-r^2)}} = \frac{(0.5767\sqrt{(22-2)})}{\sqrt{(1-(-0.5767^2))}} = 5.11$$

Probability $\qquad p < 0.0001$

complete set of Burger King (or any other type of fast-food restaurant) in a large city such as Pittsburgh will be influenced by a number of factors in determining which one to frequent. Therefore, it is to be expected that our analysis will not have revealed the whole story: there are likely to be a number of other variables that should also be considered and the analysis has treated each restaurant as having the same floor space and capacity to serve its clientele.

One further important aspect of Pearson's Product Moment Correlation Coefficient r should be considered before exploring alternative correlation analysis techniques. The square of r, which will inevitably be a smaller positive value than the coefficient itself, provides a quantity known as the **coefficient of determination** (r^2). This acts as a shorthand indication of the amount of the total variance that is explained by the correlation between the two variables. The unexplained percentage of the variance is attributed to other variables that were not included in the analysis. Such an explanation (or lack of it) is of course statistically defined in terms of the particular numerical values from which the statistic r has been calculated. It would thus be entirely feasible to perform a correlation analysis between two theoretically unrelated variables, calculate a strong positive or negative r, obtain a very high (r^2) and reach potentially rather ludicrous conclusions about their relationship.

8.2.2 Correlating ordinal variables

Some investigators take a rather relaxed view of the assumptions associated with the Pearson's Product Moment Correlation Coefficient and apply the technique in situations where they are known to be violated or simply disregarded. Such a cavalier approach is prone to criticism not only from statisticians but also from others working in the same subject area in the course of the process of peer review for publication or, more significantly in the context of this text, in the assessment of students' dissertations. Fortunately, there are alternative nonparametric techniques that are rather less demanding and may be used with ranked ratio/interval scale variables or simply with ordinal data. The following sections examine two of these correlation analysis techniques.

8.2.2.1 Spearman's Rank Correlation

Spearman's Rank Correlation is regularly used in geographical investigations and produces a correlation coefficient referred to by the symbol r_s. The underlying idea behind the technique is that if two variables, X and Y, measured in respect of a set of observations have a relationship with each other (i.e. are correlated), their rank scores in respect of these variables will be similar (positive correlation) or opposite (negative correlation). So, the questions asked in respect of the data values shown in Table 8.1 can be restated to reflect the rank scores of the variables:

1. Do the ranks of X and Y for the observations differ in parallel with each other, so that lower ranks for X are matched with lower Y ranks, and vice versa?

2. Do the ranks of X and Y for the observations differ in an irregular way, so that those ranked high on X stand just as much chance of being matched with upper, middle or lower ranks for Y, and vice versa?

3. Do the ranks of X and Y for the observations differ in opposite directions to each other, so that lower ranks for X are matched with upper ranks for Y, and vice versa?

Some of the preprocessing of data values involved in calculating Spearman's Rank Correlation coefficient resembles the preparation of data in order to apply univariate statistical tests to ordinal data values. The rank scores for a sample of observations in respect of X and Y are determined taking into account tied cases by assigning the mean (average) rank in these instances. The correlation coefficient is calculated from the squared differences between these scores (see Box 8.2) rather than the raw data values.

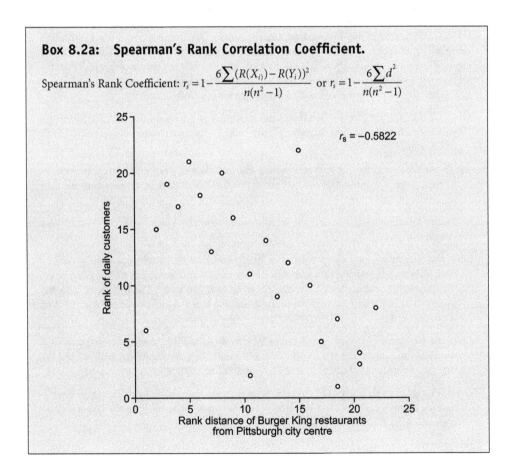

Box 8.2a: Spearman's Rank Correlation Coefficient.

Spearman's Rank Coefficient: $r_s = 1 - \dfrac{6\sum (R(X_i) - R(Y_i))^2}{n(n^2 - 1)}$ or $r_s = 1 - \dfrac{6\sum d^2}{n(n^2 - 1)}$

$r_s = -0.5822$

Rank of daily customers (y-axis)

Rank distance of Burger King restaurants from Pittsburgh city centre (x-axis)

Box 8.2b: Application of Spearman's Rank Correlation Coefficient.

The basis of Spearman's Rank Correlation Coefficient is that it examines the difference in rank scores for a set of observations in respect of two variables. These ranks are obtained by a process of sorting the data values for X and Y into ascending order and assigning a rank according to their position in the sequence taking due account of ties. The observations are then resorted back to their original order so the difference (d) in ranks can be determined. The subscripts i in the first of the definitional equations refers to an individual observation and R to its rank scores in respect of X and Y. The term n refers to the number of observations in the usual way in both equations. Testing the significance of Spearman's Rank Correlation Coefficient helps to determine whether the value of r_s obtained in respect of a sample of observations is significantly different from what might have been expected to arise through random sampling error.

The Spearman's rank coefficient has been applied to the rank scores of the distance of each Burger King restaurant from the centre of Pittsburgh and the number of daily customers (hypothetical data values). The distribution of the rank scores in the scatter graph above shows a similar general trend to the raw data examined in Box 8.1a. The rank scores associated with greater distance from the city centre seem to be associated with lower ranks for the daily number of customers. The Spearman's ran correlation coefficient is -0.5822, which is similar to the value for the Pearson's coefficient. Converting the Spearman's coefficient into a t statistic reveals that there is a very small chance of having obtained this value through sampling error, which provides support for rejecting the Null Hypothesis. Note that in this example because the number of observations is <100 the significance graph/table for Spearman's Rank Correlation Coefficient could be used to find the probability.

The key steps in applying and testing Spearman's Rank Correlation are:

Obtain and assign rank scores: determine the rank of X and Y data values by independently sorting each variable into ascending order and assign a corresponding rank to each observation.

Match correct rank pairs for each observation and calculate difference: re-sort observations by their sequence or identifier number to match up correct rank score pairs and calculate difference in ranks (d).

Calculate r_s: Insert relevant values into definitional equation for Spearman's Rank Correlation Coefficient.

State Null Hypothesis and significance level: the correlation coefficient has been produced from data values relating to a sample of observations and is subject to sampling error. The population parameter is 0.0 or some other specified value. The difference between these values has arisen through sampling error and is not significant at the 0.05 level of significance.

Calculate test statistic (t) and its probability or refer to standard r_s significance graph/table if $n < 100$: the sample size is 22 and plotting the sample r_s on Figure 8.6 with $n - 2$ (20) degrees of freedom reveals its probability is <0.05 but >0.01.

Select whether to apply a one- or two-tailed test: typically there is no prior reason to believe that the sample correlation coefficient would be larger or small than the hypothesized population coefficient.

Accept or reject the Null Hypothesis: the probability of t and the observed value of the coefficient are both <0.05, therefore reject the Null Hypothesis. This decision implies that the Alternative Hypothesis is accepted.

Box 8.2c: Assumptions of the Spearman's Rank Correlation Coefficient.

There are two main assumptions:

Ideally random sampling with replacement should be used with respect to a finite population, although random sampling without replacement is allowed from an infinite population. This restriction can be relaxed if the sample constitutes no more than 10 per cent of the population and is sufficiently large.

The technique ignores the size of differences between the X and Y data values for the observations and focuses on the simpler issue of their rank position.

Box 8.2d: Calculation of the Spearman's Rank Correlation Coefficient.

No.	x	y	$R(x_i)$	$R(y_i)$	d^2
1	13.54	1200.00	6	18	144
2	19.14	7040.00	1	6	25
3	11.67	1550.00	9	16	49
4	6.26	4670.00	16	10	36
5	9.66	2340.00	12	14	4
6	10.31	8755.00	11	2	81
7	2.75	9430.00	19	1	324
8	7.16	3465.00	14	12	4
9	2.50	8660.00	20	3	289
10	6.18	7255.00	17	5	144
11	2.40	7430.00	21	4	289
12	1.80	5555.00	22	8	196
13	10.00	4330.00	10	11	1
14	2.69	6755.00	18	7	121
15	6.50	1005.00	15	22	49
16	7.50	4760.00	13	9	16
17	15.91	1675.00	2	15	169
18	14.60	1090.00	3	19	256
19	13.87	1450.00	4	17	169
20	13.12	2500.00	7	13	36
21	12.66	1070.00	8	20	144
22	13.65	1040.00	5	21	256

$$\sum d^2 = 2802$$

r_s (Spearman's rank correlation coefficient)

$$r^s = 1 - \frac{6\sum d^2}{n(n^2-1)} \qquad 1 - \frac{6(2802)}{22(484-1)} = -0.5822$$

t statistic

$$t = r_s\sqrt{\frac{n-2}{1-r_s^2}} = -0.5822\sqrt{\frac{22-2}{1-0.339}} = -3.2023$$

Probability

$$p = 0.0045$$

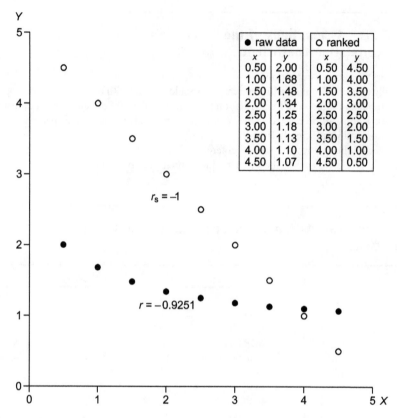

The table shown in the figure:

● raw data		○ ranked	
x	y	x	y
0.50	2.00	0.50	4.50
1.00	1.68	1.00	4.00
1.50	1.48	1.50	3.50
2.00	1.34	2.00	3.00
2.50	1.25	2.50	2.50
3.00	1.18	3.00	2.00
3.50	1.13	3.50	1.50
4.00	1.10	4.00	1.00
4.50	1.07	4.50	0.50

$r_s = -1$

$r = -0.9251$

Figure 8.5 Contrast between Pearson's Product Moment and Spearman's Rank Correlation analyses.

One of the problems associated with using Spearman's Rank Correlation is that converting the raw data values into rank scores can mask an underlying nonlinear relationship between the variables that might more appropriately be explored by means of correlation techniques more suited to curvilinear types of relationship. Figure 8.5 illustrates this by a series of data values that have a curved form, which have a Pearson's correlation coefficient of −0.9251. In order that the rank scores can be shown on a similar numerical scale to the raw data values, they have been recorded as the ascending ranks 0.5, 1.0, 1.5, etc. for X and 4.5, 4.0, 3.5, etc. as the corresponding descending ones for Y. These reveal a second issue associated with the Spearman's Rank Correlation Coefficient, namely that they display perfect negative correlation $(r_s = -1.000)$ because the ascending and descending rank scores are mirror images of each other. Such an extreme case is unusual and has been included to emphasize the need for caution when applying correlation analysis. If there is a genuine a linear relationship between the two variables, Spearman's Rank Correlation will

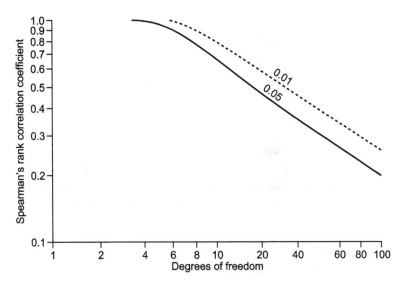

Figure 8.6 Selected Spearman's Rank Correlation coefficients' significance and degrees of freedom.

usually produce a slightly less-robust outcome that Pearson's product moment correlation.

There are two approaches to testing the significance of Spearman's Rank Correlation Coefficient depending on the number of sampled observations. If there are less than 100 observations in the sample, the significance of Spearman's correlation coefficient can be obtained by consulting Figure 8.6. This shows the critical values of r_s at selected levels of significance (0.05 and 0.01) for different degrees of freedom. The degrees of freedom are defined as $n - 2$, where n is the number of observations in the sample. If the sample coefficient when plotted on the chart falls below the 0.05 significance line for the known degrees of freedom, then it is possible that the correlation has arisen by chance and the Null Hypothesis should be accepted. Whereas if the Spearman's coefficient falls on or above the line, you can be 95 per cent confident that the result is statistically significant and you can accept the Alternative Hypothesis. If there is a larger sample, r_s can be converted into a t statistic and its probability be determined with $n - 2$ degrees of freedom.

Analysis the distance of the Pittsburgh Burger King restaurants from the city centre and their daily customer footfall in Box 8.1 tacitly assumed that these variables complied with the assumptions of Pearson's Product Moment correlation procedure, in particular that they followed the bivariate Normal Distribution. Box 8.2 presents the alternative Spearman's Rank Correlation analysis of these variables. The analysis is based on the idea that the rank scores of the raw data values will display some form of regular pattern if there is a strong relationship between them. In other words, if the restaurants were ranked in order of size in terms of their daily customer footfall

(1 being the first or most visited restaurant, 2 the second, 3 the third, and so on), then smaller ranks would be for outlets close to the city centre and larger ranks for those further away.

How could this pattern be expressed in the reverse way with respect to the ranked distance variable?

8.2.2.2 Kendall's Tau Correlation Coefficient

One of the advantages of **Kendall's τ (tau) Rank Correlation Coefficient** is that its significance can be determined without reference to a specific probability distribution (i.e. it is distribution free). Another advantage is that it measures the direction and strength of the relationship between two variables in a more direct and simpler fashion that Spearman's Rank Correlation. Both work with the rank scores of the observations rather than the raw data values of two variables. However, the Spearman's coefficient is essentially the Pearson's product moment correlation coefficient calculated for these ranks. Kendall's τ works out how jumbled up the two sets of ranks are in relation to the ideals of perfect concordance (conformity) or total discordance (disagreement) between the set of ranks produced by the two variables. However, one disadvantage of the technique is that calculation of the coefficient is more complicated if there are ties in the rank scores of the data values. Mathematically simple but relatively tedious, the technique proceeds by independently ordering the pairs of data values for two variables, X and Y, and assigning their ranks. Each pair is examined to determine whether the rank for X is greater than the rank for Y, in which case a pair is labelled as concordant otherwise they are referred to as discordant. The difference between the number of concordant n_c and discordant n_d pairs yields the quantity S. This quantity is referenced in the alternative definitional equation, which comes into play if there are any ties in rank scores (see Box 8.3).

The technique can also be thought of as identifying the number of neighbour swaps required to convert one set of rank scores into another. Suppose there is a set of four observations ranked as 4, 2, 1, 3 in respect of variable X and 1, 2, 3, 4 for variable Y. How many swaps are required to turn the first sequence into the second? We could start by swapping X ranks 1 and 2 to produce 4, 1, 2, 3; then swap 4 and 1 to obtain 1, 4, 2, 3; and then 4 and 2 to give 1, 2, 4, 3 and finally 4 and 3 to yield the desired 1, 2, 3, 4: a total of four swaps is required to convert from one rank sequence into another. This value constitutes a measure of how 'mixed up' or different are the two sets of rank scores. Recalling that two sets of perfectly in-sequence or out-of-sequence ranks would, respectively, have correlation coefficients of +1.000 and −1.000, Kendall's tau may be thought of a measuring how the rank scores in respect of two variables for a set of observations approach either of these extreme forms of relationship.

The probability of the particular value of Kendall's τ obtained for a sample of observations may not be directly found by reference to one of the standard probability distributions such as Z or t, but nevertheless it needs to be determined in order to

---

Box 8.3a: Kendall's Tau Rank Correlation Coefficient.

Kendal's Rank Correlation Coefficient: $\tau = \dfrac{n_c - n_d}{n(n-1)/2}$

Alternative Kendall's Rank Correlation Coefficient if there are ties:

$$\tau_b = \dfrac{S}{\sqrt{[n(n-1)/2 - \sum_{i=1}^{t} t_i(t_i-1)/2]\,[n(n-1)/2 - \sum_{i=1}^{u} u_i(u_i-1)/2]}}$$

X \ Y	1	2	3	4	5	6	7	8	9	10	11	12	13	14	15	16	17	18	19	20	21
2	C																				
3	C	C																			
4	C	C	D																		
5	C	C	C	C																	
6	C	C	D	C	D																
7	C	D	D	D	D	D															
8	C	C	C	C	D	C	C														
9	C	C	D	D	D	D	C	D													
10	C	D	D	D	D	D	D	D	D												
11	D	D	D	D	D	D	D	D	D	D											
12	C	D	D	D	D	D	C	D	D	C	C										
13	C	D	D	D	D	D	D	D	D	D	C	D									
14	C	D	D	D	D	D	D	D	D	C	C	D	C								
15	C	C	C	C	C	C	C	C	C	C	C	C	C	C							
16	C	D	D	D	D	D	D	D	D	D	C	D	C	D	D						
17	D	D	D	D	D	D	D	D	D	D	C	D	D	D	D	D					
18	C	D	D	D	D	D	D	D	D	D	C	D	D	D	D	D	C				
19	D	D	D	D	D	D	D	D	D	D	D	D	D	D	D	D	D	D			
20	D	D	D	D	D	D	D	D	D	D	C	D	D	D	D	D	D	D	C		
21	D	D	D	D	D	D	D	D	D	D	C	D	D	D	D	D	D	D	C	C	
22	C	D	D	D	D	D	D	D	D	D	C	D	D	D	D	D	D	C	C	C	C

test whether the coefficient is statistically significant. The approach to determining the probability varies depending upon whether there are ties in the rank scores for some of the observations and therefore on the version of the coefficient that is calculated (see Box 8.3). The τ coefficient may be viewed as representing the probability that ordering a set of observations by means of two variables will produce exactly the same or opposite outcome. A random ranking of the observations would have a 0.5 or 50 per cent probability of occurrence. The curves in Figure 8.7 trace the critical values of the τ coefficient that are associated with the 0.05, 0.01 and 0.005 (5%, 1% and 0.5%) levels of significance. If the value of the coefficient is equal to or greater than the graphed value for a sample with n observations, the result is significant. The lines on the chart in Figure 8.7 also represent the critical values for a one-tailed

Box 8.3b: Application of Kendall's Tau correlation coefficient.

The simpler version of Kendall's τ coefficient, used when none of the observations are tied with the same rank score for the two variables, is calculated from the number of concordant n_c and discordant n_d pairs of ranks. The denominator of this equation ($n(n-1)/2$) represents the total number of possible pairings of X with Y ranks when there are n observations. In the equation that is used to calculate τ_b when there are ties, the subscript i refers to the number of tied observations at a certain rank score for X and u to the number of tied observations at a certain rank score for Y.

Kendall's Rank Correlation Coefficient has been applied to variables X and Y denoting the distances of the Burger King restaurants from the centre of Pittsburgh and their daily customer footfall in order to provide some consistency with the worked examples for Pearson's and Spearman's correlation techniques. Kendall's rank correlation untangles the complexity of two sets of paired rank scores for a set of observations. The Cs and Ds in the grid (matrix) above represent the concordant and discordant connections between the 22 Burger King restaurants in Pittsburgh (numbered 1 to 22) in respect of the ranks for X and Y. Reading up the columns of the grid, where each number identifies a different restaurant, if the cell contains a C then reading along the row there is a concordant link with the numbered restaurant shown on the Y-axis, indicating that the Y rank is larger than the X rank. In contrast a D indicates that Y rank is lower than the X one. There are no tied ranks (see Box 8.3d) and the simpler, standard equation for τ can be used. The ranks for the observations are shown in ascending order for variable X (distance from city centre). The Kendall's τ coefficient is -0.429, which suggests a weaker relationship between the two variables than either the Pearson's or Spearman's correlation analyses. As usual, the probability of having obtained this value by chance should be determined in order to discover whether the sample data can be regarded as providing a statistically significant result.

The key stages in carrying out a Kendall's rank correlation are:

Calculate correlation coefficient τ: this quantity indicates the extent to which two sets of ranks are different and thus the direction and strength of the relationship between the ranked variables. Calculations for the quantity are given below.

State Null Hypothesis and significance level: the two sets of ranks are mutually independent of each other and are not different from what would have occurred if they had been randomly assigned; the correlation coefficient is not significant at the 0.05 level.

Select whether to apply a one- or two-tailed test: in most cases a two-tailed test is used since there is rarely a prior reason for assuming a concordant or discordant relationship should exist.

Determine the probability of the calculated τ: the probability of obtaining $\tau = -0.429$ is $p < 0.005$.

Accept or reject the Null Hypothesis: the probability of τ is <0.05, therefore reject the Null Hypothesis. By implication the Alternative Hypothesis is accepted.

Box 8.3c: Assumptions of Kendall's Tau correlation coefficient.

There is only one main assumption:
  Random sampling should be applied.

Box 8.3d: Calculation of Kendall's Tau correlation coefficient.

No.	$R(x_i)$	$R(y_i)$	Concordant (n_c)	Discordant (n_d)
2	1	6	16	5
17	2	15	7	13
18	3	19	3	16
19	4	17	4	14
22	5	21	1	16
1	6	18	2	14
20	7	13	4	11
21	8	20	1	13
3	9	16	1	12
13	10	11	3	9
6	11	2	10	1
5	12	14	1	9
16	13	9	3	6
8	14	12	1	7
15	15	22	0	7
4	16	10	0	6
10	17	5	2	3
14	18	7	1	3
7	19	1	3	0
9	20	3	2	0
11	21	4	1	0
12	22	8	0	0
			$\sum n_d = 6$	$\sum n_d = 165$

τ (Kendall's tau Rank Correlation Coefficient)	$\tau = \dfrac{n_c - n_d}{n(n-1)/2}$	$\tau = \dfrac{66 - 165}{22(22-1)/2} = -0.429$
Probability	$p < 0.005$	

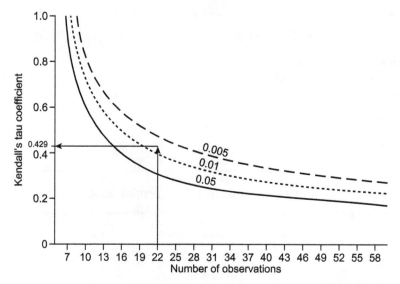

Figure 8.7 Two-tailed five and one per cent critical values of Kendall's Rank Correlation Coefficient according to sample size.

test of significance for Kendall's Rank Correlation Coefficient at 0.025, 0.0025 and levels of significance, since the probabilities are halved. Box 8.3 provides a worked example of the Kendall's tau Rank Correlation Coefficient with respect to the Burger King restaurants in Pittsburgh. The coefficient in this case has been plotted on Figure 8.7 with 22 observations and shows that the result is significant, since it falls between the curves representing the 0.01 and 0.005 probability levels.

8.2.3 Correlating nominal variables

The links between the different scales of measurement (interval/ratio, ordinal and nominal) and the likelihood that research investigations will involve variables from each of these scales has been explored in other chapters. There is a group of rather neglected correlation techniques that enable the direction and strength of the relationship between nominal attributes to be examined. Several of these correlation techniques use the χ^2 distribution in order to assess the probability of the coefficient when testing for significance. These correlation techniques may not be as powerful as Pearson's, Spearman's and Kendall's coefficients, nevertheless, they do allow the direction and strength of the relationship between nominal attributes to be quantified.

8.2.3.1 Phi correlation coefficient
The **Phi ϕ Coefficient** measures the degree of association between two nominal attributes or ratio/interval scale variables both of which have been classified into dichotomous (binary) categories. The coefficient is therefore used with data expressed

as a frequency distribution and measures the extent to which the observations are distributed into the cells along one of the diagonals of the crosstabulation or are 'off the diagonal' and allocated among the four cells. The 2 × 2 crosstabulations on the left and right of Table 8.2 represent the extremes of negative and positive relationships with all of the observations falling along the diagonal cells. The central crosstabulation depicts the distribution of observations across all four cells in an uneven fashion. The ϕ coefficient can in principle be applied to larger contingency tables than 2 × 2, but this is rare since its value can exceed +1.000 in these circumstances and its interpretation is less straightforward.

How many observations would there need to be in each of the cells in Table 8.2 for the calculated correlation coefficient to indicate the complete absence of a relationship (i.e. 0.0000)?

The calculation of the phi coefficient refers to the four cells in a 2 × 2 crosstabulation and the marginal totals of the rows and columns by letters. The cells are labelled A, B, C and D, and the marginal totals E, F, G and H as shown in Table 8.3 together with N for the total number of observations. The ϕ coefficient shares the common feature of all correlation analysis techniques that its value is constrained to fall with the range −1.000 to +1.000, thus making its interpretation comparable with other forms of correlation coefficient. Thus, values towards the extremes beyond −0.700 and +0.700 are generally accepted as indicating a strong relationship (negative or positive), −0.300 (to −0.700) and +0.300 (to +0.700) a moderate one (negative or positive) and −0.300 to +0.300 little discernible relationship. The definitional equation for the ϕ coefficient given in Box 8.4 is really a simplified version of the Pearson's product moment equation reflecting the dichotomous nature of the two variables, although the extreme values (−1.000 and +1.000) will only be achieved if both the

Table 8.2   Two-by-two crosstabulations illustrating different types of relationship between two attributes.

Strong positive relationship			Weak relationship			Strong negative relationship		
	A	B		A	B		A	B
C	0	47	47	C	21	24	45	C	53	0	53
D	53	0	53	D	33	22	55	D	0	47	47
	53	47	100		54	46	100		53	47	100

Table 8.3   Standard labelling of cells and marginal totals of a two-by-two crosstabulation for the phi correlation coefficient calculations.

	Attribute Y class 1	Attribute Y class 2	Marginal row totals
Attribute X class 1	A	B	E
Attribute X class 2	C	D	F
Marginal column totals	G	H	N

Box 8.4a: Phi Correlation Coefficient.

Phi Correlation Coefficient: $\phi = \dfrac{AD-BC}{\sqrt{((A+B)(B+C)(A+C)(B+D))}}$ or $\phi = \dfrac{AD-BC}{\sqrt{(EFGH)}}$

Box 8.4b: Application of the Phi Correlation Coefficient.

Most applications of the Phi Correlation Coefficient require that the observations are cross-tabulated into a 2×2 contingency data using two dichotomous attributes (or binomially classified) variables. The calculations for the coefficient are relatively straightforward requiring that the difference between the products resulting from multiplying the frequency counts in specific cells in the table are divided by the square root of the product of the marginal totals. Testing the significance of the ϕ coefficient is achieved by converting it to χ^2 and determining the probability of obtaining this value for comparison with the chosen level of significance, typically 0.05. This probability guides the investigator towards deciding whether to accept or reject the Null Hypothesis.

The worked application of the coefficient relates to the sample survey of households in Mid-Wales, whose shopping habits have been explored previously. The sample survey of randomly selected households in the communities were also asked whether there was a PC available in the household and the gender of the person who was designated as the 'head of household'. The 188 households have been distributed across the four cells of a 2×2 cross-tabulation of these two nominal attributes in Box 8.4d below. The majority of household heads were male and about a third of households had a PC. Does gender of head of household influence whether or not there is a PC, or vice versa? The Null Hypothesis states that there is no relationship between the attributes and any indication to the contrary in the sample data has occurred as a result of sampling error. The ϕ coefficient equals +0.089 and converting this into χ^2 to test its significance with 1 degree of freedom indicates that the Null Hypothesis should be retained. This result suggests that at the time the survey was carried out in mid-Wales there was no relationship between gender of head of household and the presence or absence of a PC. Clearly the results could be different now and/or in another locality.

The key stages in applying and testing the significance of the ϕ coefficient are:

Prepare the 2×2 crosstabulation and calculate the ϕ coefficient: The crosstabulation is given below.

State Null Hypothesis and significance level: the direction and strength of the relationship between gender of head of household and the presence or absence of a PC from the sample survey has arisen by chance and is not significant at the 0.05 level of significance.

Calculate test statistic (χ^2): the calculations for the test statistic are given below.

Determine the probability of the calculated χ^2: the probability of obtaining $\chi^2 = 1.497$ is $p = 0.221$.

Accept or reject the Null Hypothesis: the probability of χ^2 is >0.05, therefore accept the Null Hypothesis. By implication the Alternative Hypothesis is rejected.

Box 8.4c: Assumptions of the Phi Correlation Coefficient.

There are three main assumptions:

Observations should be chosen at random from the parent population.

The value of the coefficient will fall within a limited range if the marginal totals of the contingency table are very dissimilar.

The ordering the nominal categories for each attribute affects the sign of the coefficient and thus the direction of any relationship. By convention, dichotomous categories such as Yes and No, Present and Absent, etc. are usually given in this order. If the opposite order is used for one or both there will be an effect on the sign of the coefficient.

Box 8.4d: Calculation of the Phi Correlation Coefficient.

		PC in household					PC in household		
		Yes	No				Yes	No	
Gender	Male	A	B	E	Gender	Male	58	114	172
	Female	C	D	F		Female	3	13	16
		G	H	N			61	127	188

ϕ coefficient	$\phi = \dfrac{AD - BC}{\sqrt{(EFGH)}}$	$\phi = \dfrac{(58)(13) - (114)(3)}{\sqrt{(172)(16)(61)(127)}} = +0.089$
χ^2 statistic	$\chi^2 = \phi^2 n =$	$+0.089^2(188) = 1.497$
Probability		$p = 0.221$

---

row totals and the column totals are equal. The examples of strong negative and positive relationships in the contingency tables in Table 8.2 will not result in a ϕ coefficient of -1.000 and $+1.000$, since the pairs of row and column totals are very nearly but not exactly equal.

The ϕ coefficient is closely related to χ^2 (Chi-square) and another way of defining the coefficient that is equivalent to the one given on Box 8.4 is to say that it is the number of observations divided by the square root of χ^2. This gives a clue as to how the statistical significance of the phi coefficient can be assessed. The coefficient is converted into χ^2 by multiplying the squared coefficient by the total number of observations. The number of degrees freedom used when assessing the probability of χ^2 is always the same, since the correlation technique is normally only applied to a 2×2 crosstabulation and so $(c - 1)(r - 1)$ equals 1.

8.2.3.2 Cramer's V correlation coefficient and the Kappa Index of agreement

The **Cramer's V Coefficient** provides a further tool for analysing nominal attributes since it produces a quantity that normally lies within the range 0.000 and +1.000 and unlike the Phi Coefficient is not restricted to 2×2 contingency tables. Cramer's V is one of a number of measures that are linked to χ^2 and it is defined as the square root of χ^2 divided by nm, where n denotes sample size and m is the smaller of the number of rows minus 1 or columns minus 1. The application of Cramer's V therefore simply entails calculating χ^2 for a contingency table and then applying this adjustment to produce the required statistic. One way of interpreting Cramer's V is to view it as measuring the maximum possible variation of two variables. The value of the coefficient is influenced by the degree of equality between the marginal totals, which was noted as one of the assumptions of the Phi coefficient. If the marginal totals are unequal then the extreme value of 1.00 is unattainable. The Cramer's V is often used with grid-based spatial datasets, particularly where two classification schemes with unequal number of classes in each are being compared. The coefficient provides a measure of agreement between the two schemes.

The Kappa Index is another technique for comparing two classifications of spatial data that have been assembled as a frequency distribution in a contingency table. The counts in the cells of the contingency table are the number of spatial units allocated to particular categories according to classification schemes A and B. If both classification schemes allocate observations to the same categories then the spatial units would all occur in the cells forming the diagonal of the table. The Kappa Index is calculated from the difference between the observed and expected proportions of units thus matched along the diagonal $(O - E)$. This provides the numerator for the definitional Kappa Index equation that is divided by $1 - E$ as the denominator. Both the Kappa Index and Cramer's V can have negative values (i.e. between -1.00 and 0.00), but this is uncommon since it indicates an unusually low degree of agreement between the classification schemes.

8.3 Concluding remarks

This chapter has examined some of the most important and practically useful techniques for measuring the correlation between variables and association between attributes. One important issue that has been reserved for consideration at a later point is autocorrelation and in particular spatial autocorrelation. The worked example for three of the correlation techniques referred to the location of Burger King's restaurants in Pittsburgh in respect of their distance from the city centre and the typical daily footfall of customers. As a conclusion to this chapter and a foretaste of our later exploration of the spatial autocorrelation, consider the following question. Is it not likely that the restaurants located close together at the centre of the city are likely to have similar numbers of customers every day partly because they are located near to each other and convenient for the city centre workforce and shoppers, whereas those that are further away will be drawing on a different population of potential customers from the residential and suburban areas of the city?

9
Regression

Chapter 9 explores regression analysis, which contrasts with correlation to the extent that at least one of the variables is treated as exerting a controlling influence on the other and may be able to predict the values of the dependent variable(s). The starting point is Ordinary Least Squares regression that focuses on linear relationships between two variables with more advanced non-linear and multivariate regression techniques introduced as a basis for exploring correlation and regression of spatial data in the following chapter. The chapter also explores the two main ways of testing the significance of regression models. Simple regression techniques, such as those examined here, are rarely capable of fully explaining the complexities of geographical, geological and environmental processes and should be used with some caution by students undertaking independent projects in these areas, although an understanding of the principles of these techniques provides a fuller appreciation of the research literature.

Learning outcomes

This chapter will enable readers to:

- apply simple linear regression analysis to ratio/interval scale variables;

- specify Null Hypotheses in a format appropriate to regression analysis and undertake significance testing;

- appreciate the ways of extending these simpler regression techniques to examine relationships where there is more than one independent variable;

- recognize situations in which regression analyses could be applied in an independent research investigation in Geography, Earth Science and related disciplines.

Practical Statistics for Geographers and Earth Scientists   Nigel Walford
© 2011 John Wiley & Sons, Ltd

9.1   Specification of linear relationships

The main purpose of correlation analysis is to summarize the direction and strength of the relationship between two variables: therefore it offers a way of describing how their data values are linked. However, the data values for the X and Y variables are interchangeable when calculating correlation coefficients. In other words it does not matter whether the values collected for the variable X are 'plugged into' the equation for any of the correlation coefficients as corresponding to the terms X or Y—the outcome (i.e. the value of the coefficient) would be the same. Correlation techniques do not attempt to specify the form of the relationship in the sense of identifying which of the two variables is in control. However, this is the purpose of regression, which designates each variable as having a different role in the relationship. One, referred to as the **independent variable** (X), is treated as exerting a controlling influence on the other, which is known as the **dependent variable** (Y). What this means is that the variable X to some extent determines the value of Y. Putting it another way, if we know the value of X then we can estimate or predict the value of Y with a certain level of confidence. This raises the question, why should we want to estimate the value of Y if we already have a set of paired X and Y measurements. There are perhaps three main answers to this question:

- The X and Y data values may relate to a sample and we want to estimate the Y values of other observations in the same population for which only the X data values are known.

- The data values recorded for the independent X variable by the sampled observations may not be the ones for which estimated Y values are required.

- The form of the relationship revealed by regression analysis for one sample of observations from a specific area or time period may be applicable to comparable populations of the same type of entity in other areas or at different times.

In the case of analysing more complex relationships, the labels dependent and independent apply to two groups of variables, in other words there will be more than one independent variable and possibly more than one dependent variable as well. Whether the focus is on such complex multivariate relationships or simpler bivariate ones, which we will examine first, regression analysis seeks to express the form of the relationship mathematically by means of an equation. There are standard regression equations representing different forms of relationship that link together the dependent and independent variables in a particular way (e.g. linear, curvilinear, etc.). These equations are sometimes described as modelling the relationship. The particular form of regression analysis or model that an investigator chooses to apply depends on how the dependent and independent variables appear to be related when examined as a scatter graph. Some of the different types of scatter graph were examined in

Chapter 8 (see Figure 8.3). From a computational perspective it would be feasible to apply simple linear regression to the X and Y data values that produced either of the scatters of points shown in Figure 8.3. However, a simple straight line is clearly not the most suitable way of summarizing the two forms of the relationship, since one is curvilinear and the other cyclical.

The separation of variables into dependent and independent groups in regression analysis requires that an investigator thinks conceptually about how to model the process or relationship. What variables might cause people to emigrate from one country to another? What variables might cause water quality in rivers to decline over a period of time? What factors might increase the speed at which shoppers can evacuate a shopping centre in an emergency? These examples illustrate that regression analysis is often used to explore relationships where there is a cause and an effect, one thing resulting in another. There is often an interest in the Geographical Sciences, particularly Physical Geography and the Earth/Environmental Sciences but also those involving human interaction with the environment, in exploring how natural processes or systems work, for example how soil erosion occurs.

However, perhaps one of the simplest examples is to imagine farmers growing crops in an area where there is limited precipitation during the summer growing season (e.g. Mediterranean areas). A farmer choosing to irrigate his/her crops is likely to increase the yield in comparison with a farmer who opts to rely on precipitation. Furthermore, irrigation at one rate (e.g. 25 l/h for 10 h a day) may have a different effect on yield compared with another (e.g. 75 l/h for 4 h a day). The same total quantity of water is applied (250 l), but the duration of irrigation is different. A lower rate of application over a longer period may be more beneficial towards increasing yield compared with a higher rate over a shorter period of time. The former might help to counteract the effect of the irrigated water evaporating in high temperatures, whereas the latter might increase erosion of soil through overland flow. These two rates of irrigation consume the same quantity of water and therefore the cost of irrigation is the same or at least very similar. However, suppose two neighbouring farmers both irrigate their crops at a rate of 25 l/h, but one does so for 4 h and the other for 8 h a day: the water charges for the first farmer would be half of those of the second. The second farmer might achieve a higher increase in crop yield than the first, but unless this is at least double the yield achieved by the first farmer, some of the money paid out to the water supplier will have been wasted. We could carry on adding in other factors or variables into the analysis or adopt the strategy of some investigators and discuss the results with 'other things being equal' or ceteris paribus to use the term employed by economists. What this means is that quite possibly there are some other variables influencing crop yield that have not been measured and analysed, but these are assumed to have an equal impact (on crop yield) irrespective of the value of the independent variable(s). No matter what the irrigation rate the effect of these 'other things' on crop yield does not vary.

This description of the relationship between crop yield (dependent variable), and irrigation rate and payment for water (independent variables) has made certain unspecified or tacit assumptions. First, it has been assumed that agricultural crop plants need water in order to produce a yield (e.g. seeds in the case of cereal crops). Secondly, water supplied to crops by means of irrigation and precipitation is equivalent, in other words there is nothing particularly beneficial or detrimental to crop growth associated with either of these types of water supply. Thirdly, water is charged per litre and the rate does not vary with quantity used, whereas in reality there might be a financial penalty or incentive for high or low usage. Fourthly, all farmers are regarded as equal and it is assumed that there are no subsidies for certain types of farm/farmer (e.g. small or large farms) or for those growing certain crops. Fifthly, the agricultural crops grown in the area are assumed to have equal demand for water to achieve the same growth rate. Some leafier plants such as maize or corn and potatoes tend to have higher water requirements in the hot summer period having a larger surface area available for evapotranspiration in comparison with cereals (wheat, barley, oats, etc.) which are ripening in this period, and demand less water.

Some of the issues raised by these assumptions could be investigated by undertaking a series of regression analyses. For example, the water supply and growth-rate relationship could be explored for each of the different types of crops and even for different varieties of each crop type. The charge for irrigation water could be varied according to the amount used or farmers growing certain low-value staple crops could pay less for irrigating these rather than more exotic high-value ones. In this fashion a complex system, the relationship between yield, irrigation rate and water charges is broken down into its simpler components. Recognition that even complex multivariate relationships can be subdivided into their constituent parts provides a suitable point for introducing the next section, which examines bivariate regression techniques, in other words simple linear or ordinary least squares regression where there are just two variables in the mathematical model or equation. We will then progress to nonlinear (curvilinear) bivariate relationships and finally introduce multivariate (multiple) regression where there is more than one independent variable.

9.2 Bivariate regression

The application of regression techniques is sometimes described as fitting a regression line to a set of data points. The purpose of this exercise is to determine how closely the points fit to a particular line and to specify the relationship between the dependent and independent variables as an equation. The form or shape of the line produced by the different types of regression equation is fixed. The right side of Figure 9.1 shows the equations and form of the lines for three ideal types of regression (simple linear, quadratic and cubic) where just two variables are analysed. Each of these represents a relationship between one dependent variable (Y) and one independent variable (X).

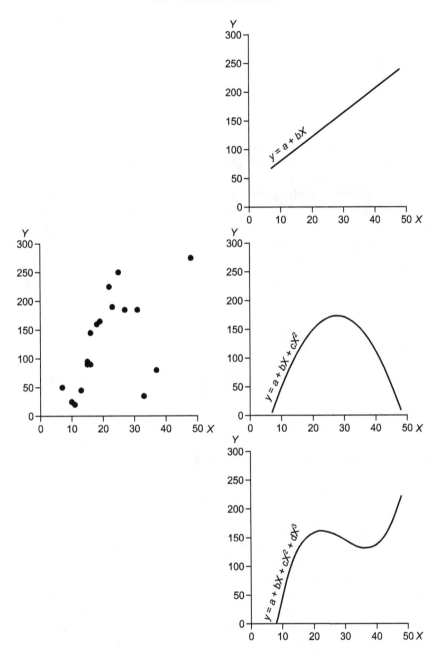

Figure 9.1   Fitting standard regression lines to a scatter of points by eye.

The left side of Figure 9.1 depicts a scatter graph of data points (observations) plotted in positions according to their paired data values on X and Y. It is clear that the point cloud does not exactly match any of the three ideal regression equations, although it is reasonably clear that the form or shape of each of the standard regression lines can reasonably be viewed as partially fitting the scatter of points.

Figure 9.1 illustrates that the process of fitting a regression line (equation) to a set of data points is about discovering the form of the relationship between the variables. Simply looking at the three regression lines and scatter plot is a rather hit and miss way of applying regression techniques. Even if you can decide which of the three standard regression lines best fits the scatter of points, it is clear that they do not fall perfectly along any of the lines. Most points measured along the X-axis (the independent variable) deviate to some extent from where they should be if they conformed exactly to the regression equation. The equations alongside the lines in Figure 9.1 define the form of the regression and are known as the linear, quadratic and cubic forms of bivariate regression. The following sections will examine how to fit a scatter of data points to these different types of bivariate regression.

9.2.1 Simple linear regression

The previous description of fitting data points to a regression line has approached the question from a graphical perspective and works from the principle of looking for the line that most closely matches the points. If the known values of X perfectly determine the corresponding values of Y, they fall along a regression line with a particular form. Each form of regression line can be represented as an equation and in the case of **simple linear** regression the general form of the equation is:

$$Y = a + bX$$

where a and b are constants, and X denotes the known values of the independent variable and Y the values of the dependent variable. If the values of the constants are given, then inserting each known X data value into the equation in turn will necessarily result in a predetermined series of Y values. This series of Y values lie along the regression line and equal the height of the line above or below the horizontal axis. These values may be viewed as the quantity that the independent variable (X) is capable of shifting the dependent variable (Y) in a positive direction above or a negative direction below the horizontal axis where Y equals zero. In some applications when $X = 0.0$ Y will also equal zero, whereas as in other cases there will always be some 'background' value of Y irrespective of the value of X. For example, there are background values of the different radioactive isotopes in the atmosphere regardless of whether there is any source in an area potentially raising this level.

This description helps to explain the role of the constants a and b in the simple linear regression equation. The constant a equals the value of Y where the regression line passes through (intercepts) the vertical axis and $X = 0.0$; it is known as the **intercept**. Suppose the equation omitted the constant b and appeared as:

$$Y = a + X$$

This would mean that Y always equalled the sum of the X value and the intercept: so with an intercept of 10.0 the series of Y values associated with $X = 1, 2, 3, 4 \ldots 10$ would equal 11, 12, 13, 14 ... 20.

Visualize the graph of this relationship between X and Y for this series of X and Y values. What is the direction of the line in relation to the X-axis? Change the value of the intercept, perhaps to 15 and recalculate the X values. What would be the direction of this line plot on the graph?

These questions illustrate the importance of the b constant in the regression equation: its value and sign (+ or −) control the **gradient or slope** of the regression line. Without the b constant the angle between line and the horizontal axis would always be 45° and this effectively means that the slope of the line is 1.0 in all cases. The b constant is the amount that Y increases per unit of X: a large value denotes a steep slope and a small one produces a gentle gradient. A positive or negative sign for the slope coefficient has the same significance as for a correlation coefficient: a positive b means that the regression line ascends from lower left to upper right, whereas a negative one denotes the reverse (it descends from upper left to lower right).

The perfect relationship defined by the simple linear regression equation is unlikely to occur when the pairs of data values forming a scatter of points on a graph have been obtained from a population or sample of observations. Figure 9.2a illustrates a

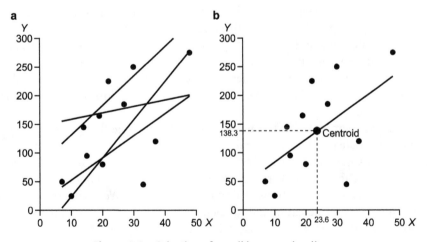

Figure 9.2 Selection of possible regression lines.

selection of the infinite number of straight lines that could be drawn through a given scatter of data. Which, if any, of these is the 'best fit' line? How do we know which is the best fit line? What does 'best' mean in this context? What criteria might be useful in deciding whether one line is a better fit than another? Obviously, the line cannot zigzag across the graph connecting all the data, since it would not be a straight line. The means of the X and Y data values measure the central tendency of the two variables, which can be plotted as the middle or centroid of the point distribution. Given this role, it is reasonable to argue that the regression line should pass through the centroid, which acts as an 'anchor point' through which pass an infinite number of straight lines. One of these is shown in Figure 9.2b, but even though this passes through the centroid it may not be the 'best fit' line. To discover whether it is, we need to remember that the values of X control the values of Y (at least to some extent) and not the other way round. We are seeking to discover the **linear regression of** Y **on** X and not the reverse. The Y values do not fall along a straight for some unknown reason, which represents the extent to which the independent variable does not fully control the dependent one.

The dependent variable will imperfectly predict the value of Y in most applications and there will differences between observed and predicted Y values for known values of X. This difference or deviation is known as the **residual** or **error.** A positive residual occurs when the regression equation overestimates the observed value of Y and a negative one arises when it is underestimated. Smaller positive or negative residuals indicate that the regression line and its equation provide a better prediction of the Y values in comparison with larger differences. The size of the residuals is therefore relative. It follows that if we were able to quantify the overall size of the total difference represented by the residuals, then this might help in determining the line that fits a given scatter of data points. Perhaps the solution is to select the line that minimizes the total sum of the residuals. However, if we simply add up the positive and negative residuals for any line that passes through the centroid of the data points, the result will equal zero. This is the same problem that arises when summing the positive and negative differences of a set of data values from their mean. This difficulty with the absolute positive and negative residuals always adding up to zero can be overcome by squaring the values before summation. This brings to mind the need to square the differences between a set of data values and their mean before dividing by the number of observations to obtain a population variance (see Chapter 4).

The 'best fit' regression line can now be defined as the line that passes through the centroid of the data points and minimizes the sum of the squared differences or deviations in Y units measured vertically from each data point to the line. However, it is unrealistic to use the trial and error method of drawing various lines and calculating the sum of the squared deviations to find the 'best fit' line. We need a more straightforward approach to calculating the coefficients (constants) a and b in the regression equation. Once these are know the equation can be solved for any value of X and the line drawn through the predicted Y values. The simple linear regression equation given previously should be slightly rewritten to indicate that the Y values are predicted and differ from the observed as:

$$\hat{Y} = a + bX$$

The overscript ^ ('hat') symbol denotes that values of the dependent variable along the regression line may differ from the observed values in a sample. The equivalent linear regression equation for a population of observations is

$$\mu = \alpha + \beta X$$

The use of μ, which normally symbolizes a population mean, in place of $\hat{Y}$ makes the important point that the regression line and the Y values lying along it are the mean or average of the relationship between the two variables in a population of observations. These mean values for the dependent variable at each known value of X and the conditional distributions associated with them lie parallel to the vertical axis at each point along the line and follow the normal probability distribution. Thus, the population mean of Y is predicted by any known value of X.

Figure 9.3 shows how the same regression line and equation can apply to more than one scatter of data points. However, it is clear that the residuals in Figure 9.3a are generally smaller than those in Figure 9.3b: the regression equation seems to act as a more accurate way of predicting the dependent variable from the independent one in the former case. However, given that the equation relates to both distributions, it would be useful to have some way of quantifying the overall variability in the residuals or errors and the extent to which the lines 'best fit' the data points. In other words, we need a statistic that measures their overall dispersion about the regression line in

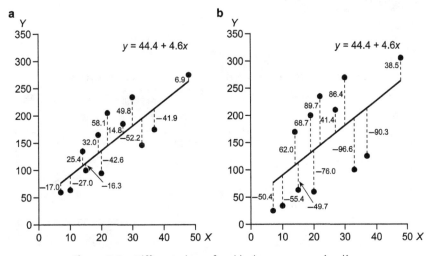

Figure 9.3 Different sizes of residuals on a regression line.

a similar way to a variance or standard deviation quantifies the dispersion of data values about their mean. The **standard error of the estimate** ($\sigma_{Y.X}$ for a population and $s_{y.x}$ for a sample) performs this function and is similar to the standard error of the mean examined in Chapter 6. It is effectively the standard deviation of the residuals from the best fit regression line, which is usually shown as ε (epilson) for a population and e for a sample and indicates that they represent the error in the estimation of the true Y values. The standard error of the estimate acts as a summary measure, similar to the mean and standard deviation and allows different samples to be compared with regard to the amount that their data points are dispersed or scattered about the regression line. It gives a measure of the overall accuracy of the predicted Y values and is the standard deviation of the sampling error.

Box 9.1 applies simple linear regression in respect of the distance of Burger King and McDonald's restaurants from the centre of Pittsburgh and illustrates how to

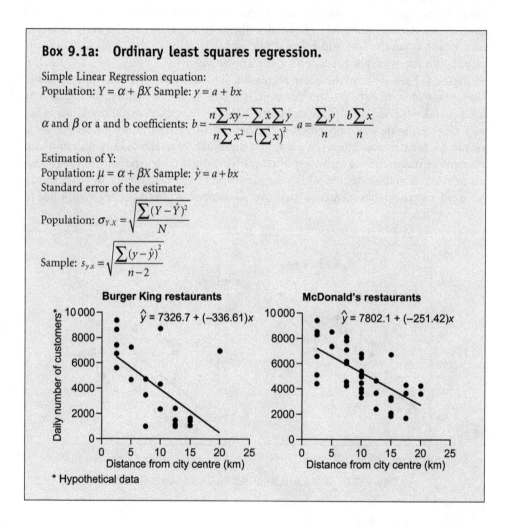

Box 9.1a: Ordinary least squares regression.

Simple Linear Regression equation:
Population: $Y = \alpha + \beta X$ Sample: $y = a + bx$

α and β or a and b coefficients: $b = \dfrac{n\sum xy - \sum x \sum y}{n\sum x^2 - \left(\sum x\right)^2}$ $a = \dfrac{\sum y}{n} - \dfrac{b\sum x}{n}$

Estimation of Y:
Population: $\mu = \alpha + \beta X$ Sample: $\hat{y} = a + bx$
Standard error of the estimate:

Population: $\sigma_{Y.X} = \sqrt{\dfrac{\sum (Y - \hat{Y})^2}{N}}$

Sample: $s_{y.x} = \sqrt{\dfrac{\sum (y - \hat{y})^2}{n-2}}$

Burger King restaurants

$\hat{y} = 7326.7 + (-336.61)x$

McDonald's restaurants

$\hat{y} = 7802.1 + (-251.42)x$

Daily number of customers*

Distance from city centre (km)

* Hypothetical data

Box 9.1b: Application of ordinary least squares regression.

Simple linear or ordinary least squares regression defines a straight line by means of a mathematical equation that connects two sets of paired data values, one measuring a dependent variable (Y) and the other an independent variable (X). The most straightforward way of obtaining the regression equation and line for a particular sample of observations is by calculating the constants a and b and inserting the resulting values into the standard equation $\hat{Y} = a + bX$. These calculations are normally carried out by statistical or spreadsheet software, although a worked example helps to clarify the technique. The series of $\hat{Y}$ values for the dependent variable lie along the regression line and are estimates or predictions of its value in the statistical population for the known values of X. The standard error of the estimate statistic ($s_{y.x}$) summarizes the overall accuracy of the regression equation in predicting Y and provides a means of comparing different samples.

Simple linear or ordinary least squares regression has been applied to investigate whether distance from the centre of Pittsburgh acts as a good predictor of daily customer footfall for two types of fast-food restaurant. The values of the independent variable (X) should not be discovered empirically but be predetermined by an investigator. This requirement has been at least partially satisfied by categorizing the distances of the Burger King and McDonald's restaurants at intervals of 2.5 km. In the interests of brevity only the calculations involved in obtaining the a and b constants in respect of the data values for the Burger King restaurants are shown in Box 9.1d, although the standard errors for both are shown. The scatter plots of the two sets of data values are reasonably similar as are the fitted regression lines, although the gradient of the McDonalds line is slightly less and the set of Burger King restaurants includes a few that have rather more customers than would be predicted by the regression equation. This visual interpretation of the two distributions is supported by the standard errors of the estimate for the two sets of data (2447.30 for Burger King and 1529.99 for McDonalds). There is rather more dispersion in the residuals for the Burger King restaurants compared with the McDonald's ones.

The key stages in applying Simple Linear Regression are:

Calculate constants a and b: these denote where the regression line intercepts the vertical Y-axis and the gradient or slope of the line;

Calculate the standard error(s) of the estimate: this statistic provides a summary of the overall fit of the data points to the regression line/equation.

Box 9.1c: Assumptions of ordinary least squares regression.

There are five main assumptions:

The values of the independent variable X should be determined by the investigator and are NOT sample values.

The dependent variable Y should be normally distributed.

The dependent variable should have the same variance for all values of the independent variable X.

The complete set of residuals or errors follow the normal distribution.

There should be no systematic pattern in the values of the residuals (i.e. no autocorrelation is present).

Box 9.1d: Calculation of ordinary least squares regression (Burger King).

No.	x	y	xy	x^2	y^2	$(y - \hat{y})^2$
1	12.5	1200.0	15 000.0	156.3	1 440 000.0	3 683 002.4
2	20.0	7040.0	140 800.0	400.0	49 561 600.0	41 543 954.6
3	12.5	1550.0	19 375.0	156.3	2 402 500.0	2 462 121.9
4	5.0	4670.0	23 350.0	25.0	21 808 900.0	948 072.2
5	10.0	2340.0	23 400.0	100.0	5 475 600.0	2 626 474.0
6	10.0	8755.0	87 550.0	100.0	76 650 025.0	22 985 887.8
7	2.5	9430.0	23 575.0	6.3	88 924 900.0	8 671 758.7
8	7.5	3465.0	25 987.50	56.3	12 006 225.0	1 788 010.2
9	2.5	8660.0	21 650.0	6.3	74 995 600.0	4 729 689.8
10	5.0	7255.0	36 275.0	25.0	52 635 025.0	2 596 319.9
11	2.5	7430.0	18 575.0	6.3	55 204 900.0	892 618.7
12	2.5	5555.0	13 887.50	6.3	30 858 025.0	865 299.9
13	10.0	4330.0	43 300.0	100.0	18 748 900.0	136 426.8
14	2.5	6755.0	16 887.50	6.3	45 630 025.0	72 783.9
15	7.5	1005.0	7537.50	56.3	1 010 025.0	14 418 462.0
16	7.5	4760.0	35 700.0	56.3	22 657 600.0	1777.9
17	15.0	1675.0	25 125.0	225.0	2 805 625.0	363 114.7

18	15.0	1090.0	16350.0	225.0	1188100.0	1410370.0
19	15.0	1450.0	21750.0	225.0	2102500.0	684905.2
20	12.5	2500.0	31250.0	156.3	6250000.0	383303.4
21	12.5	1070.0	13375.0	156.3	1144900.0	4198872.3
22	12.5	1040.0	13000.0	156.3	1081600.0	4322719.2

$$\sum x = 202.5 \qquad \sum y = 93\,025.0 \qquad \sum xy = 673\,700.0 \qquad \sum x^2 = 2406.3 \qquad \sum y^2 = 574\,582\,575.0 \qquad \sum (y-\hat{y})^2 = 11\,985\,945.6$$

Slope constant

$$\text{Burger King } b = \frac{n\sum xy - \sum x \sum y}{n\sum x^2 - \left(\sum x\right)^2} \qquad \frac{22(673\,700.00) - 202.50(93\,025.00)}{22(2406.25) - 202.50^2} = -336.61$$

Intercept constant

$$\text{Burger King } a = \frac{\sum y}{n} - \frac{b\sum x}{n} \qquad \frac{93\,025.00}{22} - \frac{-336.61(202.50)}{22} = 7326.74$$

Regression equation

$$\text{Burger King } \hat{y} = a + bx \qquad \hat{y} = 7326.74 + -336.61x$$

Standard error of the estimate

$$\text{Burger King } s_{y \cdot x} = \sqrt{\frac{\sum (y-\hat{y})^2}{n-2}} \qquad \sqrt{\frac{119\,785\,945.66}{22-2}} = 2447.30$$

$$\text{McDonald's} \qquad \sqrt{\frac{91\,294\,159.81}{40-2}} = 1529.99$$

calculate the constants a and b in order to define the specific straight line equation that fits a given set of data points. However, we shall later examine some of the difficulties associated with using a spatial variable, such as distance from city centre, in simple linear regression. The application explores whether distance from the city centre has a controlling influence over the daily footfall of customers to the two types of restaurants and whether there is a difference in the two brands. The distances have been coded to the nearest 2.5 km, which represents an attempt to address the assumption that the data values of X should be chosen by the researcher rather than be allowed to arise from the sample data. The regression lines and equations for the two distributions are reasonably similar, trending at a moderately steep angle from higher numbers of customers near the city centre to lower values in the suburban areas. The dispersion of data points for the Burger King restaurants appears to be slightly greater with two particularly well-patronized outlets at the 10 and 20 km points, the latter seemingly untypical of other suburban sites. Box 9.1 includes calculations for the standard error of the estimate for the two sets (samples) of restaurants. In comparison with data values for the variable distance from city centre used previously, the adjusted values results in the data points in the scatter plots appearing in vertical lines above the X-axis corresponding to 2.5, 5.0, 7.5, etc. km. Recoding of the original distances to the nearest 1 km or even 500 m would have produced different scatters of points that would still have been arranged in vertical lines.

What does the standard error of the estimate statistics given in Box 9.1 tell us about the dispersion of the two types of restaurant around their respective regression lines?

9.2.2 Testing the significance of simple linear regression

The differences between the predicted and observed Y values can occur for three main reasons when applying regression analysis to populations or samples of observations:

- there is an **alternative** independent variable that exerts stronger control over the dependent variable;

- there are **additional** independent variables not yet taken into account that exert stronger control over the dependent variable in combination with the currently selected independent variable;

- there is sampling error in the data.

The first two reasons suggest that further analysis is required that takes into account other independent variables, whereas the third calls for testing the significance of the regression statistics. There are two main approaches to this task: one involves applying analysis of variance (see Chapter 6) to test the whole regression model or equation and the other uses t tests to investigate the significance of its component parts (a, b and $\hat{Y}$). Confidence intervals around the regression line and the intercept, which is really the predicted value of Y when $X = 0.0$, are similar to such limits around a univariate mean. They define a range within which an investigator can be 95 or 99 per cent confident that the true population value lies.

We have seen that one-way analysis of variance can be used as a statistical test to examine whether the means of a group of independent samples are significantly different from each other and that the procedure divides the total variance into the between groups and within groups parts. When applied to simple linear regression, the total variance of the dependent variable (Y) is similarly separated into two parts: one connected with the predicted $\hat{y}$ values along the regression line and the other with the residuals of the observed values from the line. These are known, respectively, as the regression and residual variances. The purpose of dividing the total variance into these two parts is to indicate how much of the variability of the Y values is explained by the linear regression model and how much by the residuals. If a large proportion of the total variance comes from the regression equation, then it is reasonable to conclude that this model and the independent variable (X) together act as a good predictor of the dependent variable (Y). If the variance associated with the residuals accounts for a large proportion of the total, this suggests that other excluded independent variables explain more of the variability in the dependent variable than accounted for by X.

Consideration of extreme or unlikely situations sometimes helps to explain how a statistical technique works. In the case of simple linear regression it would be very unusual for a regression line to be horizontal, which would indicate that the value of Y was constant for all values of the independent variable. This would mean that there was no slope ($b = 0.0$) and $\bar{y} = \hat{y}$ for all values of X. It would follow that the regression sum of squares and variance were both zero and total variance would equal the residual variance. The regression line almost invariably slopes upwards or downwards through a scatter of points from the left of the scatter graph, which means that some of the total regression sum of squares and variance are attributed to the regression model. If the Y values all lie along the regression line, this produces another extremely unlikely situation in which the total sum of squares and variance of Y are only associated with the regression model (i.e. there is no residuals' variance). The size of F test statistic (see Box 9.2) indicates whether the amount of variability in the Y values attributed to the regression model is significant in relation to that connected with the residuals or prediction errors.

An alternative approach to testing the significance of regression analysis is to examine the sampling error that might be present in the intercept and slope constants as estimates of the corresponding population parameters α and β, and in the predicted values of Y. The starting point for testing the significance of these quantities is to recognize that the calculated values are obtained from a particular sample of observations and that this is just one of an infinite number of such samples that could have been selected from the population. This represents the standard starting point for

Box 9.2a: Significance testing of ordinary least squares regression.

Whole model or equation using analysis of variance:

Population: $F = \dfrac{\sigma_\mu^2}{\sigma_e^2}$ Sample: $F = \dfrac{s_{\hat{y}}^2}{s_e^2}$

Regression variance: $s_{\hat{y}}^2 = \dfrac{\sum(\hat{y} - \overline{y})^2}{k}$ Residual variance: $s_e^2 = \dfrac{\sum(y - \hat{y})^2}{n-2}$

Separate components of equation:
Standard errors:

Slope: $s_b = \dfrac{s_{y.x}}{\sqrt{\left(\sum x^2 - \dfrac{(\sum x)^2}{n}\right)}}$ Intercept: $s_a = s_{y.x}\sqrt{\dfrac{\sum x^2}{n\sum x^2 - (\sum x)^2}}$

Predicted Y: $s_{\hat{y}} = s_{y.x}\sqrt{\dfrac{1}{n} + \left(\dfrac{n(x - \overline{x})^2}{n\sum x^2 - (\sum x)^2}\right)}$

t test statistics:

Slope: $t = \dfrac{|b - \beta|}{s_b}$ Intercept: $t = \dfrac{|a - \alpha|}{s_a}$

Confidence limits of regression equation or line:
Slope: $b - t_{0.05}s_b$ to $b + t_{0.05}s_b$
Intercept: $a - t_{0.05}s_a$ to $a + t_{0.05}s_a$
Predicted Y: $\hat{y} - t_{0.05}s_{\hat{y}}$ to $\hat{y} + t_{0.05}s_{\hat{y}}$

Box 9.2b: Application of significance testing to ordinary least squares regression.

Using analysis of variance to test the significance of the whole regression model involves calculating the regression variance ($s_{\hat{y}}^2$) and the residual variance (s_e^2). The F test statistic is obtained by dividing the regression variance by the residual variance and its probability determined with k degrees of freedom for the regression variance and $n - k - 1$ for the residual variance. n and k are respectively, the number of observations and the number of independent variables, which is always 1 in the case of simple linear regression. In general terms the Null Hypothesis states that the variability in the dependent variable (Y) is not explained by differences in the value of the independent variable (X). The fate of the Null Hypothesis is judged by specifying a level of significance (e.g. 0.05) and determining the probability of having obtained the value of the F statistic. If the probability is less than or equal to the chosen significance threshold, the Null Hypothesis would be rejected.

The alternative method of testing the significance of regression analysis is to focus on the individual elements in the regression equation, the slope and intercept constants and the predicted Y values, which estimate the mean of the dependent variable for each known value of X. The procedure is similar in each case and recognizes that if an infinite number of samples were to be selected from a population there would be a certain amount of dispersion in the slope, intercept and predicted Y values (for each value of X). This can be quantified by means of calculating the standard error, which is used to produce t test statistics and confidence limits in each case. The general form of the Null Hypothesis for each of these tests is that any difference between the sample-derived statistics (b, a and $\hat{y}$) and the known or more likely assumed population parameters (β, α and μ) has arisen through sampling error. There are $n - 2$ degrees of freedom when determining the probability associated with the t test statistic calculated from the sample data values. As usual, the decision on whether to accept the Null Hypothesis or the Alternative Hypothesis is based on the probability of the t test statistic in relation to the stated level of significance (e.g. 0.05).

The calculation of confidence limits for the slope, intercept and predicted Y values provide a range of values within which the corresponding population parameter is likely to lie with a certain degree of confidence (e.g. 95% or 99%). In each case the standard error is multiplied by the t statistic value for the chosen level of confidence with $n - 2$ degrees of freedom and the result is added to and subtracted from the corresponding sample derived statistic in order to define upper and lower limits. In the case of confidence limits for the predicted Y values, these can be calculated for several known values of X and shown on the scatter plot to indicate a zone within which the corresponding population parameter is expected to fall.

The different types of significance testing have been applied to the simple linear regression equations in Box 9.1 relating to the Burger King and McDonald's restaurants in Pittsburgh. The Null Hypothesis relating to the regression model states that the distance from the centre of the city does not explain the daily customer footfall in respect of either type of restaurant. Each of the Null Hypotheses in the second group of tests asserts that the slope and intercept coefficients are not significantly different from what would occur by chance. The probabilities associated with $F = 10.26$ (Burger King) and $F = 27.84$ (McDonald's) respectively with 1/20 and 1/39 degrees of freedom in the ANOVA tests in Box 9.2c clearly suggest that the regression models for both types of restaurant provide evidence that there is a very small probability that distance from the city centre does not partly explain daily customer footfall. The individual t tests in Box 9.2d produce statistically significant results for both types of restaurant in respect of the slope and intercept coefficients. The probability of the t test statistic for b ($t = 3.20$) is 0.004 and for a ($t = 6.67$) is < 0.000 for Burger King restaurants and the equivalent figures for McDonald's outlets are $t = 5.28$ and $p < 0.000$, and $t = 14.71$ and $p < 0.000$. The critical t distribution values at the 0.05 significance level with 20 and 39 degrees of freedom respectively for the Burger King and McDonald's restaurants are 2.086 and 2.021. These have been inserted into the calculations for the confidence limits in Box 9.2e below. Calculation of the standard error for the predicted Y values is illustrated in Box 9.2e for when x = 15.0. It also includes the calculations for the upper and lower confidence limits of the predicted Y values for both types of restaurant.

The key stages in testing the significance in Simple Linear Regression are:

Choose type of testing procedure: it is possible to test the whole regression model using analysis of variance or the slope, intercept and predicted Y values, although in practice statistical software allows both to be carried easily.

State Null Hypothesis and significance level: the regression model does not indicate that the independent variable (X) explains any variation in the dependent variable (Y) beyond that which might be expected to occur through sampling error at the 0.05 level of significance OR the slope, intercept and predicted Y values are not significantly different from what would occur by chance at the same level of significance.

Calculate the appropriate test statistic (F or t): the calculations for the test statistics are given below.

Select whether to apply a one- or two-tailed test: in most cases there is no reason to believe the difference between the sample statistics would be smaller or larger than the hypothesized population values.

Determine the probability of the calculated test statistics: the probabilities of obtaining the individual test statistics by chance can be obtained from tables of the t and F distributions or are provided by statistical software using the appropriate degrees of freedom.

Accept or reject the Null Hypothesis: these probabilities are all <0.05 for the test statistics in respect of the Burger King and McDonald's restaurants and therefore the Null Hypothesis should be rejected and the Alternative Hypothesis is accepted.

Box 9.2c: Calculation of analysis of variance for ordinary least squares regression.

Regression variance	Burger King $s_{\hat{y}}^2 = \dfrac{\sum(\hat{y}-\bar{y})^2}{k}$	$\dfrac{61448873.67}{1} = 61448873.67$
	McDonald's	$s_{\hat{y}}^2 = 65179528.15$
Residual variance	Burger King $s_e^2 = \dfrac{\sum(y-\hat{y})^2}{n-2}$	$\dfrac{119785945.66}{22-2} = 5989597.28$
	McDonald's	$s_e^2 = 2340875.89$
Analysis of variance and probability	Burger King $F = \dfrac{s_{\hat{y}}^2}{s_e^2}$	$\dfrac{61448873.67}{5989597.28} = 10.26$
	$p=$	0.004
	McDonald's	$F = 27.84$
	$p<$	0.000

Box 9.2d: Calculation of t tests for ordinary least squares regression.

Standard error of the slope	Burger King $s_b = \dfrac{s_{y.x}}{\sqrt{\left(\sum x^s - \dfrac{(\sum x)^2}{n}\right)}}$	$\dfrac{2447.30}{\sqrt{\left(2406.25 - \dfrac{202.50^2}{22}\right)}} = 105.09$
	McDonald's	$s_b = 47.65$
T test statistic for slope and probability	Burger King $t = \dfrac{|b-\beta|}{s_b}$	$\dfrac{|-336.61-0|}{105.09} = 3.20$
	$p=$	0.004
	McDonald's	$t = 5.28$
	$p<$	0.000
Standard error of the intercept	Burger King $s_a = s_{y.x}\sqrt{\dfrac{\sum x^2}{n\sum x^2 - (\sum x)^2}}$	$2447.3\sqrt{\dfrac{2406.25}{22(2406.25)-202.5^2}} = 1099.45$
	McDonald's	$s_a = 530.46$
T test statistic for intercept and probability	Burger King $t = \dfrac{|a-\alpha|}{s_a}$	$\dfrac{|7326.74-0|}{1099.45} = 6.67$
	$p<$	0.000
	McDonald's	$t = 14.71$
	$p<$	0.000

Box 9.2e: Calculation of confidence limits for ordinary least squares regression.

Confidence limits of slope coefficient	Burger King $b - t_{0.05}s_b$ to $b + t_{0.05}s_b$	$-336.61 - 2.086(105.09)$ to $-336.61 + 2.086(105.09)$ -555.82 to -117.39
	McDonald's	$-251.42 - 2.021(47.65)$ to $-251.42 + 2.021(47.65)$ -347.72 to -155.12
Confidence limits of intercept coefficient	Burger King $a - t_{0.05}s_a$ to $a + t_{0.05}s_a$	$7326.74 - 2.086(1099.45)$ to $7326.74 + 2.086(1099.4)$ 5033.29 to 9620.19
	McDonald's	$7802.10 - 2.021(530.46)$ to $7802.10 + 2.021(530.46)$ 6730.04 to 8874.16
Standard error of the predicted Y	Burger King $s_{\hat{y}} =$ $s_{y.x}\sqrt{\dfrac{1}{n} + \left(\dfrac{n(x-\bar{x})^2}{n\sum x^2 - \left(\sum x\right)^2}\right)}$	Burger King example with x = 15.0 $2447.3\sqrt{\dfrac{1}{22} + \left(\dfrac{22(15.0-9.20)^2}{22(2406.25)-(202.5^2)}\right)}$ $= 12.32$

	x	$\hat{y}$	$s_{\hat{y}}$	$\hat{y} - t_{0.05}s_{\hat{y}}$ (lower)	$\hat{y} + t_{0.05}s_{\hat{y}}$ (upper)
Confidence limits of the predicted Y (Burger King)	2.5	14.23	6484.77	6456.01	6513.54
	5.0	8.92	5643.25	5625.21	5661.29
	7.5	3.62	4801.72	4794.41	4809.04
	10.0	1.71	3960.20	3956.74	3963.66
	12.5	7.01	3118.67	3104.50	3132.85
	15.0	12.32	2277.15	2252.25	2302.05
	17.5	17.63	1435.62	1399.99	1471.26
	20.0	22.94	594.10	547.73	640.47

	x	$\hat{y}$	$s_{\hat{y}}$	$\hat{y} - t_{0.05}s_{\hat{y}}$ (lower)	$\hat{y} + t_{0.05}s_{\hat{y}}$ (upper)
Confidence limits of the predicted Y (McDonald's)	2.5	9.06	7173.55	7155.23	7191.87
	5.0	6.02	6545.00	6532.83	6557.17
	7.5	2.98	5916.45	5910.43	5922.47
	10.0	0.17	5287.90	5287.55	5288.25
	12.5	3.12	4659.35	4653.04	4665.66
	15.0	6.17	4030.80	4018.34	4043.26
	17.5	9.21	3402.25	3383.64	3420.86
	20.0	12.26	2773.70	2748.93	2798.47

---

most statistical tests, namely that we do not know for certain whether the statistics obtained from our sample are typical or unusual in relation to the whole population of observations. Theoretically, if regression analysis in respect of X and Y variables was applied to a series of samples selected from a population, each slope coefficient and the mean of all the slope coefficients would be an estimate of the population parameter, β. This set of coefficients would have a degree of dispersion that can be measured by their standard error (s_b), which be estimated from a single sample (see Box 9.2). This quantity can be used in two ways:

- to calculate a t test statistic that examines the Null Hypothesis that the differences between the sample slope coefficient and zero is not significant;

- to obtain confidence limits within which the population parameter might be expected to fall.

Given earlier comments about the improbability that the regression line will be horizontal, it might seem strange to assign the value zero to β in order to test whether the difference between b and β is significant. However, if the slope coefficient was zero, this would imply that the relationship between the dependent and independent variables was fixed irrespective of the value of X. Testing whether the sample slope coefficient is significantly different from zero seeks to discover whether the independent variable exerts a controlling influence. If the value of Y was constant for all values of X, the latter would clearly not be exerting any control. There are some occasions when a value other than zero might be assigned to β in the Null Hypothesis, for example when a complete statistical population of observations had been analysed previously and the current investigations seeks to determine whether the regression of Y on X has changed over time.

The theory behind testing the significance of the intercept follows a similar line of argument. If an infinite number of samples containing observations whose X values equalled zero was selected from a given population, then the mean of all the intercepts would be expected to equal α, the population parameter. It would not be necessary for the intercepts of all of the samples to equal α and so there would be some degree of dispersion that could be measured by their standard error (s_a). This can again be estimated from any single sample of the observations (see Box 9.2). The standard error can be used to test the significance of the intercept in a particular sample of observations and to determine confidence intervals. In most cases the Null Hypothesis assumes that population parameter (α) equals zero, since the only evidence available to suggest that the regression does not pass through the origin of the scatter plot (i.e. $X = 0.0$ and $Y = 0.0$) comes from the sampled set of observations and these might be untypical of the population. *A priori* reasoning often leads to the conclusion that Y must equal zero when X equals zero. For example an investigation into the relationship between wind speed (independent) and energy generated by wind turbines (dependent) might reasonably assume that when there is no wind there will also be no energy output. Equally there are other occasions when *a priori* reasoning might lead to a Null Hypothesis that assumes Y equals some value more or less than zero. For instance, an examination of the relationship between an increase in tax per litre of petrol or diesel (independent) and change in people's monthly expenditure on this item (dependent) might assume that when $X = 0.0$ (no tax increase) Y will be > 0.0 because other factors affect people's consumption patterns. Simple linear regression focuses on one independent variable and there may be others that lead to a positive or negative value of Y when X equals zero, therefore the best option is to assume α is zero.

The predicted values of the dependent variable ($\hat{y}$) are estimates of the population means (μ) for each of the normally distributed conditional distributions that are associated with the different values of X. If a series of samples was drawn from a population, the means of their Y values for each value of X would be expected to equal μ or $\hat{y}$, the predicted value of the dependent variable in the population. It is virtually certain the sampled Y values and quite possibly those within any individual sample would display some degree of dispersion that could be measured by the standard error ($s_{\hat{y}}$). For example the scatter plots in Box 9.1 show that several fast-food restaurants at the same distance from the centre of Pittsburgh had different average daily customer footfalls. It is possible to carry out t tests in respect of Y for each of the known values of X, although it is perhaps more useful to gain an overall impression of the predictive strength of the regression equation by calculating the confidence limits around the regression line. Chapter 6 examined how a sample mean ($\bar{x}$) can be used to determine confidence limits within which a population mean (μ) is likely to fall in a certain percentage of randomly selected samples (e.g. 95% or 99%) by reference to the probabilities of the t distribution. A relatively wide confidence interval indicates that the specific value of the sample mean ($\bar{x}$) is a rather poor estimate of the population mean (μ), whereas a narrower interval suggests the sample statistic is a good estimate. These principles can be applied to produce confidence intervals (or limits) for the predicted values of the dependent variable. The procedure involves calculating the standard error of Y for each value of X and inserting this into the formula shown in Box 9.2e. The confidence limits for each value of X appear as curves around the regression line because more extreme values of the independent variable are further from its mean and produce higher values of $s_{\hat{y}}$.

There are two further points to note before moving on from simple bivariate linear regression analysis into the more complex realms of nonlinear relationships. The **coefficient of determination** (r^2) was explored in relation to Pearson's product moment correlation and can be calculated for simple linear regression as the ratio between the ($s_{\hat{y}}^2$) and the variance of the observed Y values (s_y^2). It quantifies the extent to which X explains Y with values in the range 0.0 to 1.0: when expressed as a percentage by moving the decimal point two places to the right this value indicates the amount of the variance of the dependent variable accounted for by the independent or explanatory variable.

Most applications of simple linear regression generate the regression equation from sample data: as a result, the predictions for Y are based on the known values of X that have been included. Although the research may have been planned so that the full range of known X values have been covered, it is quite possible for the current range to expand or contract over time. In both cases the analysis could give rise to misleading results and an incorrectly specified regression equation connecting the two variables which produce unreliable predictions. If a narrower range of X values is required then some of the original more extreme values where the confidence limits are wider may have produced a weaker relationship than is actually the case. In contrast, if a wider range of X values extending beyond the original set is needed, then simple linear regression may produce a stronger relationship than genuinely exists. Figure 9.4a includes an original set of data points (solid circles) with X values in the range 17 to 33, which produces $r^2 = 0.288$ and a regression of $y = 67.89 + 3.89x$. However, suppose the range of X values changes over time and narrows to 23 to 27 (crossed solid circles): now the regression equation (and line) is virtually identical to the previous one

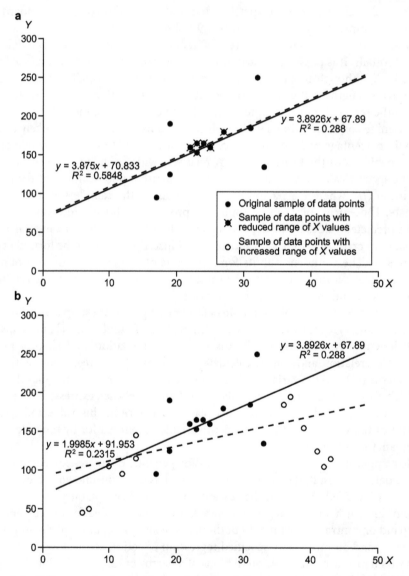

Figure 9.4 Differences in the 'goodness of fit' in relation to range of values of independent variable.

($y = 70.83 + 3.88x$) but with a higher coefficient of determination ($r^2 = 0.585$) indicating the independent variable explains a higher percentage of the variance. Figure 9.4b shows the case where the range of X values increases (6 to 43) with the original set of data points retained. The coefficient of determination now reduces to 23.1 per cent and the regression equation becomes $y = 91.95 + 2.00x$. Examination of the extra data points in Figure 9.4b (open circles) suggests that the additional X values beyond the original minimum and maximum are associated with lower Y values. Thus, there is some indication that the original regression equation provides less accurate predictions of Y at the upper and to some extent lower ends of the wider range of X values. Additionally, the new regression equation for the full range of X values has a lower coefficient of determination ($r^2 = 0.231$) and indicates a weaker relationship between the variables.

9.2.3   *Nonlinear bivariate relationships*

The previous example reveals there is no guarantee that the linear relationship measured by a regression equation applies beyond the values of the independent variable recorded by a particular sample of observations: possibly a different linear or a **curvilinear regression** equation better fits the data points. Inspection of the scatter plot should be sufficient to indicate whether a nonlinear relationship might exist and that curvilinear bivariate regression might be applied. The purpose of applying curvilinear regression is essentially the same as ordinary least squares regression, namely to measure and specify the form of the relationship between a pair of ratio or interval scale variables and, when based on sample data, to test hypotheses about the explanatory and predictive power of the independent variable over the dependent one. The main difference is that the curvilinear regression equation defines a curved rather than a straight line. What this means is that the way in which the value of the dependent variable is connected with the independent variable differs across the range of X values. The slope of the regression line is constant throughout its length in simple linear regression, whereas the slope of a curved line varies along its length. One approach to dealing with apparently curved trends in a set of data points is to transform the original data values of either X or Y, or possibly both variables, for example by turning them into logarithms, and then applying simple linear regression to the transformed values. A major drawback with this approach is that it may be difficult to interpret the results in terms of the real measurement scale of the variables.

Once we admit the possibility that a bivariate relationship might vary along the range of values of the independent variable, it is realistic to imagine that the line may have not one but a series of curves (see lower right part of Figure 9.1). We will explore curvilinear regression using **polynomial** equations of the general form shown in Table 9.1. Polynomial means that the equation has many terms, although it is evident that the simple linear regression equation is the starting point for these. Each equation seeks to predict the value of Y and includes an intercept and a series of one, two, three, four or more (up to n) terms where the X values raised to a certain power are mul-

Table 9.1 Specific and generic form of linear and curvilinear bivariate regression equations.

Polynomial equation	Alternative form of polynomial equation	Name	Number of curves
$\hat{y} = a + b_1 x_1$	$\hat{y} = a + bx$	Linear	Straight line – no curve
$\hat{y} = a + b_1 x_1 + b_2 x_1^2$	$\hat{y} = a + bx + cx^2$	Quadratic	One curve
$\hat{y} = a + b_1 x_1 + b_2 x_1^2 + b_3 x_1^3$	$\hat{y} = a + bx + cx^2 + dx^3$	Cubic	Two curves
$\hat{y} = a + b_1 x_1 + b_2 x_1^2 + b_3 x_1^3 + b_4 x_1^4$	$\hat{y} = a + bx + cx^2 + dx^3 + ex^4$	Quartic	Three curves
......	......	......	......
$\hat{y} = a + b_1 x_1 + b_2 x_1^2 + b_3 x_1^3 + b_n x_1^n$	$\hat{y} = a + bx + cx^2 + dx^3 + mx^n$	Generic equation	$n-1$ curves

tiplied by a different slope coefficient. Moving through the equations in Table 9.1 from linear through quadratic, cubic and quartic x refers to the same independent variable throughout and the terms b_1, b_2, b_3, b_4, ... b_n or b, c, d, e, ... m represent alternative ways of representing the slope coefficients. The different types of polynomial equation are differentiated by their degree, which refers to the largest exponential power (e.g. cubic degree denotes the cube of X (raised to the power 3)). These power terms introduce curves into the regression line as indicated in Table 9.1.

Curvilinear regression represents a potentially more successful way of determining the bivariate regression equation and line that best fits a scatter of data points for a sample of observations. It explicitly recognizes the possibility that the best-fit line may not be straight but curved. The process of applying curvilinear regression analysis is essentially the same as that used with simple linear regression, namely to determine the slope coefficients and the intercept. Box 9.3 illustrates the application of curvilinear regression to a subgroup of 20 farms that has been randomly selected from a larger sample in order to keep the dataset to a reasonable size. The dataset relates to these farms in 1941 and the analysis explores whether farm size measured in terms of hectares acts as an explanatory variable in respect of the total horsepower of wheeled tractors on the farms. Such motive power was a comparatively new feature on British farms at this time and one aspect of the investigation focuses on whether larger farms tended to have invested more in this type of machinery in comparison with smaller ones. The first four orders or degrees of bivariate polynomial regression analysis are illustrated using by this example in Box 9.3.

9.2.4 Testing the significance of bivariate polynomial regression

Polynomial regression adds successive terms to the simple linear regression equation in order to determine whether a line with one, two, three or more curves provides a better fit to the data points. Unless visual inspection of the scatter plot clearly suggests the form of the curved line (i.e. how many curves), the application of polynomial

Box 9.3a: Curvilinear (polynomial) bivariate regression.

Bivariate Polynomial Regression equation:

Population: $Y = \alpha + \beta_1 X_1 + \beta_2 X_1^2 + \beta_3 X_1^3 + \beta_n X_1^n$

Sample: $y = a + b_1 x_1 + b_2 x_1^2 + b_3 x_1^3 + b_n x_1^n$

Estimation of Y:

Population: $\mu = \alpha + \beta_1 X_1 + \beta_2 X_1^2 + \beta_3 X_1^3 + \beta_n X_1^n$

Sample: $\hat{y} = a + b_1 x_1 + b_2 x_1^2 + b_3 x_1^3 + b_n x_1^n$

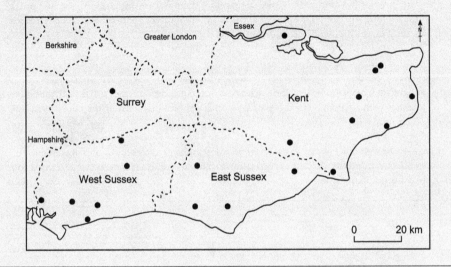

Box 9.3b: Application of curvilinear (polynomial) bivariate regression.

The bivariate polynomial regression equation is an extension of ordinary least squares regression and the latter is usually carried out first. Further terms are added that incrementally increase the power by which the values of the independent variable (X) are raised. It is usually only necessary to progress to the cubic or quartic level in order to discover the curvilinear regression line that best fits the data values of the dependent variable (Y). Polynomial regression is typically found by calculating the constants a and $b_1, b_2, \ldots b_n$ and inserting these into the appropriate definitional equation given above. The predicted values of the dependent variable ($\hat{y}$) lie along the regression line, which has zero, one, two, three, etc curves according to the order of the polynomial equation. These are estimates or predictions of the dependent variable's value in the statistical population for known values of X. The overall accuracy of these predictions can be assessed by means of the standard error of the estimate ($s_{Y.X}$), which is calculated for each polynomial regression equation in a series (i.e. 1st order, 2nd order, 3rd order, etc.). This provides a way of comparing their explanatory power.

Bivariate polynomial regression has been applied to a randomly selected subsample of farms in the South East of England that were being farmed in the early years of the Second World War. Ideally the values of the independent variable should not have been discovered empirically (i.e. simply from the sample data), but should have determined by randomly selecting farms of particular sizes. The independent variable is total farm area measured in hectares and the dependent variable is total horsepower of wheeled farm tractors, which were relatively new in British farming at the time. The slope and intercept constants together with the standard error of the estimates for the first four orders of the polynomial regression series have been included in Box 9.3d below, although details of the calculations are omitted. The standard error of the estimate statistics suggest that there is least dispersion in the first order or ordinary least squares regression equation, although it might appear from the scatter plots that second or third polynomial equations are a better fit.

The key stages in applying Bivariate Polynomial Regression are:

Calculate the constant a and the series of b coefficients: the former is the value of the dependent variable at the point where the regression line cuts through the Y-axis and the latter together with their sign are the gradients or slopes and direction of the curved sections of the line;

Calculate the standard error(s) of the estimate: these indicate the amount of dispersion in the residuals around the sections of regression lines and therefore suggest which line/equation best fits the data values.

Box 9.3c: Assumptions of curvilinear (polynomial) Bivariate regression.

There are three main assumptions:

Random sampling with replacement should be employed or without replacement from an infinite population provided that the sample is estimated at less than 10 per cent of the population.

In principle, the data values for the variables X and Y should conform to the bivariate normal distribution: however, such an assumption may be difficult, if not impossible, to state in some circumstances. It is generally accepted that provided the sample size is sufficiently large, which is itself often difficult to judge, the technique is sufficiently robust as to accept some unknown deviation from this assumption.

The observations in the population and the sample should be independent of each other in respect of variables X and Y, which signifies an absence of autocorrelation.

Box 9.3d: Calculation of curvilinear (polynomial) regression.

No.	x	y	No.	x	y
1	292.41	34	11	341.82	80
2	442.26	45	12	104.49	48
3	178.00	48	13	512.63	146
4	768.20	50	14	230.89	15
5	114.41	0	15	301.32	48
6	67.84	27	16	97.20	20
7	1549.73	50	17	343.44	22
8	53.46	30	18	196.59	86
9	129.60	0	19	50.90	20
10	130.41	26	20	65.41	0

Polynomial regression scatter plots and equations

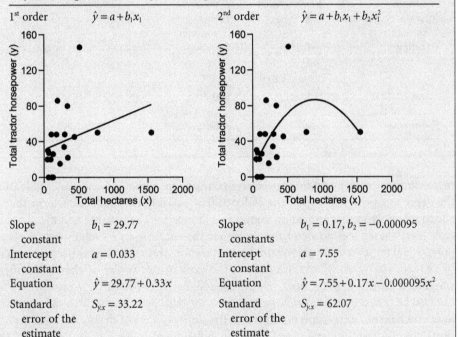

1st order $\hat{y} = a + b_1 x_1$

2nd order $\hat{y} = a + b_1 x_1 + b_2 x_1^2$

Slope constant	$b_1 = 29.77$	Slope constants	$b_1 = 0.17,\ b_2 = -0.000095$
Intercept constant	$a = 0.033$	Intercept constant	$a = 7.55$
Equation	$\hat{y} = 29.77 + 0.33x$	Equation	$\hat{y} = 7.55 + 0.17x - 0.000095x^2$
Standard error of the estimate	$S_{y.x} = 33.22$	Standard error of the estimate	$S_{y.x} = 62.07$

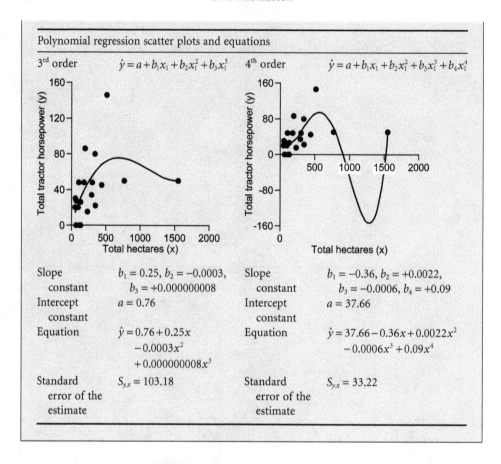

Polynomial regression scatter plots and equations

3rd order	$\hat{y} = a + b_1 x_1 + b_2 x_1^2 + b_3 x_1^3$		4th order	$\hat{y} = a + b_1 x_1 + b_2 x_1^2 + b_3 x_1^3 + b_4 x_1^4$
Slope constant	$b_1 = 0.25$, $b_2 = -0.0003$, $b_3 = +0.000000008$		Slope constant	$b_1 = -0.36$, $b_2 = +0.0022$, $b_3 = -0.0006$, $b_4 = +0.09$
Intercept constant	$a = 0.76$		Intercept constant	$a = 37.66$
Equation	$\hat{y} = 0.76 + 0.25x$ $- 0.0003x^2$ $+ 0.000000008x^3$		Equation	$\hat{y} = 37.66 - 0.36x + 0.0022x^2$ $- 0.0006x^3 + 0.09x^4$
Standard error of the estimate	$S_{y.x} = 103.18$		Standard error of the estimate	$S_{y.x} = 33.22$

regression proceeds by adding successive terms to the equation, each one increasing the degree or power to which the independent variable is raised. Each term that is added to a polynomial regression equation will increase the r^2 value and the series of Null Hypotheses seek to determine whether the increase in r^2 values between one polynomial regression equation and the next is more than might be expected to occur by chance. The series of r^2 values allows the explanatory power of the independent variable in the different equations to be assessed. Each curvilinear regression equation can also be tested to determine whether it fits the data points better than a horizontal line, which is similar to applying a t test to the slope coefficient in simple linear regression. Box 9.4 tests the results of applying polynomial regression in respect of the dependent and independent variables examined in Box 9.3, namely the subsample of farms in South-East England.

9.2.5 Complex relationships and multivariate (multiple) regression

This text has followed the accepted norm by introducing bivariate linear and curvilinear regression analysis before progressing on to multivariate or multiple

---

Box 9.4a: Significance testing of bivariate polynomial regression.

Coefficient of determination: $r^2 = \dfrac{s_{\hat{y}}^2}{s^2}$

Confidence limits of predicted Y: $\hat{y} - t_{0.05}s_{\hat{y}}$ to $\hat{y} + t_{0.05}s_{\hat{y}}$

Box 9.4b: Application of significance testing to bivariate polynomial regression.

The most straightforward way of testing significance in polynomial regression is to examine the incremental increase in the value of r^2 (the coefficient of determination) that occurs with each additional term in the equation. The starting point is that the simple linear regression equation is entirely adequate and the series of Null Hypotheses examine whether the difference in r^2 values between one equation and the next in the sequence is more than might be expected to occur by chance. There are separate degrees of freedom for the numerator and denominator of the F test statistic: in the case of the numerator these are the order of polynomial regression (i.e. 1, 2, 3, etc); and for the denominator they are $n - 2 - i$, where n is the number of observations and i is the power to which the independent variable is raised in the polynomial regression equation (i.e. 1, ,2, 3, ... k). A decision on whether to accept or reject each Null Hypothesis is made by reference to the probability of the difference having occurred by chance in relation to the level of significance, typically 0.05 or 0.01: if the probability is less than or equal to the stated level, the Null Hypothesis should be rejected.

This form of significance testing with bivariate polynomial regression has been applied to farm survey data introduced in Box 9.3. The separate r^2 statistics associated with the four polynomial regression equations are distinguished from each other by their subscripts 1, 2, 3 and 4. The numerator and denominator degrees of freedom for the 1st-order (the simple linear equation) are 1 and 20-1; 2 and 20-2-1 for the 2nd-order polynomial equation; 3 and 20-2-2 for the 3rd; and 4 and 20-2-3 for the 4th. These F statistic corresponding to these combinations of degrees of freedom for the numerator and denominator and their probabilities given below indicate that the 2nd-order polynomial regression equation provides a statistically significant increase in the value of r^2 compared with the linear, cubic and quartic versions. Whether there is any logical reasoning that can be adduced for **why** the total horsepower of tractors on farms in the South East of England in 1941 should be related to and at least partly explained by $\hat{y} = 7.55 + 0.17x - 0.000095x^2$ is another matter.

The confidence limits around the curvilinear regression can also be calculated and plotted. These indicate a zone or range around the regression line within which the population parameter (μ) of the predicted dependent variable will fall with a certain degree of confidence for known values of the independent variable. The standard error of the estimate is multiplied by the t statistic value at the specified confidence level (e.g. 95%) with $n - 2$ degrees of freedom and the resultant value is added to and subtracted from $\hat{y}$ for any known value of x. In this application the 95% confidence limits are relatively wide, which suggests that the polynomial regression equation provides only a moderately reliable prediction of the population parameter for any given value of x.

The key stages in testing the significance in Bivariate Polynomial Regression are:

Calculate the F test statistics: the F statistics are usually calculated for the individual r^2 values and the difference between these for successive polynomial regression equations. The sequence of F statistics for these differences in this application is 1.935, 5.399, 0.224 and 2.057.

State Null Hypotheses and significance level: the general form of the Null Hypothesis is that the difference between the successive pairs of r^2 values is not significantly greater than would occur by chance at the 0.05 level of significance.

Determine the probability of the calculated test statistics: the probabilities associated with the F statistics with 2 and 17 degrees of freedom is 0.015, with 3 and 16 degrees of freedom is 0.879 and with 4 and 15 degrees of freedom is 0.138.

Accept or reject the Null Hypothesis: these probabilities indicate that the increase in r^2 achieved between the 1st- and 2nd-order polynomial regression equations is more than might be expected at the 0.05 level of significance and therefore the Null Hypothesis associated with this difference is rejected and the other two Null Hypotheses are accepted.

Box 9.4c: Calculation of significance testing statistics for bivariate polynomial regression.

1st-order coefficient of determination	$r_1^2 = \dfrac{s_{\hat{y}}^2}{s^2}$	$\dfrac{134.09}{1179.88} = .1138$
F test statistic for r_1^2 and its probability	$F = \dfrac{n-2(r_1^2)}{1-r_1^2}$	$\dfrac{20-2(0.1138)}{1-0.1138} = 1.935$
	$p =$	0.146
2nd-order coefficient of determination	$r_2^2 = \dfrac{s_{\hat{y}}^2}{s^2}$	$\dfrac{134.09}{384.79} = .3274$
F test statistic and its probability	$F = \dfrac{n-2-1(r_2^2)}{1-r_2^2}$	$\dfrac{20-2-1(0.3274)}{1-0.3274} = 5.239$
	$p =$	0.034
F test statistic for difference between r_1^2 and r_2^2 and its probability	$F = \dfrac{n-2-1(r_1^2-r_1^3)}{1-r_2^2}$	$\dfrac{20-2-2 1 0.3274 - 0.1138)}{1-0.3274} = 5.399$
	$p =$	0.015
3rd-order coefficient of determination	$r_3^2 = \dfrac{s_{\hat{y}}^2}{s^2}$	$\dfrac{134.09}{397.25} = .3367$
F test statistic and its probability	$F = \dfrac{n-2-2(r_3^2)}{1-r_3^2}$	$\dfrac{20-2-2(0.3367)}{1-0.3367} = 5.050$
	$p =$	0.080

F test statistic for difference between r_2^2 and r_3^2 and its probability	$F = \dfrac{n-2-2(r_3^2 - r_2^3)}{1-r_3^2}$ $p =$	$\dfrac{20-2-2(0.3367-0.3274)}{1-0.3367} = 0.224$ 0.879
4th-order coefficient of determination	$r_4^2 = \dfrac{s_{\hat{y}}^2}{s^2}$	$\dfrac{134.09}{491.657} = .4167$
F test statistic and its probability	$F = \dfrac{n-2-3(r_4^2)}{1-r_4^2}$ $p =$	$\dfrac{20-2-3(0.4167)}{1-0.4167} = 6.251$ 0.031
F test statistic for difference between r_3^2 and r_4^2 and its probability.	$F = \dfrac{n-2-3(r_4^2 - r_3^3)}{1-r_4^2}$ $p =$	$\dfrac{20-2-3(0.4167-0.3367)}{1-0.4167} = 2.057$ 0.138

	x	$\hat{y}$	$s_{\hat{y}}$	$\hat{y} - t_{0.05}s_{\hat{y}}$ (lower)	$\hat{y} + t_{0.05}s_{\hat{y}}$ (upper)
Confidence	0	7.55	12.10	−17.87	35.56
limits of the	200	38.23	22.74	−9.54	37.54
predicted Y	400	61.31	30.91	−3.63	39.15
	600	76.79	36.48	0.15	40.31
	800	84.67	39.34	2.01	40.93
	1000	84.95	39.45	2.07	40.95
	1200	77.63	36.78	0.35	40.38
	1400	62.71	31.67	−3.82	39.27

regression. However, in practice the latter techniques are more likely to reflect the complexity of 'real-world' research questions since they include more than one independent variable in order to explain differences in the value of the dependent variable. This means that the dependent and independent variables are connected in such a way that members of the latter group are able to affect the values of the dependent variable in different ways. For example, suppose there are three independent variables, A, B and C:

- Variable A is positively correlated with the dependent variable so that low values of A are associated with low values of Y and vice versa;

- Variable B is negatively correlated with the dependent variable so that low values of B are linked to high values of Y and vice versa;

- Mid-range values of variable C are associated with high values of Y, and low and high values of C with low Y values.

It is perhaps useful to think of the separate independent variables simultaneously pushing and pulling the values of the dependent variable in different directions. This potentially offers a more satisfying way of investigating 'real-world' events, since they rarely seem to have only one cause. Multiple regression still refers to the dependent variable as Y, while the separate independent variables are usually labelled X, but with different subscripts to distinguish one from the other $x_1, x_2, x_3, \ldots x_n$. Computationally multivariate regression is similar to polynomial regression in so far as different variables are added into the regression equation to discover if they increase the r^2 value significantly. The overall purpose is to discover which combination of independent variables provides the equation with the highest explanatory power, which is normally assessed by examining the r^2 values and testing the significance of different combinations. A bivariate scatter plot provides a relatively easy way of visually assessing whether a simple linear or curvilinear relationship is present, since any trends in the data points are usually evident. However, it is more difficult to see a pattern in a single multivariate scatter plot where there are two independent and one dependent variable, and virtually impossible if there are more than two X variables. Thus, the decision on which independent variables to include in the regression equation and the order in which they should be entered is a little more problematic. There are two main ways of resolving this issue: either to introduce all of the independent variables into the analysis together or to enter them one after the other. The inclusion of more than one independent variable in regression analysis is only appropriate if this improves the predictive power of the regression equation or to test theoretical models relating to complex concepts, such as a landscape's attractiveness.

Visualization of such multivariate relationships by means of a scatter plot is not only very difficult but also would not easily enable the predictions to be made. Therefore, we usually define a multivariate relationship by means of an equation, which has the generic form:

$$\hat{y} = a + b_1 x_1 + b_2 x_2 + b_3 x_3 + \ldots b_n x_n$$

Each of the independent variables $(x_1, x_2, x_3, \ldots x_n)$ has its own slope coefficient $(b_1, b_2, b_3, \ldots b_n)$ as a multiplier, although there remains only one intercept (a). Calculation of the slope coefficients and intercept for multivariate regression is more complex than is the case for simple bivariate linear regression and the details are not examined here. However, it is worth considering why these computations are more complex. Simple linear regression assumes that all of the variance of the dependent variable is accounted for by one independent variable and so two or more independent variables might be expected to account for this variance jointly. However, some or all of these independent variables may be correlated with each other, and so they exert an overlapping or duplicate effect on the values of the dependent variable. Figure 9.5 illustrates the impact of this statement in the hypothetical situation of where there are two independent variables $(x_1$ and $x_2)$ and one dependent variable (y). The upper part of Figure 9.5 separately shows the bivariate scatter plots and linear regression equations between each independent variable and the dependent variable, which has the same values in both, where x_1 and x_2 are virtually uncorrelated with each other

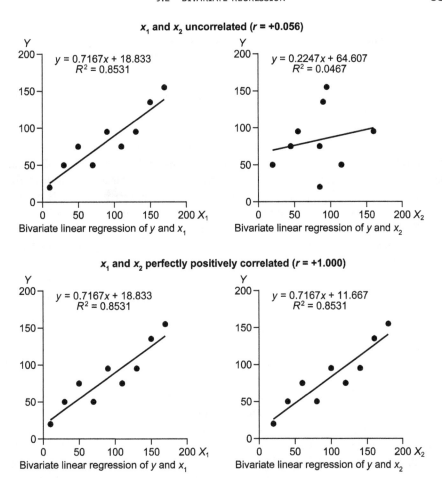

Figure 9.5 Bivariate scatter plots between one dependent and two independent variables contrasting the effect of different strengths of correlation between x_1 and x_2.

(Pearson's $r = +0.056$). The results of bivariate regression analysis indicate that x_1 has a strong explanatory relationship with y, since the slope coefficient equals 0.72 and $r^2 = 0.85$, indicating that it accounts for some 85 per cent of the variability in the dependent variable. In contrast x_2 has much weaker explanatory power with $r^2 = 0.05$. The lower pair of bivariate scatter plots and regression analyses show the comparable situation for when the two independent variables are perfectly positively correlated with each other (Pearson's $r = +1.000$). In this instance, the visual appearance of the scatter plots seems identical on first sight, although close inspection reveals that the x_2 values are each 10 units less than those for x_1, hence their perfect positive correlation. The slope coefficients and the r^2 values for the two regression equations are nevertheless identical.

Why are the values of the intercept different in the bivariate regression equations in the lower part of Figure 9.5? Describe what has happened to the regression line.

Perfect positive or negative correlation between any pair of independent variables means that one them is unnecessary in the regression analysis and could be omitted without any loss of explanation for or capacity to predict the values of the dependent variable. There is an equivalent statistic to the coefficient of determination that is used in multivariate regression analysis, known as the **multiple coefficient of determination** (r^2), that indicates the percentage of the dependent variable's variance that is accounted for by the independent variables. The r^2 value for the multiple regression of x_1 and x_2 with y when applied to the data values in Figure 9.5 equals 0.88 (88%). This indicates that the addition of the second independent variable marginally improves on accounting for the variance of y compared with using the first independent variable on its own. One consequence of these issues is that when applying multivariate regression investigators should select independent variables that are not only weakly or uncorrelated with each other but that are strongly correlated with the dependent variable.

Multivariate regression potentially offers an investigator a more comprehensive statistical explanation of the causes of variability in a dependent variable, since it explicitly recognizes that outcomes can have multiple causes. The calculations involved with applying multivariate regression analysis are not examined in detail, since statistical and spreadsheet software is readily available for the task. However, Box 9.5 illustrates the application of multiple regression to the same subsample of farms in South-East England. However, in this case we have the total areas of the 20 farms from three separate surveys carried out in 1941, 1978 and 1998. The underlying research question aims to determine whether farm size as measured by land area in the first two years provides a statistical explanation of where these farms were located in the overall spectrum of farm size in the latest survey year.

9.2.6   Testing the significance of multivariate regression

The application of multivariate regression analysis to a particular research problem produces a multiple regression equation that best fits the sample data and that predicts the values of the dependent variable in the statistical population. The elements in this equation, the intercept and slope coefficients, which are multipliers of the independent variables, predict $\hat{y}$ values that lie on a sloping plane within the multidimensional space defined by the total number of variables. However, just as with bivariate regression the observed values of y in the sample data are unlikely to fall exactly on this regression plane. In other words, there will be an error or residual difference between the predicted $\hat{y}$ value and the observed y value. The regression plane summarizes the relationship between the variables so that the total sum of the residuals between the

Box 9.5a: Multivariate (multiple) regression.

Multiple Regression equation:
Population: $Y = \alpha + \beta_1 X_1 + \beta_2 X_2 + \beta_3 X_3 + \ldots \beta_n X_n$
Sample: $y = a + b_1 X_1 + b_2 X_2 + b_3 X_3 + \ldots b_n X_n$
Population: $\mu = \alpha + \beta_1 X_1 + \beta_2 X_2 + \beta_3 X_3 + \ldots \beta_n X_n$
Sample: $\hat{y} = a + b_1 x_1 + b_2 x_2 + b_3 x_3 \ldots b_n x_n$

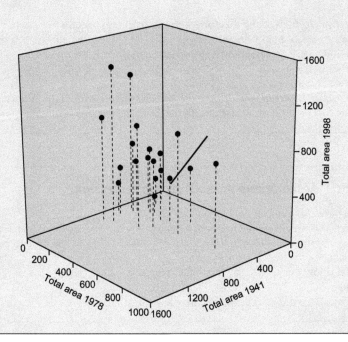

Box 9.5b: Application of multivariate regression.

Multivariate linear regression represents an extension of ordinary least squares regression to include two or more independent variables labelled $x_1, x_2, x_3, \ldots x_n$. Each of these has a slope coefficient ($b_1, b_2, b_3, \ldots b_n$) that denotes the gradient of its relationship with the dependent variable: in other words the amount that the value of y (the dependent variable) increases per unit of the corresponding x. The multivariate linear regression equation includes the intercept (a), which represents the point where the regression plane passes through the axis of the dependent variable. The multiple coefficient of determination (r^2), which can easily be converted into a percentage by moving the decimal point two spaces to the right, indicates the explanatory power of all or some of the independent variables.

Multivariate linear regression has been applied to a sample of farms in the South East of England (Kent, Surrey, East and West Sussex). Box 9.5d below includes a subsample of 20 of these farms in the interests of limiting the amount of data presented. The total area (ha) of the sampled farms has been obtained by means of a questionnaire survey on three separate occasions: the agricultural census on 4th June 1941 and author-conducted interview surveys in 1978 and 1998. The independent variables x_1 and x_2, respectively the 1941 and 1978 areas, and the regression model seek to explore whether these accurately predict farm size in 1998 (y). It is difficult to determine from the 3D scatter plot how well the independent variables account for the variability in the dependent variable, which can be assessed by means of the standard error of the estimate, which equals 272.46 in this case.

The key stages in applying Multivariate Linear Regression are:

Calculate a and b_1, b_2, b_3 ... b_n: these are constants in the regression equation and the slope coefficients act as multipliers for known values of the independent variables that define the multidimensional regression plane, which intercepts the vertical Y-axis and the gradient or slope of the plane;

Calculate the standard error(s) of the estimate: a small standard error of the estimate indicates that the data points have a close fit with the linear multivariate regression equation.

Box 9.5c: Assumptions of multivariate regression.

There are three main assumptions:

Sampled observations should be selected randomly with replacement or without replacement from an infinite population if the sample is considered to be less than 10 per cent of the population.

The deviations or residuals from the regression line (plane) follow a normal distribution.

The residuals from the regression line have a uniform variance

---

Box 9.5d: Calculation of multivariate regression.

No.	x_1	x_2	y	No.	x_1	x_2	y
1	292.41	725.00	930.00	11	341.82	617.10	607.04
2	442.26	329.70	512.00	12	104.49	530.10	647.51
3	178.00	423.40	600.00	13	512.63	572.00	910.57
4	768.20	532.10	748.68	14	230.89	369.80	404.69
5	114.41	259.20	250.91	15	301.32	475.90	627.28
6	67.84	817.60	930.79	16	97.20	661.80	444.00
7	1549.73	749.00	748.68	17	343.44	422.80	404.69
8	53.46	411.50	424.93	18	196.59	396.20	141.64
9	129.60	379.80	485.63	19	50.90	625.80	283.19
10	130.41	789.10	1401.05	20	65.41	486.00	1254.17

Slope constants	$b_1 =$	$b_1 = -0.034$
	$b_2 =$	$b_2 = 1.228$
Intercept constant	$a =$	$a = -1.098$
Regression equation	$y = a + b_1x_1 + b_2x_2 + b_3x_3...b_nx_n$	$y = -1.098 - 0.34x_1 + 1.228x_2$
Standard error of the estimate	$s_{y.x_1x_2} =$	$= 272.46$

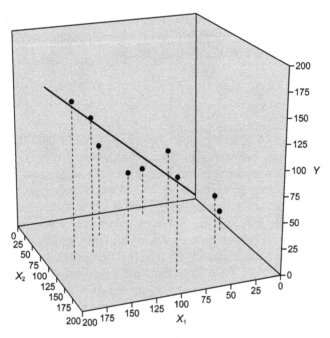

Figure 9.6   3D scatter plot of x_1 and x_2 with y showing multiple regression plane.

data points and the plane is minimized. Figure 9.6 shows the data values associated with a 3-dimensional scatter plot.

There are again two main approaches to testing the significance of multivariate regression with sample data: one involves testing the whole regression model and the other focuses on the individual parts or elements in the equation. The partitioning of the total variability of the dependent variable between the regression and residual effects that we came across with respect to bivariate regression also works with multivariate regression. The total variability or sum of squares of the dependent variable is defined as $\sum(y - \bar{y})^2$, which is divided between the regression sum of squares $\left(\sum(\hat{y} - \bar{y})^2\right)$ and the residual sum of squares $\left(\sum(y - \hat{y})^2\right)$. Significance testing involves using analysis of variance to examine whether the regression sum of squares is significantly different from zero. However, the application of ANOVA to test the significance of the multivariate regression equation is not quite as simple as in bivariate regression where there is only one independent variable and a significant difference between the regression sum of squares and zero also indicates that the slope coefficient (b) and correlation between the variables ($r_{y.x}$) are significant at the chosen level. The situation is slightly more complicated with multivariate regression because there is more than one independent variable. A significant difference between the regression sum of squares and zero in these circumstances does not indicate which of the independent variables is significantly related to the dependent variable.

This issue can be resolved by testing the significance of the slope coefficients with the Null Hypotheses that $b = 0.0$ using a series of t tests. Each of the t values is calculated by dividing the slope coefficient by the corresponding standard error, for example in respect of the first independent variable:

$$t = \frac{b_1}{s_{b_1}}$$

The t values are assessed by reference to the critical value in the t distribution corresponding to the chosen level of significance with the degrees of freedom equal to the total number of observations minus the number of independent variables minus 1 $(n - k - 1)$. An investigator examines each t test result to determine which of the independent variables plays a statistically significant role in the multivariate relationship. This examination may indicate that all, some or none are significant and each possible outcome can arise with both significant and nonsignificant ANOVA results. This can occur because the common variance (i.e. that associated with all the independent variables) does not form part of the individual t tests for the slope coefficients. The rather surprising outcome of a significant ANOVA result without any significant independent variables occurs when the latter share a large amount of the common variance but they each have a very small unique variance. Any independent variable with a nonsignificant slope coefficient does not provide an accurate prediction of the dependent variable and so can realistically be ignored. Box 9.6 provides the results of testing the significance of the multiple regression equation produced by applying to the subsample of farms in South-East England in order to investigate the possible explanatory relationship between farm area at different times.

9.3 Concluding remarks

This chapter has given a comprehensive exploration of simple linear regression and provided an introduction to the more complex techniques of polynomial and multiple or multivariate regression. However, in many respects it has only scratched the surface of these statistical techniques and a full account is beyond the scope of an

Box 9.6a: Significance testing of multivariate (multiple) regression.

Whole model or equation using analysis of variance:

Population: $F = \dfrac{\sigma_\mu^2}{\sigma_\varepsilon^2}$ Sample: $F = \dfrac{s_{\hat{y}}^2}{s_e^2}$

Regression variance: $s_{\hat{y}}^2 = \dfrac{\sum(\hat{y}-\bar{y})^2}{n}$ Residual variance: $s_e^2 = \dfrac{\sum(y-\hat{y})^2}{n}$

Separate components of equation:

Slope: $t = \dfrac{|b_{1...k} - \beta_{1...k}|}{s_{b...k}}$ Intercept: $t = \dfrac{|a - \alpha|}{s_a}$ Predicted Y

Confidence limits of regression equation or line:

Predicted Y: $\hat{y} - t_{0.05}s_{\hat{y}}$ to $\hat{y} + t_{0.05}s_{\hat{y}}$

Box 9.6b: Application of significance testing to multivariate (multiple) regression.

Testing the significance of multiple regression proceeds in a similar way to simple linear regression by using analysis of variance to focus on the whole regression model by calculating an F statistic by dividing the regression variance ($s_{\hat{y}}^2$) by the residual variance (s_e^2) and/or by computing t test statistics for the individual coefficients in the regression equation (a, b_1, b_2, ... b_k), where k is the number of independent variables and the predicted values of the dependent variable ($\hat{y}$). Calculation of the t test statistics and the confidence limits requires that the relevant standard error is computed first since this is used as the divisor in the t statistic equation. These calculations are not shown below, since although they follow the principle outlined previously for simple linear regression they need to take into account the covariance between the independent variables.

When using ANOVA there are degrees of freedom relating to the regression variance (k) and the residual variance ($n - k - 1$), which are used to determine the probability of F. The degrees of freedom for the different t tests on the coefficients are $n - 2$. The ANOVA Null Hypothesis states that the particular set of independent variables included in the analysis do not provide a statistically significant explanation for the variability in the dependent variable at a particular level of significance (e.g. 0.05 or 95%). The Null Hypotheses relating to each of the slope coefficients assert that they are not significantly different from zero (i.e. constant for all values of the corresponding independent variable). The Null Hypothesis for the intercept (a) also states that the 'true' value is zero and the difference between the observed value and zero has only arisen through sampling error. If the probability associated with these test statistics is less than or equal to the chosen significance (e.g. 0.05 or 0.01), the Null Hypothesis would be rejected.

The confidence limits for the slope, intercept and predicted Y values are also usually calculated in order to define a range of values within which the corresponding population parameters (β_1, β_2, ... β_k, α and μ) are likely to lie with a certain degree of confidence (e.g. 95% or 99%). In each case the standard error is multiplied by the t statistic value for the chosen level of confidence with $n - k - 1$ degrees of freedom and the result is added to and subtracted from the corresponding sample derived value in order to define upper and lower limits.

Both types of significance testing have been applied to the farm survey data introduced in Box 9.3, where the dependent variable is total farm area in 1998 and the two independent variables are farm area in 1941 and 1978. It is difficult to determine from the 3D scatter plot given above whether the independent variables provide a strong explanation for farm size in 1998. The significance test results calculated in Box 9.6c indicate that historical farm size does provide some explanation for subsequent total area, but that the area 20 years previously has a stronger relationship than this combined with the area nearly 60 years before. The confidence limits for the predicted Y values have also been calculated. These define a relatively wide zone around the predicted 1998 farm areas.

The key stages in testing the significance in Multiple Regression are:

Choose type of testing procedures: most statistical software will carry out both types of significance testing and the investigator's task is to interpret the information provided correctly.

State Null Hypothesis and significance level: the H_0 with the ANOVA procedure states the combination of independent variables do not explain any variability in the dependent variable beyond what might be expected through sampling error at the 0.05 level of significance OR the slope and intercept coefficients are not significantly different from zero at the same level of significance.

Calculate the appropriate test statistic (F or t): the calculations for the test statistics are given below.

Determine the probability of the calculated test statistics: the probability of obtaining $F = 4.89$ is 0.021 when there are 2 (k) and 17 ($n - k - 1$,) degrees of freedom, respectively, for the regression and residual variances; the probability of the t test statistic for b_1 ($t = 0.183$) is 0.857, for b_2 ($t = 3.079$) is 0.007 and for the intercept (a) ($t = 0.005$) is 0.996.

Accept or reject the Null Hypothesis: these test results and probabilities indicate that the Null Hypothesis for the ANOVA procedure would be rejected, whereas only the b_2 slope coefficient is significantly different from zero at the 0.05 level. These results suggest that it was 1978 total farm area that had greater explanatory power in respect of influencing 1998 farm area.

Box 9.6c: Calculation of significance testing statistics for multivariate (multiple) regression.

Regression variance	$s_{\hat{y}}^2 = \dfrac{\sum(\hat{y}-\overline{y})^2}{k}$	$\dfrac{725\,423.711}{2} = 362\,711.86$
Residual variance	$s_e^2 = \dfrac{\sum(y-\hat{y})^2}{n-k-1}$	$\dfrac{1\,261\,976.09}{20-2-1} = 74\,233.87$
Analysis of variance and probability of F	$F = \dfrac{s_{\hat{y}}^2}{s_e^2}$	$\dfrac{362\,711.89}{74\,233.89} = 4.89$
	$p =$	0.021
Test statistics for slope coefficients b_1 and b_2, and probabilities of their t values	$t = \dfrac{|b_1-\beta_1|}{s_{b_1}} =$	$\dfrac{|-0.034-0|}{0.19} = 0.183$
	$p =$	0.857
	$t = \dfrac{|b_2-\beta_2|}{s_{b_2}} =$	$\dfrac{|1.228-0|}{0.40} = 3.079$
	$p =$	0.07
Test statistic for intercept and probability of t	$t = \dfrac{|a-\alpha|}{s_a} =$	$\dfrac{|-1.08-0|}{219.60} = 0.005$
	$p =$	0.996

	x_1	x_2	$\hat{y}$	$s_{\hat{y}}$	$\hat{y}-t_{0.05}s_{\hat{y}}$ (lower)	$\hat{y}+t_{0.05}s_{\hat{y}}$ (upper)
Confidence	0	0	−1.10	215.01	−452.83	450.63
limits of the	200	200	237.70	142.43	−61.60	536.95
predicted Y	400	400	476.50	84.90	298.13	654.76
	600	600	715.30	83.50	539.86	890.56
	800	800	954.10	139.98	660.00	1247.96
	1000	1000	1192.90	212.30	746.86	1638.65
	1200	1200	1431.70	288.80	824.94	2038.11
	1400	1400	1670.50	366.86	899.72	2440.87

introductory statistical analysis text such as this. We could for example have moved on to consider polynomial multiple regression where the plane on which the predicted values of the dependent variable is warped and curved in different directions. This chapter has mentioned that there are different ways of entering independent variables into a multivariate regression analysis, but detailed examples have not been provided. The intention has been to provide sufficient information to enable students to apply regression analysis in their project in a sensible and meaningful way, even if more complex techniques could have been tried.

Multiple regression analysis explicitly recognizes that the set of independent variables included in the final model after all the candidate variables have been tried will not predict the dependent variable perfectly–something will be missing. This is acknowledged by the inclusion of an error term in the definitional equation for multiple regression (see Box 9.4). The presence of this term in the population and sample equations (respectively (ε) and (e) represents the combined residuals of the independent variables.

However, there is one further important point that has so far been glossed over in the discussion that relates to difficulties that often arise with applying correlation and regression to spatially distributed phenomena. These are examined in the next chapter, but by way of introduction it is worth recalling Tobler's first law of geography that was stated in Chapter 2: 'Everything is related to everything else, but near things are more related than distant things' (Tobler, 1970, p. 236). The implication is that phenomena of the same type (i.e. from the same population) that are spatially close to each other are likely to be more similar to each other than they are to occurrences of the same of phenomena that are located further away. This seems to contravene the assumption of independence of the data values for variables in a set of observations that is the fundamental basis of most of the correlation and all the regression techniques examined in this and the previous chapter. The following chapter will explore this problem of **spatial autocorrelation** in more detail and examine geospatial analysis techniques that can be applied to harness the benefits of Tobler's first law of geography and to measure spatial autocorrelation.

10
Correlation and regression of spatial data

The issues that can arise when applying correlation and regression analysis techniques to data relating to observations that are located at specific places in space or occur on fixed occasions in time are examined in this chapter. Indices for quantifying global spatial autocorrelation and local spatial association are explored together with an introduction to the relatively advanced techniques of trend surface analysis and geographically weighted regression. Students often develop a level of confidence with the correlation and regression techniques covered in previous chapters, but the issues associated with applying these to spatially autocorrelated data are sometimes neglected. This chapter shows how relatively simple measures can be calculated and in some cases tested statistically to avoid the pitfalls of unwittingly ignoring the lack of independence in spatial data by students and researchers in Geography, Earth and Environmental Science and related disciplines.

Learning outcomes

This chapter will enable readers to:

- describe the characteristics and implications of spatial autocorrelation;

- calculate and apply suitable indices to measure the global and local effects of spatial autocorrelation;

- consider how to incorporate these measures when analysing geographical datasets in an independent research investigation in Geography, Earth Science and related disciplines.

Practical Statistics for Geographers and Earth Scientists   Nigel Walford
© 2011 John Wiley & Sons, Ltd

10.1   Issues with correlation and regression of spatial data

Correlation and regression analysis are often used to investigate research questions in the geographical sciences, although there are some important issues that need to be considered when the variables and attributes relate to spatial entities. Some applications of correlation and regression may be carried out in a particular geographical context, such as with respect to businesses operating in a certain city region in Human Geography or to the concentration of pollutants in a particular river system in Environmental Science. Provided that the spatial distribution of the population and sample of observations are only of incidental interest and they are independent of each other, both types of statistical analysis can be carried out with relative ease. However, once the spatial location of the entities starts to be regarded as relevant to the investigation, for example the distribution of businesses in relation to each other or to some other place, such as the centre of the city or the sites along the river channels where water samples are selected in relation to land use, then some issues overshadow the application of correlation and regression as described in the previous chapters.

The origin of these problems arises from the fact the individual entities that make up a given collection of spatial units (points, lines and areas) are rarely, if ever, entirely independent of each other. Yet a fundamental assumption of correlation and regression is that the values possessed by each observation in respect of the variables and attributes being analysed should be independent. If any dependence between the entities is ignored then its effect on the results of the correlation and regression will be undetected. For example, it might have artificially increased or decreased the value of the correlation coefficient, thus indicating a stronger or weaker relationship than is really present. Similarly, it might have affected the form of the regression equation and could lead to unreliable predicted values for the dependent variable. The possible problems that might arise from a lack of independence between spatial features is illustrated in Box 10.1 with respect to a section of the moraine where the material has emerged from Les Bossons Glacier near Chamonix in France and been transported and been deposited on the sandur plain. There is a mixture of sizes of material in the area of moraine shown and it clear from a superficial examination that the different-sized material is not randomly distributed. There are clumps of individual boulders, stones and pebbles together with finer material not visible in the image. The upward facing surface of a random sample of these boulders, stones and pebbles has been digitized and shown on a 'map' superimposed on the image. The sampled items have been measured in respect of their surface area and the length of their long axis.

Regression and correlation analyses have been carried out on a subsample of these items (in the interests of limiting the calculations shown) with surface area as the independent and axis length as the dependent variables. The results (Pearson's correlation coefficient and linear regression equation) are shown on the scatter plot in Box 10.1. These suggest a very strong positive relationship between the variables

Box 10.1:   Spatial autocorrelation.

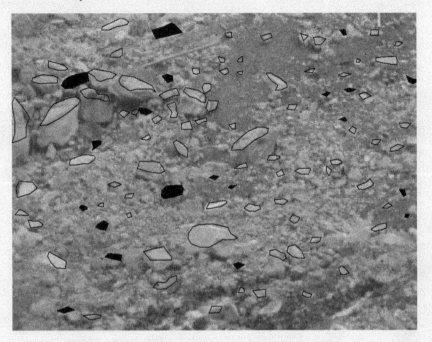

Subsample of debris shown with dark shading.

A lack of independence in the data values for a collection of n observations is likely to mean that there is some systematic pattern in the size of the residuals along the regression line. It could be that the lower and upper ends of the range of values for the variable x produce larger residuals and so a poorer prediction of the dependent variable in regression, or perhaps there is a repeating pattern of large and small residuals along the range x values: either way, these patterns indicate the presence of autocorrelation in the data.

The image of part of the moraine of Les Bossons Glacier suggests that the size and long axis length of debris material is not distributed randomly. For example, there seems to be a group of large boulders towards the upper left and a relatively larger number of small items in the upper and central right areas. The 20 debris items shaded black have been randomly selected as a subsample of the full 100 boulders, stones and pebbles in the full sample. Their surface area has been measured and simple linear regression analysis has been applied to examine the supposed relationship that hypothesizes area as an explanatory variable in respect of axis length. The calculations for the regression analysis have not been included since the standard procedures discussed in Chapter 9 have been followed. The regression equation is $\hat{y} = 7.873 + 0.0001x$ and with $r^2 = 0.905$ there is a strong indication that surface area has significant explanatory power in respect of the long axis length. The residuals from this regression analysis seem to display some systematic pattern along the regression line with smaller residuals at the lower end of the range of x values. The residuals seem to become progressively larger towards the upper end.

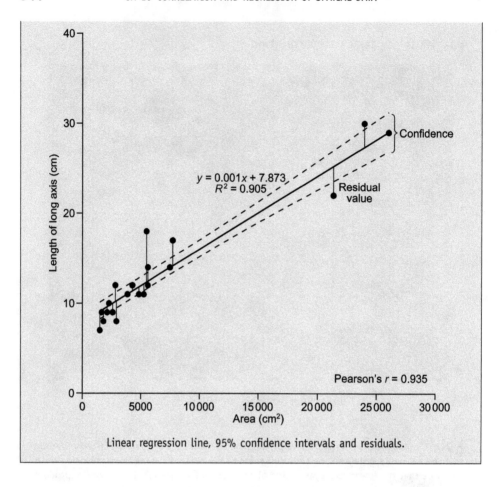

Linear regression line, 95% confidence intervals and residuals.

(+0.951) and with r^2 equal to 0.905, there is some indication that surface area explains 90.5 per cent of the variability in axis length. The scatter plot also shows the residuals of the sampled data points as vertical lines connected to the regression line, which represents the predicted value of the dependent variable for the known values of surface area (dependent). These reveal an interesting feature: generally speaking the residuals are smaller for sampled stones that were towards the lower end of the surface area scale. The confidence limits support this notion since they are curving away from the regression lines towards the upper end of the independent variable axis. In other words, there appears to be a relationship between successive values of the residuals along the regression line and they vary in a systematic way. This might not be a problem if the different sizes of moraine material were randomly distributed across the area, but the image clearly shows this is not the case. Separate subsamples of material from the different parts of the moraine could potentially produce contrasting and even contradictory results from their respective correlation and regression analyses.

10.2 Spatial and temporal autocorrelation

The example in Box 10.1 illustrates a problem known as **spatial autocorrelation**. Correlation analysis as outlined previously concentrates on the strength and direction of the relationship between two variables for either a population or a sample of observations, but it does not take into account the relationship between the individual entities. So far, we have ignored the possibility that one observation possessing a certain value for the X variable might have some bearing on the value of X (or Y) of other observations. Rather than the observations being independent they might be **interdependent**. Autocorrelation occurs when some or all of the observations are related to each other. Spatial autocorrelation arises when it is locational proximity that results in observations being related and temporal autocorrelation when closeness together in time is the cause. Spatial and temporal autocorrelation are most commonly positive in nature in the sense that the observations possess similar values for attributes and variables, where as negative autocorrelation, when spatially or temporally close observations have dissimilar values, is rarer but by no means unknown. Many geographical phenomena display positive spatial autocorrelation, for example people living in housing on the same street and soil samples taken from the same field, are likely to be more similar to each other than they are to the same types of observation from locations that are further apart.

The underlying reason why this might be a problem is illustrated in Figure 10.1, which shows scatter plots for the complete sample of boulders, stones and pebbles on the Bossons Glacier moraine. Rather than plotting the dependent variable (length of long axis) against the independent one (area), the upper and lower pairs of plots, respectively show these plotted against the X and Y coordinates of the locations of the sampled debris. There are a number of important features to note from these scatter plots. First, the r^2 values are relatively low, which indicates that the X and Y coordinates do not provide a strong explanation for variability in area or length. Secondly, the relationships are all negative, although the slope of the regression line is much higher in the case of the X coordinates. However, perhaps the most striking feature is that there are some clumps of data points where there are groups of observations that have very similar coordinates and area or length values. One clear example of this is to be found just above the centre of the horizontal axis of the upper-left plot where there is a group of 11 observations with low area values and X coordinates around 200. Spatial autocorrelation extends the general concept of autocorrelation in two ways: first that adjacent values are strongly related and second that randomly arranged values indicate the absence of autocorrelation.

Where else are there clumps of data points in Figure 10.1? What are the combinations of variable and coordinate values at these locations?

Understanding of spatial autocorrelation owes much to earlier work concerned with **time-series analysis** and the fact that geographical investigations are often focused not

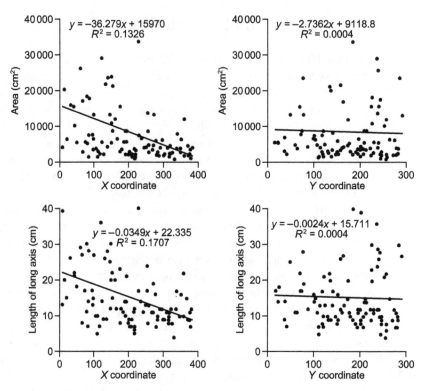

Figure 10.1 Scatter plots of full sample of moraine material by area and length of long axis against X and Y spatial coordinates.

only on spatial occurrences of phenomena but also the measurement of variables as they change over time. For example, human geographers might be interested in how deprivation is distributed spatially **and** temporally between different census areas. The concept of covariance, the way in which two independent pairs of data values for variables X and Y vary jointly, was introduced in Chapter 8 as the starting point for understanding correlation. Dividing the covariance by the product of the squares roots of the variances of X and Y produces the Pearson's correlation coefficient (r). This effectively standardizes the value of the coefficient to lie within the range -1.0 to $+1.0$. In Box 10.1 we focused on the relationship between the area and long axis length of boulders, stones and pebbles on part of the Les Bossons Glacier moraine, but suppose we were interested in a set of n values for one of the variables, say surface area, measured in respect of the spatially contiguous debris over the surface of the moraine. Box 10.2 illustrates the effects of spatial autocorrelation by examining the **spatial contiguity** with respect to the subset of all 19 items (boulders, pebbles and stones) lying partly or wholly within a transect across the surface. The series of four scatter plots in Box 10.2b are known as h-scatter plots, where h refers to the spatial lag between data values. When such lags are used in time-series analysis the length of time periods or intervals is often constant throughout the sequence, for example daily amounts of precipitation,

Box 10.2a: Spatial lags

Serial correlation coefficient for lag 1: $r_{,1} = \dfrac{\displaystyle\sum_{h=1}^{n-1}(x_h - \overline{x}_{,1})(x_{h+1} - \overline{x}_{,2})}{\sqrt{\displaystyle\sum_{h=1}^{n-1}(x_h - \overline{x}_{,1})^2}\sqrt{\displaystyle\sum_{h=1}^{n-1}(x_{h+1} - \overline{x}_{,2})^2}}$

Serial correlation coefficient for k lags: $r_k = \dfrac{\displaystyle\sum_{h=1}^{n-k}(x_h - \overline{x})(x_{h+k} - \overline{x})}{\displaystyle\sum_{h=1}^{n}(x_h - \overline{x})^2}$

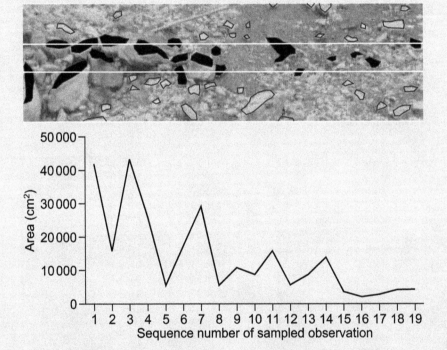

Transect through sample of debris on Les Bossons Glacier moraine.

The data values for variables measured in respect of observations that are located in space may be related to each other and display positive or negative autocorrelation. A transect has been superimposed on the top of the image representing the part of Les Bossons Glacier's moraine and the boulders, pebbles and stones intersecting with this area have been identified and numbered 1 to 19 in sequence from left to right. This example deals with objects located irregularly in space, but the procedure could as easily be applied to regularly spaced features, for example items that are a fixed distance apart.

Autocorrelation can be examined by means of the serial correlation coefficient where there are k lags and each lag is identified as h units (e.g. $h = 1,2,3$ up to k) or t time periods in the case of time-series analysis. The 19 values for the variable measuring the surface area of these objects, denoted as x, have been tabulated in Box 10.2b and labelled as x_1 to x_{19}. In the second column of data values they have been shifted up by one row, thus pairing the data value for one object with the next in the sequence. Lag 2 works in a similar way, but pairs one data value with the next but one in the sequence, and so on for however many lags are required. Once the data values have been paired in this way the Pearson's Correlation coefficients are calculated and these have been shown in h-scatter plots for spatial lags 1 to 4.

The r coefficients show an increase through lags 1, 2 and 3 (0.3341, 0.3471 and 0.4246) and them decline to 0.1039 for lag 4. This indicates that spatial autocorrelation in respect of area for these observations starts to reduce after spatial lag 3.

Box 10.2b: Linking data values by spatial lags.

		$x_{1...n} + h$	$x_{1...n} + h$	$x_{1...n} + h$	$x_{1...n} + h$
		Lag 1 ($h = 1$)	Lag 2 ($h = 1$)	Lag 3 ($h = 1$)	Lag 4 ($h = 1$)
x_1	41 928.88	15 889.19	43 323.43	26 015.02	5536.74
x_2	15 889.19	43 323.43	26 015.02	5536.74	17 865.73
x_3	43 323.43	26 015.02	5536.74	17 865.73	29 164.62
x_4	26 015.02	5536.74	17 865.73	29 164.62	5620.59
x_5	5536.74	17 865.73	29 164.62	5620.59	10 889.42
x_6	17 865.73	29 164.62	5620.59	10 889.42	8805.97
x_7	29 164.62	5620.59	10 889.42	8805.97	15 814.79
x_8	5620.59	10 889.42	8805.97	15 814.79	5703.54
x_9	10 889.42	8805.97	15 814.79	5703.54	8 95.05
x_{10}	8805.97	15 814.79	5703.54	8795.05	13 891.54
x_{11}	15 814.79	5703.54	8795.05	13 891.54	3669.44
x_{12}	5703.54	8 95.05	13 891.54	3669.44	2116.38
x_{13}	8795.05	13 891.54	3669.44	2116.38	2824.25
x_{14}	13 891.54	3669.44	2116.38	2824.25	4148.04
x_{15}	3669.44	2116.38	2824.25	4148.04	4276.09
x_{16}	2116.38	2824.25	4148.04	4276.09	
x_{17}	2824.25	4148.04	4276.09		
x_{18}	4148.04	4276.09			
x_{19}	4276.09				

---

whereas spatial lags can be regular or irregular. In Box 10.2 the separate items of moraine debris in the transect are not located at a regular distance apart, but are lagged according to their sequential spatial contiguity or neighbourliness.

The series of correlation coefficients for the lagged variable should be approximately zero if they were calculated for a random set of data values and plotting a **correlogram** is a useful way of examining whether this is the case when the spacing of the observations is equal. Although the observations in our example are not spaced

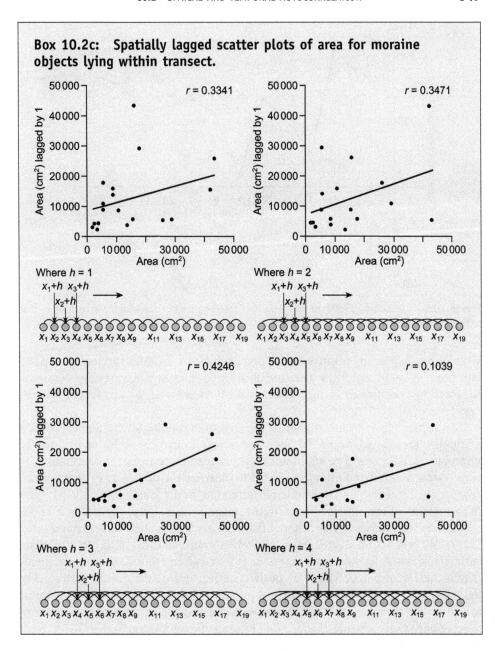

Box 10.2c: Spatially lagged scatter plots of area for moraine objects lying within transect.

at a regular distance apart across the transect, they are in a unitary sequence relating to the first, second, third and so on up to $n - 1$ nearest neighbours. It is therefore not entirely inappropriate to plot the series of correlation coefficients for the lags as a correlogram. Figure 10.2 shows the correlograms for the area and axis length variables

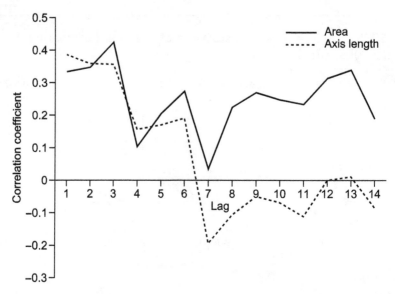

Figure 10.2 Correlograms for area and length of the long axis of moraine material intersecting with transect.

with moderately strong positive correlation coefficients for both variables for spatial lags 1 to 3 followed mainly by a decline until lag 7. Thereafter, the area line continues to record low positive correlation coefficient values, whereas for axis length they are very low negative ones.

This section has introduced some of the ways of examining spatial autocorrelation as though they could simply be migrated across from time-series analysis in an unproblematic fashion. The addition of a transect to the image of the moraine is to some extent an artefact simply being used to illustrate the principles of spatial auto-correlation. There may be some underlying trend in the data values not only in respect of the portion of the moraine shown in the image but also across the area as a whole, which may be connected with distance from the glacier snout, slope angle and other factors. We will return to this issue later when examining the application of trend surface analysis. A further important issue is that a given sequence of measurements may include some rogue values or **outliers**, which distort the overall pattern. The following sections will examine a range of procedures available for examining patterns in spatial data starting with those dealing with global spatial autocorrelation and then moving onto those capable of indicating local spatial association.

10.2.1 Global spatial autocorrelation

Indices of global spatial autocorrelation summarize the extent of this characteristic across the whole of the area under study. There are a number of measures available that are suited for use with different types of data. The starting point for many of the techniques is that the area or region of interest can be covered by a regular grid of squares or by a set of irregular-shaped polygons. A further factor influencing the choice of technique concerns whether the values are numerical measurements or counts of nominal attributes. The data type presented by the Bossons Glacier moraine example where there are data values for 100 randomly distributed points is also covered. The essential purpose of all the techniques outlined in the following sections is to explore the correlation between the units (areas or points) at different degrees of spatial separation and to produce a measure that is comparable to the serial correlation coefficient used in time-series analysis.

10.2.1.1 Join counts statistics

Join count statistics (JCS) focus on the patterns produced by sets of spatial units that have nominal data values by counting the number of joins or shared boundaries between areal units in different nominal categories. Most applications relate to binary data values, for example the absence or presence of a particular characteristic, although data with more than two classes can be regrouped into a binary form. Perhaps the simplest place to start with exploring JCS is the case where a regular grid of squares has been superimposed over the study area and these squares have been coded with a value of 0 and 1 to denote the binary categories. Chapter 6 discussed three 'standard' ways in which spatial features could be arranged, clustered, equidistant and random. Figure 10.3 illustrates the three situations with respect to a regular grid of 100 squares that belong to binary classes, here shown as either black or white. In Figure 10.3a the 100 squares are split equally in half with all the white ones at the top and the black ones in the bottom. The middle grid has a systematic pattern of white and black squares, rather similar to a chess board, with none of the same coloured squares sharing a boundary and only meeting at the corners. Figure 10.3c, again with half of the squares shaded black and the other half white, shows a random distribution with some same coloured squares sharing edges and others meeting at corners.

These comments have already given a clue as to how we might analyse the different patterns and to decide whether a given pattern is likely to have occurred by chance or randomly. First, consider the situation in time-series analysis, where time periods are usually assumed to form a linear sequence so that one period of 24 hours (a day) is followed by another and so on and each period has one join with its predecessor and one with its successor, apart from those at the arbitrary start and end of the series. If these time periods were classified in a binary fashion (e.g. absence or presence of President Obama's name on the front page of the New York Times over a period of 10 days) they could be represented as a series of black and white squares in one dimension, such as those down the right-hand side of Figure 10.3. The three linear sequences correspond to their grid square counterparts on the left-hand side. The joins between the spatial units (squares in this case) work in two dimensions rather than the one dimension of the time series. Joins between squares in the grid occur in two ways edge to edge and corner to corner, and in an analogy with chess the former are referred to

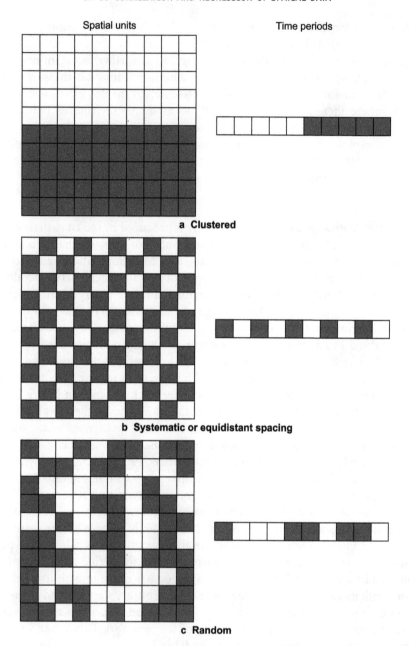

Figure 10.3 Binary join count patterns for regular grid squares and linear time series.

as rook's and the latter as queen's moves. This produces up to eight possible joins for each square with those around the edge and at the corners of the overall grid covering the study area having less. The edge effects can be disproportionately important if the size of the study area or the number of grid squares is relatively small, although this obviously raises the question of how small is small. Given that each square has the binary codes 0 or 1, where they join there are four possible combinations: 1–1, 0–0, 1–0 and 0–1. Counting the number of joins of these different types indicates the origin of join count statistics.

How many rook's and queen's joins does each corner square have in the grids down the left-hand side of Figure 10.3? How many of both types of join does each of the four squares at the centre of these grids have?

Each corner square has two adjacent squares, all of the other squares along the boundaries or sides of the grid have three rook adjacencies and all of the remaining squares have four. A 10×10 grid of 100 squares will therefore have 4 corner squares (8 joins), 32 side squares (96 joins) and 64 inner squares (256 joins) summing to 360, but because this has double counted joins between adjacent squares the sum is halved to give a total of 180. The 100 squares in the grids in Figure 10.3 are equally divided between black and white, therefore each join combination (1–1, 0–0, 1–0 and 0–1) has an equal probability of occurrence and we would expect there to be 45 joins of each type (180/4). However, we are interested in the deviation from the two extreme situations of perfect separation (Figure 10.3a) and regularity (Figure 10.3b), which respectively have 10 and 180 0–1 and 1–0 joins. The random pattern shown in Figure 10.3c has 92 0–1 and 1–0 joins, which is slightly more than the expected total of 90, but is the difference more or less than might have occurred through chance or sampling error. These comments indicate that the empirical count of each adjacency combination, with 0–1 and 1–0 being taken together since they are equally indicative of a mixed pattern, should be tested for their significance. This can be achieved by converting the difference between the observed and expected frequency into a Z score having calculated the standard deviation of the expected number of counts corresponding to each combination.

One complicating factor should be noted before examining the application of JCS, which relates to whether the data has been obtained by means of free sampling with replacement or nonfree sampling without replacement. Mention of sampling might seem a little odd, since the regular grids shown in Figure 10.3 have squares that cover all of the study area. So in what way have these data been sampled? In Chapter 2 we saw that the main difference between sampling with and without replacement when using nonspatial statistics is that the probability of an item being selected changes as each additional entity enters the sample from the population. Here, the issue concerns whether the probability that any particular square in the grid will be black or white. If this probability can be determined *a priori*, for example from published figures for another location, in other words without reference to the empirical data for the study

Box 10.3a: Join count statistics

Free sampling with replacement

Expected number of B–B joins: $E_{BB} = Jp^2$

Expected number of B–W joins: $E_{BW} = 2Jpq$

Expected number of W–W joins: $E_{WW} = Jq^2$

Standard deviation of expected B–B joins: $\sigma_{BB} = \sqrt{Jp^2 + 2Kp^3 - (J + 2K)p^4}$

Standard deviation of expected B–W joins: $\sigma_{BW} = \sqrt{(2J + \sum L(L+1)pq - 4(J + \sum L(L-1)p^2q^2}$

Standard deviation of expected W–W joins: $\sigma_{WW} = \sqrt{Jq^2 + 2Kq^3 - (J + 2K)q^4}$

Free sample without replacement

Expected number of B–B joins: $E_{BB} = J\dfrac{n_W(n_W - 1)}{n(n-1)}$

Expected number of B–W joins: $E_{BW} = 2J\dfrac{n_B n_W}{n(n-1)}$

Expected number of W–W joins: $E_{WW} = J\dfrac{n_B(n_B - 1)}{n(n-1)}$

Standard deviation of expected B–B joins:

$$\sigma_{BB} = \sqrt{E_{BB} + 2K\frac{n_B(n_B-1)(n_B-2)}{n(n-1)(n-2)} + [J(J-1) - 2K]\frac{n_B(n_B-1)(n_B-2)(n_B-3)}{n(n-1)(n-2)(n-3)} - (E_{BB}^2)}$$

Standard deviation of expected B–W joins:

$$\sigma_{BW} = \sqrt{\frac{2(J+K)n_B n_W}{n(n-1)} + 4[J(J-1) - 2K]\left(\frac{n_B(n_B-1)n_W(n_W-1)}{n(n-1)(n-2)(n-3)}\right) - 4\left(\frac{Jn_B n_W}{n(n-1)}\right)^2}$$

Standard deviation of expected W–W joins:

$$\sigma_{WW} = \sqrt{E_{WW} + 2K\frac{n_W(n_W-1)(n_W-2)}{n(n-1)(n-2)} + [J(J-1) - 2K]\frac{n_W(n_W-1)(n_W-2)(n_W-3)}{n(n-1)(n-2)(n-3)} - (E_{WW}^2)}$$

Z test statistic: $Z = (O_{BW} - E_{BW})/\sigma_{BW}$

W–W	B–B	W–B/B–W
2	3	4
0	7	2
0	7	2
2	6	1
3	2	4
3	5	1
5	1	3
1	7	1
0	5	4
5	0	4

W–W	5	5	2	1	0	0	0	0	1	4	39	
B–B	0	2	3	6	6	4	5	5	5	3		82
W–B/B–W	4	2	4	2	3	5	4	4	3	2			59

Box 10.3b: Application of join counts statistics.

Join Count Statistics work by examining the amount of separation between individual spatial units in respect of the nominal categories to which they have been assigned as the result of the distribution of some phenomenon. The procedure involves counting the number of joins between spatial units (grid squares in this example) that fall into the different possible combinations (0–0, 1–1 and 1–0/0–1). 0 denotes the absence of the phenomenon and 1 its presence, which are here represented as White and Black squares. The total number of cells in the grid (n) divides between n_W and n_B, where the subscripts denote the type of square. The total number of joins is identified as J and K is defined as $K = \sum J_i(J_i - 1)/2$ where the subscript i refers to individual squares from 1 to n. The probabilities of presence and absence of the phenomenon in any individual cells are referred to as p and q, respectively. These probabilities, which are used to calculate the expected numbers of joins in the different combinations, are determined in one of two ways depending upon whether free (with replacement) or nonfree (without replacement) sampling is used. The difference between these relates to whether the probabilities are defined by *a priori* reasoning or by *a posteriori* empirical evidence. The present application is typical in so far as nonfree sampling is assumed.

This application of JCS concerns the distribution of *dianthus gratianopoltanus* on part of the slope of Mont Cantal in the Auvergne. A 10×10 square grid has been superimposed over the area and the presence or absence of the species is shown by black and white shading. There were 59 black and 41 white squares. The calculations in Box 10.3d show that the expected number of BB joins was 29.82, of BW or WB was 87.96 and 62.22 for WW. The observed numbers were, respectively, 82, 39 and 59 (see above). The Null Hypothesis when testing JCS is that the spatial pattern of the phenomena in the grid squares is random, while the Alternative Hypothesis states it is either clustered or dispersed. Given the differences between these figures it is not surprising that the Z tests indicate that they are significant at the 0.05 level and the spatial pattern is not likely to have occurred by chance. It is reasonable to conclude that there is significant spatial autocorrelation in the distribution of *dianthus gratianopoltanus* in this area.

The key stages in applying Join Count Statistics are:

Tabulate the individual squares in the grid and count the different types of join for each: this can be a laborious process, since it involves inspecting each square and determining the code (0/1 or B/W) of all of its neighbours;

Calculate the values J and K, and count the numbers of observed BB, BW/WB and WW joins: these calculations are illustrated below;

Calculate the counts and standard deviations of the expected number of BB, BW/WB and WW joins: these are obtained by applying the equations appropriate to free or nonfree sampling

Calculate the Z test statistics for each type of join: the Z test statistics are calculated in a similar way to other tests and in this application are 2.54 (BB), 4.49 (BW/WB) and 6.22 (WW);

State Null Hypotheses and significance level: each Null Hypothesis for the three types of join states that the difference between the observed and expected counts is not significantly greater than would occur by chance at the 0.05 level of significance.

Determine the probabilities of the Z test statistics: the probabilities are equal to or less than 0.01;

Accept or reject the Null Hypothesis: each of the Null Hypotheses should be rejected at the 0.05 level of significance and the Alternative Hypotheses are therefore accepted leading to the conclusion that the spatial pattern tends toward being clustered.

Box 10.3c: Calculation of Join Count Statistics and significance testing.

$i = 1\ldots n$	WW	BB	WB/BW	J_i	$J_i(J_i - 1)$
1	2			2	2
2	2		1	3	6
3	1		2	3	6
4		2	1	3	6
5		3		3	6
6		2	1	3	6
7			3	3	6
8		2	1	3	6
9		2	1	3	6
10	1		1	2	2
11	2		1	3	6
12		2	2	4	12
13		3	1	4	12
14		4		4	12
15		4		4	12
16		4		4	12
17		3	1	4	12
18		4	1	4	12
19		3	1	4	12
20	2		1	3	6
21	1		2	3	6
22		3	1	4	12
23		4		4	12
24		4		4	12
25		4		4	12
26		4		4	12
27		4		4	12
28		3	1	4	12
29	2	2	2	4	12
30			1	3	6
31		1	2	3	6
32		3	1	4	12

33	12	4	1	3	
34	12	4		4	
35	12	4	1	3	
36	12	4	1	3	
37	12	4	1	3	
38	12	4	3		1
39	12	4	2		2
40	6	3			3
41	6	3	1		2
42	12	4	1		3
43	12	4	2		2
44	12	4	3	1	
45	12	4	3		1
46	12	4	3		1
47	12	4	1	3	
48	12	4	1	3	
49	12	4	2	2	1
50	6	3	2		1
51	6	3			3
52	12	4			4
53	12	4	2	2	4
54	12	4	2	2	2
55	12	4	2	3	
56	12	4	1	3	
57	12	4	1	4	
58	12	4		2	
59	12	4	1		
60	6	3			
61	6	3			3
62	12	4	1		4
63	12	4	2	2	3
64	12	4	2		2
65	12	4	3		1
66	12	4	2		2
67	12	4	3		1
68	12	4			

$i = 1 \ldots n$	WW	BB	WB/BW	J_i	$J_i(J_i - 1)$
69		3	1	4	12
70		3		3	6
71	2		1	3	6
72	3		1	4	12
73		2	2	4	12
74		3	1	4	12
75		4		4	12
76		3	1	4	12
77		3	1	4	12
78		3	1	4	12
79		3	1	4	12
80		3		3	6
81			3	3	6
82	2		2	4	12
83		2	2	4	12
84		4		4	12
85		3	1	4	12
86		3	1	4	12
87		3	1	4	12
88		3	1	4	12
89	1		3	4	12
90		1	2	3	6
91	1		1	2	2
92	3			3	6
93	1		2	3	6
94		1	2	3	6
95	1		2	3	6
96	2		1	3	6
97	1		2	3	6
98		1	2	3	6
99	2		1	3	6
100	1		1	2	2

$$\sum J_i / 2 = 180$$

$$180$$

$$K = \sum_i J_i(J_i - 1)/2 = 484$$

Expected number of B–B joins

$$E_{BB} = J \frac{n_W(n_W - 1)}{n(n-1)} \qquad 180\frac{41(40)}{100(99)} = 29.82$$

Standard deviation of expected B–B joins

$$\sigma_{BB} = \sqrt{E_{BB} + 2K\frac{n_B(n_B-1)(n_B-2)}{n(n-1)(n-2)} + \frac{[J(J-1)-2K]}{n(n-1)(n-2)(n-3)}\,n_B(n_B-1)(n_B-2)(n_B-3) - (E_{BB}^2)}$$

$$\sqrt{29.83 + 2(484)\frac{59(n58)(57)}{100(99)(98)} + \frac{[180(179)-2(484)]}{100(99)(98)(97)}\,59(58)(57)(56) - 29.82^2} = 3.61$$

Z test statistic of B–B joins

$$Z = (O_{BB} - E_{BB})/\sigma_{BB} \qquad (39 - 29.82)/3.61 = 2.54$$

Probability $\qquad p = 0.0111$

Expected number of W–B/B–W joins

$$E_{BW} = 2J\frac{n_B n_W}{n(n-1)} \qquad \frac{2(180)(59)(41)}{100(100-1)} = 87.96$$

Standard deviation of expected B–W joins

$$\sigma_{BW} = \sqrt{\frac{2(J+K)n_B n_W}{n(n-1)} + 4[J(J-1)-2K]\left(\frac{n_B(n_B-1)n_W(n_W-1)}{n(n-1)(n-2)(n-3)}\right) - 4\left(\frac{Jn_B n_W}{n(n-1)}\right)^2}$$

$$\sqrt{\frac{2(180+484)59(41)}{100(99)} + 4[180(179)-2(484)]\left(\frac{59(58)41(40)}{100(99)(98)(97)}\right) - 4\left(\frac{180(59)(58)n_B n_W}{100(99)}\right)^2} = 6.45$$

Z test statistic of B–W joins

$$Z = (O_{BW} - E_{BW})/\sigma_{BW} \qquad (59 - 87.96)/6.45 = 4.49$$

Probability $\qquad p < 0.000$

Expected number of W–W joins

$$E_{WW} = J\frac{n_B(n_B-1)}{n(n-1)} \qquad 180\frac{59(58)}{100(99)} = 62.22$$

Standard deviation of expected W–W joins

$$\sigma_{WW} = \sqrt{E_{WW} + 2K\frac{n_W(n_W-1)(n_W-2)}{n(n-1)(n-2)} + \frac{[J(J-1)-2K]}{n(n-1)(n-2)(n-3)}\,n_W(n_W-1)(n_W-2)(n_W-3) - (E_{WW}^2)}$$

$$\sqrt{62.22 + 2(484)\frac{41(40)(39)}{100(99)(98)} + \frac{41(40)(39)(39)}{100(99)(98)(97)}\,[180(179)-2(484)] - 62.22^2} = 3.41$$

Z test statistic of W–W joins

$$Z = (O_{WW} - E_{WW})/\sigma_{WW} \qquad (41 - 62.22)/3.41 = 6.22$$

Probability $\qquad p < 0.000$

area, then free sampling applies. Sampling without replacement is much more common and its effect is to alter the expected number of joins in each combination (0–0, 1–1 and 1–0/0–1) from an equal distribution or some other hypothesized values.

The application in Box 10.3 examines the application of JCS in respect of the presence or absence of *dianthus gratianopoltanus* on the side of the Mont Cantal in the southern Auvergne region of France. The expected counts are calculated under the assumption of nonfree sampling, since there is no *a priori* reason to assign specific values to p and q, respectively the probabilities of presence and absence of the species in a square. The data values in this example are nominal codes (1 and 0) relating to the presence and absence of *dianthus gratianopoltanus* and there is no indication of the number of individual plants, whereas examination of the image of the slope in Box 10.3a hints at some variation in the density of occurrence. The observed numbers of B–B, B–W/W–B and W–W joins are all significantly different from what would be expected by chance, therefore it is reasonable to conclude that the spatial pattern is not random, but indicates an underlying process in relation to the distribution of *dianthus gratianopoltanus* in this area.

It should be noted that the spatial lag in this example is 1 (i.e. adjacent grid squares), whereas further analyses could be carried out where the comparison was made between 2nd-order neighbours, this would mean that the counts of B–B, B–W/W–B and W–W combinations were made by 'jumping over' adjacent squares to the next but one. Similarly, queen's move adjacencies could also be included. Finally, it should be noted that grids such as the one used in this example are often placed over a study in a relatively arbitrary fashion and the size and number of grid squares may be chosen for convenience rather than in a more rigorous way. There is no reason why a study area should be constrained so that it is covered by a square or rectangular grid. Suppose our study area is bounded on one or more sides by coastline or river, it is highly unlikely that such natural features of the environment will be delimited by straight lines and some of the cells in the grid or lattice are likely to overlap the coast or river. These units would have a reduced chance of including or excluding the phenomenon under investigation. Examination of the image and superimposed grid in Box 10.3a shows that the size of each flower is relatively small in relation to the size of a grid square. Thus, some squares contain just one occurrence, whereas others have many, yet both are counted as presences of the phenomenon. Smaller grid squares closer to the size of each flower head would perhaps give a more realistic impression of the species' distribution, since isolated occurrences may have distorted the situation.

10.2.1.2 *Moran's* I *Index*

Some of the issues mentioned at the end of the previous section arise from the rather artificial superimposition of a regular grid or lattice over a study area and that JCS applies to nominal data values. **Moran's** *I* is a widely available technique that can be used when the study is covered by an incomplete regular grid or a set of planar polygons and the data values are real numbers rather than counts of units in nominal dichotomous categories. The difference between the presentation of the raw data for Moran's I compared with JCS is that rather than tabulating count statistics the data are organized in a three-column format where the first two columns contain X and Y coordinates relating to row and column numbers, the grid references of points features or the centroids of polygons. Figure 10.4 illustrates the procedure with respect to an incomplete lattice, irregular polygons and points. The data tables shows the X and Y coordinates or row and column numbers and the third column contains the Z data values corresponding to these locations, which may be decimal values or integer counts, as in Figure 10.4. Although this process retains references to the spatial

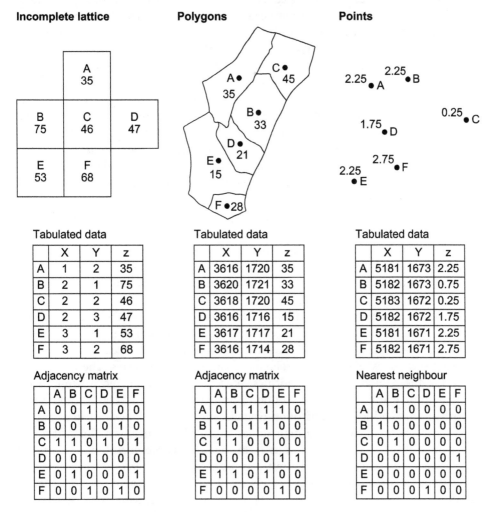

Figure 10.4 Tabular representation of irregular lattice, polygon and point feature data and weights matrices.

location of the features it discards information about their topological connections, in other words whether one unit is adjacent to another. Since such information is a vital component in the analysis of spatial patterns, it is necessary to create a **weights matrix** (W) that records which units share a common boundary in the case of area data or are nearest neighbours to each other in the case of point feature data. Figure 10.4 includes the first-order weights matrices for the three types of spatial data. Normally these weights matrices are obtained for 1st-order neighbours, but 2nd-, 3rd- or higher-order neighbours can also be used. The weights matrix can incorporate rook's and queen's adjacencies, which can be weighted to denote their relative importance if required.

Pairs of adjacent features in a spatial pattern displaying or possessing spatial autocorrelation will both have positive **or** negative values for the variable under investigation and if these values are relatively large, it will indicate stronger rather than weaker spatial autocorrelation. In contrast, pairs of neighbouring features where one has a positive and the other a negative value suggest the absence of spatial autocorrelation. These statements are effectively another way of describing Tobler's first law of Geography. Moran's I encapsulates the essence of these statements in a single index value. The sequence of calculations to compute Moran's I is somewhat protracted and involves the manipulation of data in matrix format. The first stage is subtraction of the overall mean of the variable (Z) from the data values of each spatial feature and then multiplying the result for each pair of features. The spatial weights are used to select which pairs of features are included and excluded from the final calculation of the index: those with a 0 in the weights matrix (denoting nonadjacency) are omitted, whereas those with a 1 are included. Summing these values and dividing by the sum of the weights produces a covariance. This forms the numerator in the equation for Moran's I, which is divided by the variance of the data to produce the index value that lies within the range −1.0 to +1.0. Values towards the extremes of this range, respectively, indicate negative and positive spatial autocorrelation, whereas a value around zero generally its absence.

Box 10.4 illustrates the application of Moran's I in relation to the mean scores on the UK government's 2007 Index of Multiple Deprivation for the 33 London Boroughs. The IMD is computed from a series of different variables within domains covering such areas as income, employment and social conditions. The map of the IMD for the London local authorities suggests that higher mean scores were computed for many of the inner London Boroughs, whereas several of those more suburban ones had lower deprivation overall. Some degree of positive spatial autocorrelation seems to exist with high values clustered together near the centre and lower ones in outer areas. The results from the analysis provide moderate support for this claim, since Moran's I computes as 0.305. One possible explanation for this outcome is the presence of one relatively small authority in the centre, the City of London, with a very low index value (12.54) in comparison with its seven adjacent neighbours, which apart from one have IMD values over 25.00. The Moran's I index has been recomputed leaving out the City of London Borough and the effect is to

Box 10.4a: Moran's I Index.

Moran's I: $$I = \frac{1}{p} \frac{\displaystyle\sum_{I=1}^{n} \sum_{j=1, j \neq i}^{n} w_{i,j}(z_i - \bar{z})(z_j - \bar{z})}{\displaystyle\sum_{i=1}^{n}(z_i - \bar{z})^2}$$

where $p = \displaystyle\sum_i \sum_j w_{ij} / n$

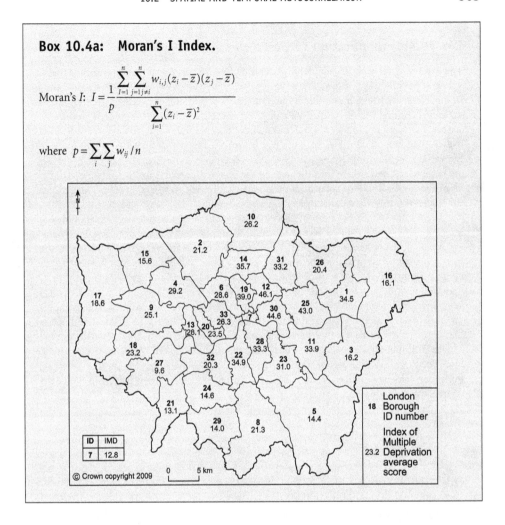

© Crown copyright 2009

raise the value to 0.479, which seems to confirm the initial visual impression of strong positive spatial autocorrelation.

There are two approaches to testing the significance of Moran's I and similar global spatial statistics such as **Geary's C** and the Getis and Ord's **G statistic**. One approach, following the standard assumption of nonspatial statistics, is to assume that the calculated statistic or index value is from a Normal Distribution of such quantities computed from a series of independent and identical samples or selections. The notion of sampling with respect to spatial features was examined in Section 6.2.2. The second approach is rather more empirical in nature and views the particular set of data values that has arisen as just one of all the possible random distributions of observed data values across the set of zones. The number of random distributions equals the factorial of the number of spatial features, for example if there are four

Box 10.4b: Application of Moran's *I* index.

The Moran's *I* index is an adaptable statistic for measuring spatial autocorrelation that is widely used in a range of disciplines. The variable being analysed (Z) has subscripts i and j to distinguish between the different observations in a pair. Binary weights are used in this example denoting whether any given pair of areas share a common boundary and the subscripts in the weight term $w_{i,j}$ also refer to a pair of observations i and j in a set with n features overall. The individual values of X are usually adjusted by subtracting the overall mean of the variable before multiplying pairs of values together. Variance/covariance-like quantities are computed, the former from the sum of the squared products in the C matrix that fall along the diagonal and the latter from the nondiagonal elements where there is a join between the spatial features represented by the row and column.

The Moran's *I* index has been applied to the average score variable in the 2007 Index of Multiple Deprivation with respect to the 33 local authorities (boroughs) in London. Visual inspection of the pattern of data values for these areas on the map suggests that lower scores (less deprivation) are found in the more peripheral zones and higher ones in the centre. The tabulated data in Box 10.4c records the mean index value is 25.68 and the following adjacency matrix of 0s and 1s shows that the minimum and maximum numbers of joins between areas are 3 and 7 with a total of 164. The values along the diagonal in the matrix used to compute the variance-like quantity are unshaded and the values for those pairs of local authorities that are adjacent (i.e. have a 1 in the weights matrix) are shaded in a darker grey. The Moran's *I* index in this application is 0.305, which indicates a moderate positive spatial autocorrelation. The randomization significance testing procedure has been applied and indicates that this index value is significant at the 0.05 level with 9999 permutations used to produce the reference distribution.

areas the number of permutations is 24 (4x3x2x1), whereas there are 8,683,317,618, 811,890,000,000,000,000,000,000,000,000 (33!) ways in which the 33 values of the 2007 Index of Multiple Deprivation could be arranged across the London Boroughs.

Both methods are implemented in various software packages that carry out spatial statistics and in the case of the randomization approach, users are normally asked to specify the number of random permutations to be generated and the spatial index is then calculated for each of these to produce a pseudo-probability distribution. The index computed from the observed data values is compared with this distribution. Both approaches to testing the significance of spatial indices involve converting the observed index value into a Z score in the standard way using the hypothesized or empirically derived population mean and standard deviation. For example, in the randomization approach the mean and standard deviation of the index in the pseudo-probability distribution are used. Box 10.4 includes the results of testing the significance of the observed Moran's *I* value, 0.305, using one of these statistical packages. The probability of the Moran's *I* obtained from the data values can be compared with a significance level that relates to the number of permutations used to generate the

Box 10.4c: Calculation of Moran's I.

Id No.	z	Id No.	z	Id No.	z	Id No.	z	Id No.	z
1	34.49	8	21.31	15	15.59	22	34.94	29	13.98
2	21.16	9	25.10	16	16.07	23	31.04	30	44.64
3	16.21	10	26.19	17	18.56	24	14.62	31	33.19
4	29.22	11	33.94	18	23.20	25	42.95	32	20.34
5	14.36	12	46.10	19	38.96	26	20.36	33	26.30
6	28.62	13	28.07	20	23.51	27	9.55		
7	12.84	14	35.73	21	13.10	28	33.33		

Mean $\sum x/n = 847.44/33 = 25.68$

Adjacency matrix.

i\j	1	2	3	4	5	6	7	8	9	10	11	12	13	14	15	16	17
1	0	0	1	0	0	0	0	0	0	0	1	0	0	0	0	0	0
2	0	0	0	1	0	1	0	0	0	1	0	0	0	1	0	1	0
3	1	0	0	0	1	0	0	0	0	0	1	0	0	0	0	0	0
4	0	1	0	0	0	1	0	1	1	0	0	0	1	0	0	0	0
5	0	0	1	0	0	0	0	0	0	0	1	0	0	0	0	0	0
6	0	1	0	1	0	0	1	0	0	0	0	0	0	0	0	0	0
7	0	0	0	0	0	1	0	0	0	0	0	1	0	0	0	0	0
8	0	0	0	1	0	0	0	0	0	0	0	0	0	0	0	0	0
9	0	0	0	0	1	0	0	0	0	0	0	0	0	0	1	0	1
10	0	1	0	1	0	0	0	0	0	0	0	0	0	1	0	0	0
11	1	0	1	0	0	0	0	0	0	0	0	0	0	0	1	0	0
12	0	0	0	0	0	0	1	0	0	0	0	0	1	0	0	0	1
13	0	0	0	1	0	0	0	0	0	0	0	1	0	0	0	0	0
14	0	1	0	0	0	0	0	0	0	1	0	0	0	0	1	0	1
15	0	0	0	0	1	0	0	0	1	0	1	0	0	1	0	0	0
16	1	0	1	0	0	0	0	0	0	0	0	0	0	0	0	0	0
17	0	0	0	0	0	0	0	0	1	0	0	0	0	0	0	0	0
18	0	0	0	0	0	1	0	0	0	0	0	1	0	1	0	0	1
19	0	0	0	0	0	0	1	0	0	0	0	0	0	1	0	0	0

Adjacency matrix.

j / i	1	2	3	4	5	6	7	8	9	10	11	12	13	14	15	16	17
20	0	0	0	1	0	0	0	0	0	0	0	0	1	0	0	0	0
21	0	0	0	0	0	0	0	0	0	0	0	0	0	0	0	0	0
22	0	0	0	0	1	0	1	0	0	0	0	0	0	0	0	0	0
23	0	0	0	0	1	0	0	1	0	0	1	0	0	0	0	0	0
24	0	0	0	0	0	0	0	0	0	0	0	1	0	0	0	0	0
25	1	0	0	0	0	0	0	1	0	0	1	0	0	0	0	1	0
26	1	0	0	0	0	0	0	0	0	0	0	0	1	0	0	0	0
27	0	0	0	0	0	0	0	0	0	0	0	0	0	0	0	0	0
28	0	0	0	0	1	0	0	0	0	0	0	0	0	0	0	0	0
29	0	0	0	0	0	0	1	0	0	0	0	0	0	0	0	0	0
30	0	0	0	0	0	0	0	1	0	1	0	1	0	0	0	0	0
31	0	0	0	0	0	0	1	0	0	0	1	1	0	1	0	0	0
32	0	0	0	0	0	1	0	0	0	0	0	0	1	0	0	0	0
33	0	1	0	1	0	0	1	0	0	0	0	0	0	0	0	0	0
$\sum w_i$	5	5	4	7	6	6	7	4	5	3	6	6	6	6	4	3	3

j / i	18	19	20	21	22	23	24	25	26	27	28	29	30	31	32	33
1	0	0	0	0	0	0	0	1	1	0	0	0	0	0	0	0
2	0	0	0	0	0	0	0	0	0	0	0	0	0	0	0	0
3	0	0	0	0	0	0	0	0	0	0	0	0	0	0	0	0
4	0	0	1	0	0	1	0	0	0	0	1	0	0	0	0	1
5	0	1	0	0	1	1	0	0	0	0	0	0	0	0	0	0
6	0	1	0	0	0	0	0	0	0	0	0	1	0	0	0	1
7	0	1	0	0	1	0	0	0	0	0	1	0	0	0	0	1
8	0	0	0	0	1	0	1	0	0	0	0	0	0	0	0	0
9	1	0	0	0	0	0	0	0	0	0	0	0	0	0	0	0
10	0	0	0	0	0	1	0	0	0	0	0	0	1	1	0	0
11	0	0	0	0	0	0	0	1	0	0	0	0	0	0	0	0
12	0	1	0	0	0	0	0	1	0	0	0	0	1	1	0	0

(continued matrix, rows 13–33)

13	1	0	0	0	0	0	0	0	0	0	0	0	0	0	1	0
14	0	1	0	0	0	0	0	0	0	0	0	0	0	0	0	0
15	0	0	0	0	0	0	0	0	0	0	0	0	0	0	0	0
16	0	0	0	1	0	0	0	0	0	0	0	0	0	0	0	0
17	1	0	0	0	0	0	0	0	0	0	0	0	0	0	0	0
18	0	0	0	0	0	0	0	0	0	0	1	0	0	0	0	0
19	0	0	0	0	0	0	0	1	0	1	0	0	0	0	0	0
20	0	0	0	0	0	0	0	0	1	0	0	0	0	0	0	0
21	0	0	0	0	0	0	0	0	0	1	0	0	0	0	0	0
22	0	0	0	0	0	0	0	0	0	0	0	0	0	0	0	0
23	0	0	0	0	0	0	0	0	0	0	1	1	0	0	0	0
24	0	0	0	0	0	1	0	1	0	0	0	0	0	0	0	0
25	0	0	0	0	0	0	0	0	0	0	0	0	0	0	1	0
26	0	0	0	0	0	0	0	0	0	1	1	0	0	0	0	0
27	1	0	0	0	0	0	0	0	0	0	0	0	0	0	0	0
28	0	0	0	0	0	0	0	0	0	0	0	0	1	0	0	0
29	0	0	0	0	0	0	0	0	0	0	0	0	0	0	0	0
30	0	0	0	0	0	0	0	0	0	0	0	0	0	0	0	0
31	1	0	0	0	0	0	0	0	0	0	0	0	0	0	0	0
32	0	0	0	0	0	0	0	0	0	0	0	0	0	0	0	0
33	0	0	0	0	0	0	1	0	0	0	0	0	0	0	0	0
$\sum w_i$	4	4	4	4	7	5	4	4	5	6	4	4	3	6	5	7	6	164

Calculation of variance/covariance-like quantities.

				j							
i	z	$z-\bar z$	1	2	3	4	5	6	7	8
			8.8	−4.5	−9.5	3.5	−11.3	2.9	−12.8	−4.4
1	34.5	8.8	77.5	−39.8	−83.4	31.1	−99.7	25.9	−113.1	−38.5
2	21.2	−4.5	−39.8	20.5	42.9	−16.0	51.2	−13.3	58.1	19.8
3	16.2	−9.5	−83.4	42.9	89.8	−33.5	107.3	−27.8	121.7	41.4
4	29.2	3.5	31.1	−16.0	−33.5	12.5	−40.0	10.4	−45.4	−15.5
5	14.4	−11.3	−99.7	51.2	107.3	−40.0	128.2	−33.2	145.4	49.5
6	28.6	2.9	25.9	−13.3	−27.8	10.4	−33.2	8.6	−37.7	−12.8
7	12.8	−12.8	−113.1	58.1	121.7	−45.4	145.4	−37.7	165.0	56.2
8	21.3	−4.4	−38.5	19.8	41.4	−15.5	49.5	−12.8	56.2	19.1

Calculation of variance/covariance-like quantities.

					j					
			1	2	3	4	5	6	7	8
i	*z*	*z* − *z̄*	8.8	−4.5	−9.5	3.5	−11.3	2.9	−12.8	−4.4
9	25.1	−0.6	−5.1	2.6	5.5	−2.1	6.6	−1.7	7.5	2.6
10	26.2	0.5	4.5	−2.3	−4.8	1.8	−5.7	1.5	−6.5	−2.2
11	33.9	8.3	72.7	−37.3	−78.2	29.2	−93.5	24.2	−106.0	−36.1
12	46.1	20.4	179.8	−92.4	−193.4	72.2	−231.2	59.9	−262.2	−89.3
13	28.1	2.4	21.0	−10.8	−22.6	8.4	−27.0	7.0	−30.6	−10.4
14	35.7	10.0	88.5	−45.4	−95.2	35.5	−113.8	29.5	−129.0	−43.9
15	15.6	−10.1	−88.9	45.7	95.6	−35.7	114.3	−29.6	129.6	44.1
16	16.1	−9.6	−84.7	43.5	91.1	−34.0	108.9	−28.2	123.5	42.1
17	18.6	−7.1	−62.7	32.2	67.5	−25.2	80.7	−20.9	91.5	31.2
18	23.2	−2.5	−21.9	11.2	23.5	−8.8	28.1	−7.3	31.9	10.9
19	39.0	13.3	116.9	−60.1	−125.8	46.9	−150.3	39.0	−170.5	−58.1
20	23.5	−2.2	−19.1	9.8	20.6	−7.7	24.6	−6.4	27.9	9.5
21	13.1	−12.6	−110.8	56.9	119.2	−44.5	142.5	−36.9	161.6	55.0
22	34.9	9.3	81.5	−41.9	−87.7	32.7	−104.8	27.2	−118.9	−40.5
23	31.0	5.4	47.2	−24.2	−50.7	18.9	−60.7	15.7	−68.8	−23.4
24	14.6	−11.1	−97.4	50.1	104.8	−39.1	125.3	−32.5	142.1	48.4
25	43.0	17.3	152.0	−78.1	−163.6	61.1	−195.5	50.7	−221.8	−75.5
26	20.4	−5.3	−46.9	24.1	50.4	−18.8	60.3	−15.6	68.4	23.3
27	9.6	−16.1	−142.1	73.0	152.9	−57.1	182.7	−47.4	207.2	70.6
28	33.3	7.6	67.3	−34.6	−72.4	27.0	−86.6	22.4	−98.2	−33.4
29	14.0	−11.7	−103.1	52.9	110.9	−41.4	132.5	−34.4	150.3	51.2
30	44.6	19.0	166.9	−85.8	−179.6	67.0	−214.7	55.7	−243.5	−82.9
31	33.2	7.5	66.1	−34.0	−71.1	26.5	−85.0	22.0	−96.4	−32.8
32	20.3	−5.3	−47.1	24.2	50.6	−18.9	60.5	−15.7	68.6	23.4
33	26.3	0.6	5.4	−2.8	−5.8	2.2	−7.0	1.8	−7.9	−2.7

i	z	j $z-\bar{z}$	9	10	11	12	13	14	15	16
			−0.6	0.5	8.3	20.4	2.4	10.0	−10.1	−9.6
1	34.5	8.8	−5.1	4.5	72.7	179.8	21.0	88.5	−88.9	−84.7
2	21.2	−4.5	2.6	−2.3	−37.3	−92.4	−10.8	−45.4	45.7	43.5
3	16.2	−9.5	5.5	−4.8	−78.2	−193.4	−22.6	−95.2	95.6	91.1
4	29.2	3.5	−2.1	1.8	29.2	72.2	8.4	35.5	−35.7	−34.0
5	14.4	−11.3	6.6	−5.7	−93.5	−231.2	−27.0	−113.8	114.3	108.9
6	28.6	2.9	−1.7	1.5	24.2	59.9	7.0	29.5	−29.6	−28.2
7	12.8	−12.8	7.5	−6.5	−106.0	−262.2	−30.6	−129.0	129.6	123.5
8	21.3	−4.4	2.6	−2.2	−36.1	−89.3	−10.4	−43.9	44.1	42.1
9	25.1	−0.6	0.3	−0.3	−4.8	−11.9	−1.4	−5.9	5.9	5.6
10	26.2	0.5	−0.3	0.3	4.2	10.3	1.2	5.1	−5.1	−4.9
11	33.9	8.3	−4.8	4.2	68.2	168.6	19.7	82.9	−83.3	−79.4
12	46.1	20.4	−11.9	10.3	168.6	416.8	48.7	205.1	−206.1	−196.3
13	28.1	2.4	−1.4	1.2	19.7	48.7	5.7	24.0	−24.1	−22.9
14	35.7	10.0	−5.9	5.1	82.9	205.1	24.0	100.9	−101.4	−96.6
15	15.6	−10.1	5.9	−5.1	−83.3	−206.1	−24.1	−101.4	101.9	97.0
16	16.1	−9.6	5.6	−4.9	−79.4	−196.3	−22.9	−96.6	97.0	92.4
17	18.6	−7.1	4.2	−3.6	−58.8	−145.4	−17.0	−71.6	71.9	68.5
18	23.2	−2.5	1.5	−1.3	−20.5	−50.7	−5.9	−25.0	25.1	23.9
19	39.0	13.3	−7.8	6.7	109.6	271.0	31.7	133.4	−134.0	−127.6
20	23.5	−2.2	1.3	−1.1	−17.9	−44.4	−5.2	−21.8	21.9	20.9
21	13.1	−12.6	7.3	−6.4	−103.9	−256.9	−30.0	−126.4	127.0	121.0
22	34.9	9.3	−5.4	4.7	76.4	189.0	22.1	93.0	−93.4	−89.0
23	31.0	5.4	−3.1	2.7	44.2	109.4	12.8	53.8	−54.1	−51.5
24	14.6	−11.1	6.5	−5.6	−91.3	−225.9	−26.4	−111.1	111.7	106.4
25	43.0	17.3	−10.1	8.7	142.6	352.5	41.2	173.5	−174.3	−166.0
26	20.4	−5.3	3.1	−2.7	−44.0	−108.7	−12.7	−53.5	53.7	51.2
27	9.6	−16.1	9.4	−8.2	−133.2	−329.4	−38.5	−162.1	162.9	155.1
28	33.3	7.6	−4.5	3.9	63.1	156.1	18.2	76.8	−77.2	−73.5
29	14.0	−11.7	6.8	−5.9	−96.6	−238.9	−27.9	−117.6	118.1	112.5
30	44.6	19.0	−11.1	9.6	156.5	387.0	45.2	190.4	−191.3	−182.2
31	33.2	7.5	−4.4	3.8	62.0	153.2	17.9	75.4	−75.8	−72.2
32	20.3	−5.3	3.1	−2.7	−44.1	−109.1	−12.8	−53.7	53.9	51.4
33	26.3	0.6	−0.4	0.3	5.1	12.6	1.5	6.2	−6.2	−5.9

i	z	j $z-\bar{z}$	17 −7.1	18 −2.5	19 13.3	20 −2.2	21 −12.6	22 9.3	23 5.4	24 −11.1
1	34.5	8.8	−62.7	−21.9	116.9	−19.1	−110.8	81.5	47.2	−97.4
2	21.2	−4.5	32.2	11.2	−60.1	9.8	56.9	−41.9	−24.2	50.1
3	16.2	−9.5	67.5	23.5	−125.8	20.6	119.2	−87.7	−50.7	104.8
4	29.2	3.5	−25.2	−8.8	46.9	−7.7	−44.5	32.7	18.9	−39.1
5	14.4	−11.3	80.7	28.1	−150.3	24.6	142.5	−104.8	−60.7	125.3
6	28.6	2.9	−20.9	−7.3	39.0	−6.4	−36.9	27.2	15.7	−32.5
7	12.8	−12.8	91.5	31.9	−170.5	27.9	161.6	−118.9	−68.8	142.1
8	21.3	−4.4	31.2	10.9	−58.1	9.5	55.0	−40.5	−23.4	48.4
9	25.1	−0.6	4.2	1.5	−7.8	1.3	7.3	−5.4	−3.1	6.5
10	26.2	0.5	−3.6	−1.3	6.7	−1.1	−6.4	4.7	2.7	−5.6
11	33.9	8.3	−58.8	−20.5	109.6	−17.9	−103.9	76.4	44.2	−91.3
12	46.1	20.4	−145.4	−50.7	271.0	−44.4	−256.9	189.0	109.4	−225.9
13	28.1	2.4	−17.0	−5.9	31.7	−5.2	−30.0	22.1	12.8	−26.4
14	35.7	10.0	−71.6	−25.0	133.4	−21.8	−126.4	93.0	53.8	−111.1
15	15.6	−10.1	71.9	25.1	−134.0	21.9	127.0	−93.4	−54.1	111.7
16	16.1	−9.6	68.5	23.9	−127.6	20.9	121.0	−89.0	−51.5	106.4
17	18.6	−7.1	50.7	17.7	−94.6	15.5	89.6	−65.9	−38.2	78.8
18	23.2	−2.5	17.7	6.2	−33.0	5.4	31.3	−23.0	−13.3	27.5
19	39.0	13.3	−94.6	−33.0	176.3	−28.9	−167.1	122.9	71.1	−146.9
20	23.5	−2.2	15.5	5.4	−28.9	4.7	27.4	−20.1	−11.6	24.1
21	13.1	−12.6	89.6	31.3	−167.1	27.4	158.4	−116.5	−67.4	139.2
22	34.9	9.3	−65.9	−23.0	122.9	−20.1	−116.5	85.7	49.6	−102.4
23	31.0	5.4	−38.2	−13.3	71.1	−11.6	−67.4	49.6	28.7	−59.3
24	14.6	−11.1	78.8	27.5	−146.9	24.1	139.2	−102.4	−59.3	122.4
25	43.0	17.3	−123.0	−42.9	229.2	−37.5	−217.3	159.8	92.5	−191.0
26	20.4	−5.3	37.9	13.2	−70.7	11.6	67.0	−49.3	−28.5	58.9
27	9.6	−16.1	114.9	40.1	−214.2	35.1	203.0	−149.3	−86.4	178.5
28	33.3	7.6	−54.5	−19.0	101.5	−16.6	−96.2	70.8	41.0	−84.6
29	14.0	−11.7	83.4	29.1	−155.4	25.4	147.3	−108.3	−62.7	129.5
30	44.6	19.0	−135.0	−47.1	251.7	−41.2	−238.5	175.5	101.5	−209.7
31	33.2	7.5	−53.5	−18.6	99.7	−16.3	−94.5	69.5	40.2	−83.0
32	20.3	−5.3	38.1	13.3	−70.9	11.6	67.2	−49.5	−28.6	59.1
33	26.3	0.6	−4.4	−1.5	8.2	−1.3	−7.8	5.7	3.3	−6.8

i	z	$z - \bar{z}$	25	26	27	28	29	30	31	32
			17.3	−5.3	−16.1	7.6	−11.7	19.0	7.5	−5.3
1	34.5	8.8	152.0	−46.9	−142.1	67.3	−103.1	166.9	66.1	−47.1
2	21.2	−4.5	−78.1	24.1	73.0	−34.6	52.9	−85.8	−34.0	24.2
3	16.2	−9.5	−163.6	50.4	152.9	−72.4	110.9	−179.6	−71.1	50.6
4	29.2	3.5	61.1	−18.8	−57.1	27.0	−41.4	67.0	26.5	−18.9
5	14.4	−11.3	−195.5	60.3	182.7	−86.6	132.5	−214.7	−85.0	60.5
6	28.6	2.9	50.7	−15.6	−47.4	22.4	−34.4	55.7	22.0	−15.7
7	12.8	−12.8	−221.8	68.4	207.2	−98.2	150.3	−243.5	−96.4	68.6
8	21.3	−4.4	−75.5	23.3	70.6	−33.4	51.2	−82.9	−32.8	23.4
9	25.1	−0.6	−10.1	3.1	9.4	−4.5	6.8	−11.1	−4.4	3.1
10	26.2	0.5	8.7	−2.7	−8.2	3.9	−5.9	9.6	3.8	−2.7
11	33.9	8.3	142.6	−44.0	−133.2	63.1	−96.6	156.5	62.0	−44.1
12	46.1	20.4	352.5	−108.7	−329.4	156.1	−238.9	387.0	153.2	−109.1
13	28.1	2.4	41.2	−12.7	−38.5	18.2	−27.9	45.2	17.9	−12.8
14	35.7	10.0	173.5	−53.5	−162.1	76.8	−117.6	190.4	75.4	−53.7
15	15.6	−10.1	−174.3	53.7	162.9	−77.2	118.1	−191.3	−75.8	53.9
16	16.1	−9.6	−166.0	51.2	155.1	−73.5	112.5	−182.2	−72.2	51.4
17	18.6	−7.1	−123.0	37.9	114.9	−54.5	83.4	−135.0	−53.5	38.1
18	23.2	−2.5	−42.9	13.2	40.1	−19.0	29.1	−47.1	−18.6	13.3
19	39.0	13.3	229.2	−70.7	−214.2	101.5	−155.4	251.7	99.7	−70.9
20	23.5	−2.2	−37.5	11.6	35.1	−16.6	25.4	−41.2	−16.3	11.6
21	13.1	−12.6	−217.3	67.0	203.0	−96.2	147.3	−238.5	−94.5	67.2
22	34.9	9.3	159.8	−49.3	−149.3	70.8	−108.3	175.5	69.5	−49.5
23	31.0	5.4	92.5	−28.5	−86.4	41.0	−62.7	101.5	40.2	−28.6
24	14.6	−11.1	−191.0	58.9	178.5	−84.6	129.5	−209.7	−83.0	59.1
25	43.0	17.3	298.1	−91.9	−278.6	132.0	−202.1	327.3	129.6	−92.3
26	20.4	−5.3	−91.9	28.3	85.9	−40.7	62.3	−100.9	−40.0	28.5
27	9.6	−16.1	−278.6	85.9	260.3	−123.4	188.8	−305.8	−121.1	86.2
28	33.3	7.6	132.0	−40.7	−123.4	58.5	−89.5	144.9	57.4	−40.9

i	z	$z-\bar{z}$	j 25	26	27	28	29	30	31	32
		$z-\bar{z}$	17.3	−5.3	−16.1	7.6	−11.7	19.0	7.5	−5.3
29	14.0	−11.7	−202.1	62.3	188.8	−89.5	137.0	−221.9	−87.9	62.5
30	44.6	19.0	327.3	−100.9	−305.8	144.9	−221.9	359.3	142.3	−101.3
31	33.2	7.5	129.6	−40.0	−121.1	57.4	−87.9	142.3	56.3	−40.1
32	20.3	−5.3	−92.3	28.5	86.2	−40.9	62.5	−101.3	−40.1	28.6
33	26.3	0.6	10.6	−3.3	−9.9	4.7	−7.2	11.7	4.6	−3.3

i	z	$z-\bar{z}$	j 33 ($z-\bar{z}$ = 0.6)	$(z_i-\bar{z})^2$	Row sum $w_{i,j}(z_i-\bar{z})(z_j-\bar{z})$
1	34.5	8.8	5.4	77.5	9.78
2	21.2	−4.5	−2.8	20.5	−31.35
3	16.2	−9.5	−5.8	89.8	36.72
4	29.2	3.5	2.2	12.5	−40.44
5	14.4	−11.3	−7.0	128.2	−188.73
6	28.6	2.9	1.8	8.6	29.67
7	12.8	−12.8	−7.9	165.0	−938.92
8	21.3	−4.4	−2.7	19.1	108.63
9	25.1	−0.6	−0.4	0.3	8.05
10	26.2	0.5	0.3	0.3	6.59
11	33.9	8.3	5.1	68.2	244.27
12	46.1	20.4	12.6	416.8	1106.69
13	28.1	2.4	1.5	5.7	−55.32
14	35.7	10.0	6.2	100.9	403.02
15	15.6	−10.1	−6.2	101.9	87.77
16	16.1	−9.6	−5.9	92.4	57.60
17	18.6	−7.1	−4.4	50.7	93.76
18	23.2	−2.5	−1.5	6.2	53.29

19	39.0	13.3	8.2	176.3	272.88
20	23.5	-2.2	-1.3	4.7	-2.60
21	13.1	-12.6	-7.8	158.4	556.78
22	34.9	9.3	5.7	85.7	-339.58
23	31.0	5.4	3.3	28.7	126.05
24	14.6	-11.1	-6.8	122.4	273.83
25	43.0	17.3	10.6	298.1	1012.08
26	20.4	-5.3	-3.3	28.3	-127.58
27	9.6	-16.1	-9.9	260.3	290.82
28	33.3	7.6	4.7	58.5	71.88
29	14.0	-11.7	-7.2	137.0	327.96
30	44.6	19.0	11.7	359.3	873.81
31	33.2	7.5	4.6	56.3	322.09
32	20.3	-5.3	-3.3	28.6	158.70
33	26.3	0.6	0.4	0.4	-2.85

$$\sum_{1}^{33}(z_i - \bar{z})^2 = 3167.58 \qquad \sum_{i=1}^{33}\sum_{j=1}^{33} w_{i,j}(z_i - \bar{z})(z - \bar{z}) = 4805.35$$

Moran's I

$$I = \frac{1}{p}\frac{\sum_{l=1}^{n}\sum_{j=1,j\neq i}^{n} w_{i,j}(z_i - \bar{z})(z_j - \bar{z})}{\sum_{i=1}^{n}(z_i - \bar{z})^2} \qquad \frac{33(4805.35)}{164(3167.58)} = 0.305$$

Z test statistic of Moran's I

$$Z = (i_O - \mu_e)/\sigma_e \qquad (0.305 - (-0.0297))/0.1045 = 2.63$$

Probability   Randomization applied with 9999 permutations $p = 0.004$

reference distribution, for example 99 and 999 are associated with the 0.01 and 0.001 significance levels, respectively.

Adjacency and contiguity are clearly important concepts in understanding the principles underlying Join Counts Statistics, the Global Moran's I index and other similar measures of spatial autocorrelation. Another important concept is distance between features, both points and areas, with the centroid usually marking the location in the latter case. Features that are located further apart may be expected to exert less influence on each other in respect of contributing towards spatial autocorrelation compared with those that are closer together. This presumption leads to the use of inverse distance weighting as a way of taking into account the diminishing or decaying effect of distance on the data values of features that are further apart. Spatial weighting methods are based on contiguity (e.g. rook's and queen's adjacency for polygons) or distance using polygons' centroids or user-defined X, Y coordinate pairs. Other methods of weighting data values focus on each point (centroid) location and then average over a prespecified number of nearest neighbours. The outcome of spatial weighting is a spatially lagged variable that is an essential requirement for testing autocorrelation and carrying out spatial regression. The application of Moran's I in Box 10.4 used a contiguity weight (i.e. adjacent boundaries) in respect of irregular polygons.

A useful way of visualizing the extent of spatial autocorrelation is by means of a Moran's I scatter plot (MSP). Figure 10.5 plots the standardized and spatially lagged values of the 2007 IMD for the 33 London Boroughs. The MSP is divided into four segments centred on the mean of the two variables that can be summarized as follows:

- Low–Low: spatial units where standardized and lagged values are low;

- Low–High: spatial units where standardized value is low and lagged value is high;

- High–Low: spatial units where standardized value is high and lagged value is low;

- High–High: spatial units where lagged and standardized values are high.

In the Low–Low quadrant of Figure 10.5 the point (borough) with lowest combination of values (-1.24 for standardized IMD and -1.08 for lagged IMD) is coincidentally Kingston upon Thames. The map in Box 10.4a shows this borough to have an IMD value of 13.1 and its four contiguous neighbours (Richmond upon Thames (9.6), Merton (14.6), Sutton (14.0) and Wandsworth (20.3)) also have comparatively low values. Although Richmond upon Thames has the lowest IMD value of all the boroughs, two of its neighbours have comparatively high values (Hounslow at 23.2 and

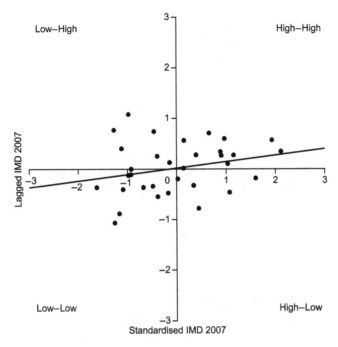

Figure 10.5   Moran's I scatter plot.

Hammersmith and Fulham at 28.1 and the influence of these outweighs its contiguity with Kingston upon Thames and Wandsworth. At the other end of the scale in the High-High quadrant are Newham and Tower Hamlets, respectively, with standardized and spatially lagged IMD values of 1.92 and 0.59, and 2.10 and 0.37.

What are the 'raw' IMD values of Newham's 6 adjacent boroughs and what are the values for the 6 contiguous boroughs of Tower Hamlets? Note: the ID numbers of Newham and Tower Hamlets are 25 and 30, respectively.

10.2.2   Local Indicators of Spatial Association (LISA)

Moran's I index is a global measure of the extent of spatial autocorrelation across a study area and quantifies the degree to which features that are spatially proximate have similar values. However, it is entirely possible for local 'pockets' of positive and negative spatial autocorrelation to exist that at least partially cancel each other out and lead to a deflated index in comparison with what might have been obtained had the study area been divided into subareas and separate indices computed for these. Such variability in the distribution of spatial autocorrelation can be quantified by **Local Indicators of Spatial Association (LISA)**, these focus on the extent to which features that are close to a specific point have similar values. The last part of the covariance/variance computation tabulation in Box 10.4c showed the row sums (S) obtained by summing the nondiagonal elements that were shaded dark grey (i.e. the spatial features are joined). If each of these S values is standardized by dividing by the sum of the squared deviations along the diagonal elements (the variance-like quantity) and the results are multiplied by the total number of spatial units (n), the figures obtained represent the local contribution of each row (feature) to the global spatial autocorrelation. These provide another way of computing the overall Moran's I index, since the sum of these local components divided by the total number of joins equals the global I computed from the definitional equation given in Box 10.4a:

$$I = 50.06/164 = 0.305$$

These local components are the LISA values, which can be mapped and tested for significant differences between areas. Rather than using these 'raw' LISA values they are often standardized by dividing by the number of joins possessed by each feature (row): thus the raw LISA values of a feature with four joins is divided by 4. When this has been done for each row in the matrix the results are in the form of a local average. Again, the row sums (S) can be divided by the sum of the squared deviations along the diagonal and then multiplied by $n - 1$ rather than n to produced standardized LISA values.

The calculation of these LISA values is shown in Box 10.5 with respect to the 2007 Index of Multiple Deprivation mean score for the 33 London Boroughs. The effect of these adjustments is to increase global Moran's I to 0.331 (0.305 previously) and when the City of London area is excluded (see above for justification) Moran's I index becomes 0.663 compared with 0.479. The results of LISA calculation can be visualized in a number of ways. Box 10.5a shows the significance and cluster maps relating to the application of Moran's I LISA for the 33 London Boroughs. Each LISA is tested for significance using a randomization process to generate a reference distribution and areas are shaded according to whether their LISA is significantly different from what would be expected at the 0.05, 0.01 and 0.001 levels. The cluster maps include those areas that are significant in the four quadrants of the MSP. Taken together these maps allow the significant combinations of positive and negative local spatial autocorrelation to be discovered. The London Boroughs used in the analyses included in Boxes 10.4 and 10.5 are in some respects rather large and have been used in order to keep the complexity of the calculations to a manageable scale. The analysis would more appropriately be carried out for smaller spatial units such as local authority wards across the whole of London. Arguably, the scale of these units is better suited to reflecting local variations that can easily become lost for relatively large areas. Nevertheless, even at the borough scale there is some evidence of local spatial autocorrelation in certain parts of the Greater London Authority's area.

Box 10.5a: Moran's *I* local indicator of spatial association.

Local Moran's I Indicator of Spatial Association: $I_i = \dfrac{\sum\limits_{j=1}^{n} w_j (z_i - \overline{z})(z_j - \overline{z})}{s_z^2 \sum\limits_{j=1}^{n} w_{ij}}$

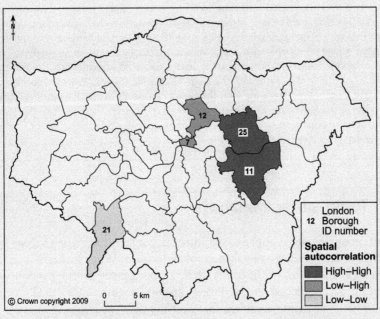

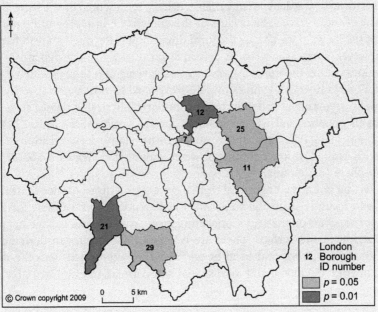

Box 10.5b: Application of the Moran's *I* local indicator of spatial association.

The Moran's *I* Local Indicator of Spatial Association focuses attention on the extent of spatial autocorrelation around individual points, including centroids as point locations for polygons, rather than providing a global summary measure. The calculations produce Moran's *I* LISA values for each spatial feature that can be tested for significance using a randomization process to generate a reference distribution. The results of the analysis are shown as two maps: the cluster map that shows combinations of high and low indices; and the significance map that shows whether a local index value is significant at certain levels. The results support the earlier global Moran's I with a pair of contiguous boroughs in South-West London (Kingston upon Thames and Sutton) having Low–Low spatial autocorrelation and Greenwich and Newham east of the centre with the High–High combination. There are two boroughs, the City of London and Hackney, with the Low–High combination. The probability of the index values associated with all of these areas is ≤0.03, which is smaller than the 'standard' 0.05 significance and therefore it is reasonable to conclude that there is significant local spatial autocorrelation in these parts of London.

10.3   Trend surface analysis

The most straightforward approach to introducing **trend surface analysis** is to recognize that it comprises a special form of regression. What makes it special is the inclusion of spatial location in terms of X, Y coordinates as independent variables recording the perpendicular spatial dimensions on a regular square grid. The purpose of the analysis is to define a surface that best fits these locations in space where the third dimension or the dependent variable, usually represented by the letter Z, denotes the height of the surface. The origins of the analysis lie in representing the physical surface of the Earth in which case the third dimension is elevation above a fixed datum level. However, surfaces can be in principle produced for any dependent variable whose values can be theoretically conceived as varying and thus producing a surface in space. For example, land values might be expected to vary across space with peaks and troughs occurring at different places. Similarly, physical variables such as measurements of pressure and water vapour in the troposphere, the lowest layer in the Earth's atmosphere, differ in a similar way. It is the relative concentration of different values, such as high or low amounts of water vapour that creates 'spatial features' in the atmosphere (e.g. clouds).

The starting point is a horizontal regular square grid in two dimensions (X and Y) and the end result is estimated values for the dependent or Z variable for the set of evenly spaced points on the grid. Connecting these estimated values together produces a visualization of the surface. There are two main approaches to deriving a trend surface known as global and local fit and these are connected with the difference between quantifying global and local spatial autocorrelation examined previously.

Box 10.5c: Calculation of local indicator of spatial association and significance testing.

Calculation of variance/covariance-like quantities.

i	z	$z - \bar{z}$	1	2	3	4	5	6	7	8
			8.8	-4.5	-9.5	3.5	-11.3	2.9	-12.8	-4.4
1	34.5	8.8	77.5	0.0	-16.7	0.0	0.0	0.0	0.0	0.0
2	21.2	-4.5	0.0	20.5	0.0	-3.2	0.0	-2.7	0.0	0.0
3	16.2	-9.5	-20.9	0.0	89.8	0.0	26.8	0.0	0.0	0.0
4	29.2	3.5	0.0	-2.3	0.0	12.5	0.0	1.5	0.0	0.0
5	14.4	-11.3	0.0	0.0	17.9	0.0	128.2	0.0	0.0	8.3
6	28.6	2.9	0.0	-2.2	0.0	1.7	0.0	8.6	-6.3	0.0
7	12.8	-12.8	0.0	0.0	0.0	0.0	0.0	-5.4	165.0	0.0
8	21.3	-4.4	0.0	0.0	0.0	0.0	12.4	0.0	0.0	19.1
9	25.1	-0.6	0.0	0.0	0.0	-0.4	0.0	0.0	0.0	0.0
10	26.2	0.5	0.0	-0.8	0.0	0.0	0.0	0.0	0.0	0.0
11	33.9	8.3	12.1	0.0	-13.0	0.0	-15.6	0.0	0.0	0.0
12	46.1	20.4	0.0	0.0	0.0	0.0	0.0	0.0	-43.7	0.0
13	28.1	2.4	0.0	0.0	0.0	1.4	0.0	0.0	0.0	0.0
14	35.7	10.0	0.0	-7.6	0.0	0.0	0.0	4.9	0.0	0.0
15	15.6	-10.1	0.0	11.4	0.0	-8.9	0.0	0.0	0.0	0.0
16	16.1	-9.6	-28.2	0.0	30.4	0.0	0.0	0.0	0.0	0.0
17	18.6	-7.1	0.0	0.0	0.0	0.0	0.0	0.0	0.0	0.0
18	23.2	-2.5	0.0	0.0	0.0	0.0	0.0	0.0	0.0	0.0
19	39.0	13.3	0.0	0.0	0.0	0.0	0.0	9.7	-42.6	0.0
20	23.5	-2.2	0.0	0.0	0.0	-1.9	0.0	0.0	0.0	0.0
21	13.1	-12.6	0.0	0.0	0.0	0.0	0.0	0.0	0.0	0.0
22	34.9	9.3	0.0	0.0	0.0	0.0	-15.0	0.0	-17.0	-5.8
23	31.0	5.4	0.0	0.0	0.0	0.0	-15.2	0.0	0.0	0.0
24	14.6	-11.1	0.0	0.0	0.0	0.0	0.0	0.0	0.0	0.0
25	43.0	17.3	25.3	0.0	0.0	0.0	0.0	0.0	0.0	9.7
26	20.4	-5.3	-11.7	0.0	0.0	0.0	0.0	0.0	0.0	0.0
27	9.6	-16.1	0.0	0.0	0.0	0.0	0.0	0.0	0.0	0.0
28	33.3	7.6	0.0	0.0	0.0	0.0	-17.3	0.0	-19.6	0.0

Top section

i	z	$z - \bar{z}$	1	2	3	4	5	6	7	8
		j → $z - \bar{z}$	8.8	−4.5	−9.5	3.5	−11.3	2.9	−12.8	−4.4
29	14.0	−11.7	0.0	0.0	0.0	0.0	0.0	0.0	0.0	17.1
30	44.6	19.0	0.0	0.0	0.0	0.0	0.0	0.0	−40.6	0.0
31	33.2	7.5	0.0	0.0	0.0	0.0	0.0	0.0	0.0	0.0
32	20.3	−5.3	0.0	0.0	0.0	0.0	0.0	0.0	0.0	0.0
33	26.3	0.6	0.0	0.0	0.0	0.4	0.0	0.3	−1.3	0.0

Bottom section

i	z	$z - \bar{z}$	9	10	11	12	13	14	15	16
		j → $z - \bar{z}$	−0.6	0.5	8.3	20.4	2.4	10.0	−10.1	−9.6
1	34.5	8.8	0.0	0.0	14.5	0.0	0.0	0.0	0.0	−16.9
2	21.2	−4.5	0.0	−0.5	0.0	0.0	0.0	−9.1	9.1	0.0
3	16.2	−9.5	0.0	0.0	−19.6	0.0	0.0	0.0	0.0	22.8
4	29.2	3.5	−0.3	0.0	0.0	0.0	1.2	0.0	−5.1	0.0
5	14.4	−11.3	0.0	0.0	−15.6	0.0	0.0	0.0	0.0	0.0
6	28.6	2.9	0.0	0.0	0.0	0.0	0.0	4.9	0.0	0.0
7	12.8	−12.8	0.0	0.0	0.0	−37.5	0.0	0.0	0.0	0.0
8	21.3	−4.4	0.0	0.0	0.0	0.0	0.0	0.0	0.0	0.0
9	25.1	−0.6	0.3	0.0	0.0	0.0	−0.3	0.0	1.2	0.0
10	26.2	0.5	0.0	0.3	0.0	0.0	0.0	1.7	0.0	0.0
11	33.9	8.3	0.0	0.0	68.2	0.0	0.0	0.0	0.0	0.0
12	46.1	20.4	0.0	0.0	0.0	416.8	5.7	34.2	0.0	0.0
13	28.1	2.4	0.0	0.0	0.0	0.0	−5.7	0.0	0.0	0.0
14	35.7	10.0	−0.2	0.8	0.0	34.2	0.0	100.9	0.0	0.0
15	15.6	−10.1	1.5	0.0	0.0	0.0	0.0	0.0	101.9	0.0
16	16.1	−9.6	0.0	0.0	0.0	0.0	0.0	0.0	0.0	92.4
17	18.6	−7.1	1.4	0.0	0.0	0.0	0.0	0.0	24.0	0.0
18	23.2	−2.5	0.4	0.0	0.0	0.0	−1.5	0.0	0.0	0.0
19	39.0	13.3	0.0	0.0	0.0	67.8	0.0	33.3	0.0	0.0
20	23.5	−2.2	0.0	0.0	0.0	0.0	−1.3	0.0	0.0	0.0
21	13.1	−12.6	0.0	0.0	0.0	0.0	0.0	0.0	0.0	0.0
22	34.9	9.3	0.0	0.0	0.0	0.0	0.0	0.0	0.0	0.0
23	31.0	5.4	0.0	0.0	11.1	0.0	0.0	0.0	0.0	0.0
24	14.6	−11.1	0.0	0.0	0.0	0.0	0.0	0.0	0.0	0.0

i	z	$z-\bar{z}$	\(j\) 17	18	19	20	21	22	23	24
			−7.1	−2.5	13.3	−2.2	−12.6	9.3	5.4	−11.1
1	34.5	8.8	0.0	0.0	0.0	0.0	0.0	0.0	0.0	0.0
2	21.2	−4.5	0.0	0.0	0.0	0.0	0.0	0.0	0.0	0.0
3	16.2	−9.5	0.0	0.0	0.0	0.0	0.0	0.0	0.0	0.0
4	29.2	3.5	0.0	0.0	0.0	−1.1	0.0	0.0	0.0	0.0
5	14.4	−11.3	0.0	0.0	0.0	0.0	0.0	−17.5	−10.1	0.0
6	28.6	2.9	0.0	0.0	6.5	0.0	0.0	0.0	0.0	0.0
7	12.8	−12.8	0.0	0.0	−24.4	0.0	0.0	−17.0	0.0	0.0
8	21.3	−4.4	0.0	0.0	0.0	0.0	0.0	−10.1	0.0	12.1
9	25.1	−0.6	0.8	0.3	0.0	0.0	0.0	0.0	0.0	0.0
10	26.2	0.5	0.0	0.0	0.0	0.0	0.0	0.0	0.0	0.0
11	33.9	8.3	0.0	0.0	0.0	0.0	0.0	0.0	7.4	0.0
12	46.1	20.4	0.0	0.0	45.2	0.0	0.0	0.0	0.0	0.0
13	28.1	2.4	0.0	−1.0	0.0	−0.9	0.0	0.0	0.0	0.0
14	35.7	10.0	0.0	0.0	22.2	0.0	0.0	0.0	0.0	0.0
15	15.6	−10.1	18.0	0.0	0.0	0.0	0.0	0.0	0.0	0.0
16	16.1	−9.6	0.0	0.0	0.0	0.0	0.0	0.0	0.0	0.0
17	18.6	−7.1	50.7	4.4	0.0	0.0	0.0	0.0	0.0	0.0
18	23.2	−2.5	4.4	6.2	0.0	0.0	0.0	0.0	0.0	0.0
19	39.0	13.3	0.0	0.0	176.3	0.0	0.0	0.0	0.0	0.0
20	23.5	−2.2	0.0	0.0	0.0	4.7	0.0	0.0	0.0	0.0
21	13.1	−12.6	0.0	0.0	0.0	0.0	158.4	0.0	0.0	34.8
25	43.0	17.3	0.0	0.0	23.8	58.8	0.0	0.0	0.0	0.0
26	20.4	−5.3	0.0	0.0	0.0	0.0	0.0	0.0	0.0	12.8
27	9.6	−16.1	0.0	0.0	0.0	0.0	−9.6	0.0	0.0	0.0
28	33.3	7.6	0.0	0.0	0.0	0.0	0.0	0.0	0.0	0.0
29	14.0	−11.7	0.0	0.0	0.0	64.5	0.0	0.0	0.0	0.0
30	44.6	19.0	0.0	0.8	26.1	30.6	0.0	15.1	0.0	0.0
31	33.2	7.5	0.0	0.0	0.0	0.0	0.0	0.0	0.0	0.0
32	20.3	−5.3	0.0	0.0	0.0	0.0	−1.8	0.0	0.0	0.0
33	26.3	0.6	0.0	0.0	0.0	0.0	0.0	0.0	0.0	0.0

i	z	$z-\bar{z}$	j 1	2	3	4	5	6	7	8
			8.8	−4.5	−9.5	3.5	−11.3	2.9	−12.8	−4.4
22	34.9	9.3	0.0	0.0	0.0	0.0	0.0	85.7	0.0	−14.6
23	31.0	5.4	0.0	0.0	0.0	0.0	0.0	0.0	28.7	0.0
24	14.6	−11.1	0.0	0.0	0.0	0.0	27.8	−20.5	0.0	122.4
25	43.0	17.3	0.0	0.0	0.0	0.0	0.0	0.0	0.0	0.0
26	20.4	−5.3	0.0	0.0	0.0	0.0	0.0	0.0	0.0	0.0
27	9.6	−16.1	0.0	10.0	0.0	0.0	50.8	0.0	0.0	0.0
28	33.3	7.6	0.0	0.0	0.0	0.0	0.0	14.2	8.2	0.0
29	14.0	−11.7	0.0	0.0	0.0	0.0	49.1	0.0	0.0	43.2
30	44.6	19.0	0.0	0.0	0.0	0.0	0.0	0.0	16.9	0.0
31	33.2	7.5	0.0	0.0	0.0	0.0	0.0	0.0	0.0	0.0
32	20.3	−5.3	0.0	0.0	0.0	1.7	9.6	−7.1	0.0	8.4
33	26.3	0.6	0.0	0.0	0.0	−0.2	0.0	1.0	0.0	0.0

i	z	$z-\bar{z}$	j 25	26	27	28	29	30	31	32
			17.3	−5.3	−16.1	7.6	−11.7	19.0	7.5	−5.3
1	34.5	8.8	30.4	−9.4	0.0	0.0	0.0	0.0	0.0	0.0
2	21.2	−4.5	0.0	0.0	0.0	0.0	0.0	0.0	0.0	0.0
3	16.2	−9.5	0.0	0.0	0.0	0.0	0.0	0.0	0.0	0.0
4	29.2	3.5	0.0	0.0	0.0	−14.4	0.0	0.0	0.0	0.0
5	14.4	−11.3	0.0	0.0	0.0	0.0	0.0	0.0	0.0	0.0
6	28.6	2.9	0.0	0.0	0.0	0.0	0.0	0.0	0.0	0.0
7	12.8	−12.8	0.0	0.0	0.0	−14.0	0.0	−34.8	0.0	0.0
8	21.3	−4.4	0.0	0.0	0.0	0.0	12.8	0.0	0.0	0.0
9	25.1	−0.6	0.0	0.0	0.0	0.0	0.0	0.0	0.0	0.0
10	26.2	0.5	0.0	0.0	0.0	0.0	0.0	0.0	1.3	0.0
11	33.9	8.3	23.8	0.0	0.0	0.0	0.0	26.1	0.0	0.0
12	46.1	20.4	58.8	0.0	0.0	0.0	0.0	64.5	25.5	0.0
13	28.1	2.4	0.0	0.0	−6.4	0.0	0.0	0.0	0.0	−2.1

i	z	$z-\bar{z}$												
14	35.7	10.0	0.0	0.0	0.0	0.0	0.0	0.0	0.0	12.6	0.0
15	15.6	−10.1	0.0	0.0	0.0	0.0	0.0	0.0	0.0	0.0	0.0
16	16.1	−9.6	17.1	0.0	0.0	0.0	0.0	0.0	0.0	0.0	0.0
17	18.6	−7.1	0.0	0.0	0.0	0.0	0.0	0.0	0.0	0.0	0.0
18	23.2	−2.5	0.0	10.0	0.0	0.0	0.0	0.0	0.0	0.0	0.0
19	39.0	13.3	0.0	0.0	0.0	0.0	0.0	0.0	0.0	0.0	0.0
20	23.5	−2.2	0.0	0.0	0.0	0.0	0.0	0.0	0.0	0.0	2.9
21	13.1	−12.6	0.0	50.8	0.0	36.8	0.0	0.0	0.0	0.0	16.8
22	34.9	9.3	0.0	0.0	10.1	0.0	0.0	0.0	0.0	0.0	−7.1
23	31.0	5.4	0.0	0.0	10.2	0.0	25.4	0.0	0.0	0.0	0.0
24	14.6	−11.1	0.0	0.0	0.0	25.9	0.0	0.0	0.0	0.0	11.8
25	43.0	17.3	298.1	0.0	0.0	0.0	54.5	0.0	21.6	0.0	0.0
26	20.4	−5.3	−23.0	0.0	0.0	0.0	0.0	0.0	−10.0	0.0	0.0
27	9.6	−16.1	0.0	260.3	0.0	0.0	0.0	0.0	0.0	0.0	21.6
28	33.3	7.6	0.0	0.0	58.5	0.0	0.0	29.0	0.0	0.0	0.0
29	14.0	−11.7	0.0	0.0	0.0	137.0	0.0	0.0	0.0	0.0	0.0
30	44.6	19.0	54.5	0.0	24.2	0.0	359.3	0.0	0.0	0.0	0.0
31	33.2	7.5	25.9	0.0	0.0	0.0	0.0	0.0	56.3	0.0	28.6
32	20.3	−5.3	0.0	12.3	0.0	0.0	0.0	0.0	0.0	0.0	28.6
33	26.3	0.6	0.0	0.0	0.0	0.0	0.0	0.0	0.0	0.0	−0.5

			j		Row sum	$\dfrac{S}{\sum (z_1-\bar{z})^2}$	$\dfrac{S}{\sum (z_1-\bar{z})^2}\, {}^{n-1}$
			33		(S)		
i	z	$z-\bar{z}$	0.6				
1	34.5	8.8	0.0		1.96	0.001	0.02
2	21.2	−4.5	0.0		−6.27	−0.002	−0.06
3	16.2	−9.5	0.0		9.18	0.003	0.09
4	29.2	3.5	0.3		−5.78	−0.002	−0.06
5	14.4	−11.3	0.0		−31.46	−0.010	−0.32
6	28.6	2.9	0.3		4.95	0.002	0.05
7	12.8	−12.8	−1.1		−134.13	−0.042	−1.36
8	21.3	−4.4	0.0		27.16	0.009	0.27

i	z	$z-\bar{z}$	1	2	3	4	5	6	7	8
			8.8	−4.5	−9.5	3.5	−11.3	2.9	−12.8	−4.4
9	25.1	−0.6	0.0	1.61	0.001			0.02		
10	26.2	0.5	0.0	2.20	0.001			0.02		
11	33.9	8.3	0.0	40.71	0.013			0.41		
12	46.1	20.4	0.0	184.45	0.058			1.86		
13	28.1	2.4	0.0	−9.22	−0.003			−0.09		
14	35.7	10.0	0.0	67.17	0.021			0.68		
15	15.6	−10.1	0.0	21.94	0.007			0.22		
16	16.1	−9.6	0.0	19.20	0.006			0.19		
17	18.6	−7.1	0.0	31.25	0.010			0.32		
18	23.2	−2.5	0.0	13.32	0.004			0.13		
19	39.0	13.3	0.0	68.22	0.022			0.69		
20	23.5	−2.2	−0.3	−0.65	0.000			−0.01		
21	13.1	−12.6	0.0	139.20	0.044			1.41		
22	34.9	9.3	0.8	−48.51	−0.015			−0.49		
23	31.0	5.4	0.0	31.51	0.010			0.32		
24	14.6	−11.1	0.0	54.77	0.017			0.55		
25	43.0	17.3	0.0	168.68	0.053			1.70		
26	20.4	−5.3	0.0	−31.90	−0.010			−0.32		
27	9.6	−16.1	0.0	72.71	0.023			0.73		
28	33.3	7.6	0.0	14.38	0.005			0.15		
29	14.0	−11.7	0.0	109.32	0.035			1.10		
30	44.6	19.0	0.0	145.64	0.046			1.47		
31	33.2	7.5	0.0	64.42	0.020			0.65		
32	20.3	−5.3	−0.5	22.67	0.007			0.23		
33	26.3	0.6	0.4	−0.48	0.000			0.00		
				1048.21	0.331			10.589		

10.3.1 *Fitting a global surface*

Global fitting produces one mathematical function that describes the entire surface of the study area on the basis of estimating Z values for the nodes on the grid in a single operation, whereas local fitting derives multiple equations based on using a subset of points around successive individual nodes in the grid. The regression analysis techniques examined in Chapter 9 produce equations that define the line providing the best fit to the known data values, although it is acknowledged that these are unlikely to provide a 100 per cent accurate prediction of the dependent variable. The same caveat applies with trend surface analysis and Figure 10.6 illustrates the link between prediction in linear and polynomial regression and trend surface analysis. The regression lines may be thought of as transecting the surface along a particular trajectory. There are known data values above and below the first-, second- and third-degree polynomial regression lines on the left of Figure 10.6, but the lines show the overall slope. The three dimensional representations on the right reveal that the known data values fall above or below (respectively solid and open circles) the surface.

The fitting of a global trend surface is achieved through a form of polynomial regression using the least squares method to minimize the sum of the squared deviations from the surface at the known, sampled locations (control points). The equation can be used to estimate the values of the dependent variable at any point and this is commonly carried out for the nodes in the regular grid. The polynomial equation obtained by means of least squares fitting provides the best approximation of the surface from the available data for the control points. The process of creating a surface from the equation is commonly known as **interpolation**, since it involves determining or interpolating previously unknown Z values for the nodes and connecting these together usually by means of a 'wire frame' in order to visualize the surface, although strictly speaking this term should be reserved for dealing with smoothing local variation. Figure 10.7 illustrates the outcome of this process with respect to an irregular sample of data points for elevation on the South Downs in South-East England. The points from which the polynomial equation for the surface was generated are identified by white markings at the peaks.

Despite the general use of polynomial regression-type equations to fit a global surface, there are some limitations to this approach. These are in some respects extensions of the same problems that were identified with respect to using regression to predict nonspatial distributed dependent variables. One of the limitations noted with respect to simple linear regression (first-order polynomial) is that it may be inappropriate to define the relationship as a straight line, especially if visualization of the empirical data suggested some curvilinear connection. It would be just as nonsensical to argue that all surfaces are flat sloping planes. However, moving to a surface based on the second-order polynomial only introduces one maximum or minimum location on the surface and the third order only provides for one peak and one trough. Thus, global fitting does not necessarily produce a very realistic surface. Another problem identified with regression in statistical analysis was that prediction of the

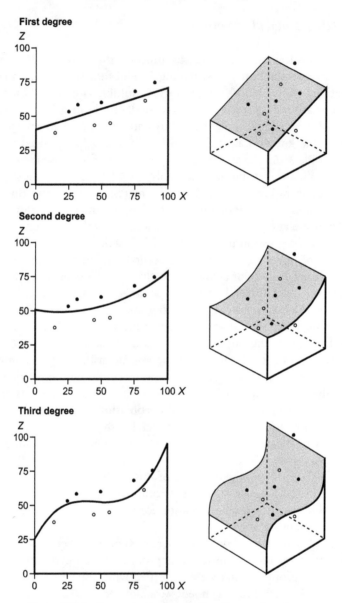

Figure 10.6 Comparison of regression lines and surfaces.

dependent variable much beyond the range of the known values of the independent variable(s) is potentially inaccurate. Similarly, the extension of a trend surface to beyond the area for which there are known data points could also be misleading. Such gaps in the observed or measured data can occur around the edges or in parts of a study area where there is a dearth of control points. More realistic surfaces may be obtained by increasing the order of the polynomial equation beyond three to four, five, six or more, although the calculations involved are computationally taxing.

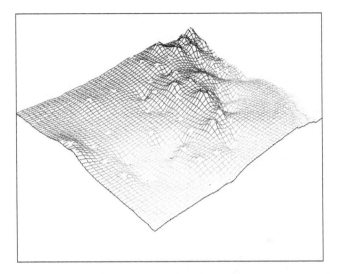

Figure 10.7 Trend surface for part of South Downs, East Sussex, England.

Box 10.6 illustrates selected aspects of the calculations involved in producing a first-order polynomial trend surface (i.e. a linear surface) in respect of the 2007 Index of Multiple Deprivation for the 33 London Boroughs. The measurement of spatial autocorrelation with Moran's I has already shown this to be high in some boroughs just to the east of the City of London and low in others towards the south west. The equation has been computed using geostatistical software and the predicted Z values and the residuals are tabulated in Box 10.6c. The percentile maps (Box 10.6a) divide the predicted figures for the linear surface and the residuals into six groups and emphasize the importance of very low and very high values. The linear surface tracks north east to south west and simplifies the spatial pattern. The residuals reveal the highest positive difference was in Hackney and three of its neighbours (Newham, Lambeth and Tower Hamlets) in the central area. The largest negative residual was in Havering on the eastern edge with the next two being the City of London and Redbridge.

10.3.2 Dealing with local variation in a surface

The global fitted trend surface relates to the concept of regional features. Returning to the example of creating a surface using atmospheric pressure and water vapour to identify features in the troposphere, the cyclones or anticyclones and clouds can be viewed as regional or large-scale features. However, at a smaller scale there will often be local variation in the variables producing highs and lows in the overall feature. Fitting the polynomial equation results in a surface, whereas the residuals, the differences between the fitted surface and known values may be thought of as local

Box 10.6a: Trend surface analysis.

Polynomial equation for fitting a trend surface: $\hat{Z}(x,y) = \sum_{i=1}^{M} a_i x^{\beta_{1i}} y^{\beta_{2i}}$

Minimization of error of estimate: $E = \sum_{i=1}^{L} (Z(x_i, y_i) - Z_i)^2 \rightarrow \text{minimum}$

Linear trend (flat dipping plane): $\hat{Z}(x,y) = a_0 + a_1 x + a_2 y$

Quadratic trend (one maximum or minimum): $\hat{Z}(x,y) = a_0 + a_1 x + a_2 y + a_3 x^2 + a_4 xy + a_5 y^2$

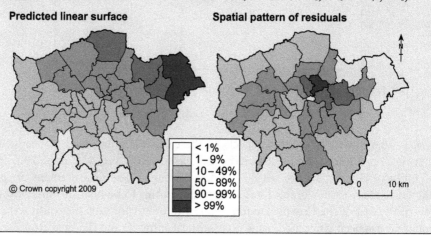

Predicted linear surface   **Spatial pattern of residuals**

© Crown copyright 2009

| < 1% |
| 1 – 9% |
| 10 – 49% |
| 50 – 89% |
| 90 – 99% |
| > 99% |

0   10 km

Box 10.6b: Application of the trend surface analysis.

The terms x and y are coordinates on the plane surface and M is the number of degrees or orders of the polynomial equation. The series of coefficients $(a_0, a_1, \ldots a_i)$ are obtained by minimizing the error of the estimation where L is the number of sample or control points. A linear trend surface has been produced using spatial analysis software (Geoda) for the London Boroughs in respect of their average score on the 2007 IMD using rook spatial contiguity weights. The percentile maps in Box 10.6a show the predicted surface and the residuals emphasizing the extreme cases. The specific 1st-order polynomial equation that best fits the empirical data is shown in Box 10.6c and this has been used to calculate the predicted values of the IMD average score for the 500 m grid squares covering the area (Box 10.6d). This simplifies the spatial pattern and provides a clear visualization of the north east to south west trending surface.

Box 10.6c: Calculation of linear trend surface.

i	X	Y	Z	$\hat{Z}$	e
1	547980	186083	34.5	30.50	−3.99
2	524392	191070	21.2	27.61	6.45
3	548572	175109	16.2	26.58	10.37
4	520575	185876	29.2	24.93	−4.29
5	542070	166316	14.4	22.04	7.68
6	527756	184528	28.6	25.87	−2.75
7	532573	181257	12.8	25.63	12.79
8	533193	164481	21.3	19.59	−1.72
9	516188	181612	25.1	22.48	−2.62
10	532197	195153	26.2	30.67	4.48
11	542484	175829	33.9	25.62	−8.32
12	533820	185439	46.1	27.42	−18.68
13	523067	179688	28.1	23.15	−4.92
14	531141	189534	35.7	28.39	−7.34
15	515258	189084	15.6	25.05	9.46
16	553902	186883	16.1	31.98	15.91
17	507523	183686	18.6	21.51	2.95
18	514876	176035	23.2	20.17	−3.03
19	531241	185097	39.0	26.78	−12.18
20	525389	180022	23.5	23.74	0.23
21	519598	167257	13.1	17.89	4.79
22	530881	173991	34.9	22.62	−12.32
23	537779	174133	31.0	24.06	−6.98
24	526431	169390	14.6	20.04	5.42
25	540804	184171	43.0	28.36	−14.59
26	543517	189417	20.4	30.83	10.47
27	517275	173187	9.6	19.60	10.05
28	533757	176088	33.3	23.97	−9.36
29	526245	164539	14.0	18.22	4.24
30	536246	181938	44.6	26.62	−18.02
31	538073	189802	33.2	29.88	−3.31
32	526925	173681	20.3	21.72	1.38
33	526765	181473	26.3	24.55	−1.75

First-order polynomial equation (linear surface)

$$\hat{Z}(x, y) = a_0 + a_1 x + a_2 y \qquad 147.72 + 0.00020x + 0.00037y$$

Box 10.6d: Calculation of linear trend surface for grid squares.

1	527500	197500	32.76	34	512500	182500	22.77
2	532500	202500	36.09	35	517500	187500	26.10
3	537500	207500	39.42	36	522500	192500	29.43
4	512500	192500	27.20	37	527500	197500	32.76
5	517500	197500	30.53	38	532500	202500	36.09
6	522500	202500	33.86	39	537500	207500	39.42
7	527500	207500	37.19	40	542500	212500	42.75
8	532500	212500	40.52	41	547500	217500	46.08
9	537500	217500	43.85	42	552500	222500	49.41
10	542500	222500	47.18	43	507500	172500	17.23
11	547500	227500	50.51	44	512500	177500	20.56
12	507500	187500	23.87	45	517500	182500	23.89
13	512500	192500	27.20	46	522500	187500	27.22
14	517500	197500	30.53	47	527500	192500	30.55
15	522500	202500	33.86	48	532500	197500	33.88
16	527500	207500	37.19	49	537500	202500	37.21
17	532500	212500	40.52	50	542500	207500	40.54
18	537500	217500	43.85	51	547500	212500	43.87
19	542500	222500	47.18	52	517500	167500	17.24
20	547500	227500	50.51	53	522500	172500	20.57
21	552500	232500	53.84	54	527500	177500	23.90
22	557500	237500	57.17	55	532500	182500	27.23
23	507500	182500	21.66	56	537500	187500	30.56
24	512500	187500	24.99	57	542500	192500	33.89
25	517500	192500	28.32	58	547500	197500	37.22
26	522500	197500	31.65	59	517500	162500	15.02
27	527500	202500	34.98	60	522500	167500	18.35
28	532500	207500	38.31	61	527500	172500	21.68
29	537500	212500	41.64	62	532500	177500	25.01
30	542500	217500	44.97	63	537500	182500	28.34
31	547500	222500	48.30	64	542500	187500	31.67
32	552500	227500	51.63	65	532500	157500	16.15
33	507500	177500	19.44	66	537500	162500	19.48

disturbances. These can be dealt with by a range of techniques that focus on groups of data points in a 'window', frame or **kernel**. These broadly divide into exact methods that produce measured values for a series of points or areas by smoothing the original data and inexact ones that estimate a local trend and include kriging and local trend surface analysis (splines).

The simplest approach is to smooth data value means of a **moving average**, which is similar to the approach often adopted with time-series data. It involves partitioning the data points into groups falling within a frame that moves along a series of data

Box 10.6e: Grid-based linear trend surface for London Boroughs.

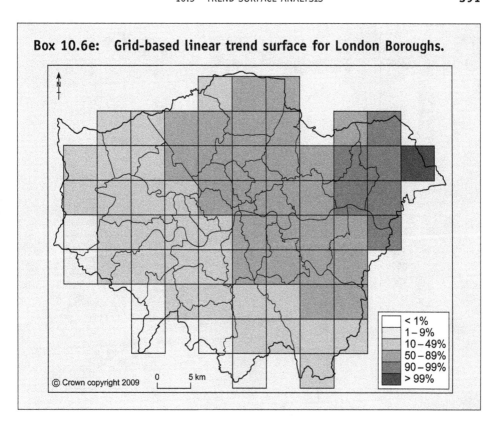

Legend:
- < 1%
- 1–9%
- 10–49%
- 50–89%
- 90–99%
- > 99%

© Crown copyright 2009   0   5 km

points and then interpolating by weighting the values usually according to the mid-point of the window. This is illustrated in Figure 10.8a, where the distances of the Burger King restaurants in Pittsburgh have been interpolated using a 5 km wide frame moving outwards from the city centre and the mean of their daily customer footfall (hypothetical) calculated for each frame (2.5, 3.5, 4.5 ... 19.5 km). The averages at these points are shown by the larger solid circles. The points representing the restaurant locations (small solid circles) have been treated as though they all lie along a single, unidirectional X-axis, but it would have been possible to use a two-dimensional frame (e.g. a square) and then calculate a weighted average within the areas formed by successive zones moving outwards (large solid circles). The smoothing of the data achieved by a moving average is highly dependent on the size of the frame. Smoothing based on **inverse distance weighting** adjusts the value of each point in an inverse relationship to its distance from the point being estimated. The moving average in Figure 10.8a interpolated values along a one-dimensional axis from the city centre and assumed all points to be located on the eastern side of a north–south line through the centre point. In contrast inverse distance smoothing weights the values according to a predetermined number of nearest neighbours to each point being estimated. The application of inverse distance weighted interpolation to the same data values

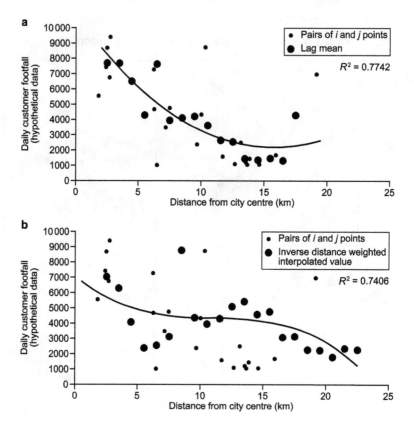

Figure 10.8 Moving average and inverse distance weighted data smoothing.

produces the interpolated values shown in Figure 10.8b (large solid circles). The inverse distance used in this example is $1/d^2$, where d is the distance between the fixed points out from the city centre (2.5, 3.5, 4.5, ... 22.5 km) and their four nearest neighbours. The best-fit polynomial regression equation has been included for both sets of smoothed data: these are, respectively, 2nd- and 3rd-degree polynomials.

Another approach to dealing with local variation in a surface is to use **splines**. This involves fitting a polynomial regression equation to discrete groups of the data points along sections of the surface. These splines are then 'tied' together to produce a smooth curve following the overall surface. The points where they connect are known as knots and the polynomial equation for each section (spline) is constrained to predict the same values where they meet. Unlike the moving average, the frames in which the splines are produced do not move across the set of data points but each has a fixed location, although they can have different widths. One similarity with the moving average is that a smaller frame will reflect the local structure of the data values, whereas a wider one will produce a smoother surface.

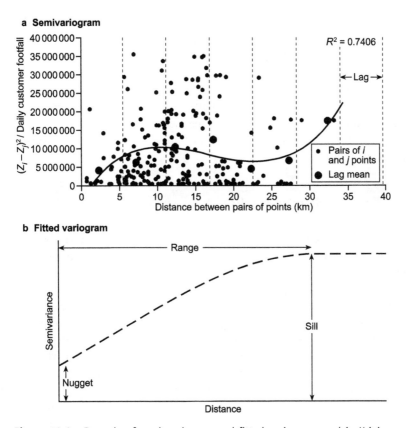

Figure 10.9 Example of semiovariogram and fitted variogram used in Kriging.

One of the most common techniques for dealing with local variation is **kriging**. Trend surface analysis may be used to identify and then remove the overall trend in a surface before using kriging to map short-range variations. Kriging is based on inverse distance weighting and uses the local spatial structure to produce weights from which to predict the values of points. The first stage involves describing the spatial structure by means of a semivariogram, which involves calculating the distance between each pair of points and the square of the difference between their values for the variable Z. These are visualized in a scatter plot that includes spatial lags and is known as a **semovariogram**: Figure 10.9 shows the semivariogram for the Burger King restaurants in Pittsburgh with 5 km lags and daily customer footfall as the variable. The mean of each group of data values within a given lag is plotted at the midpoint (large solid circles). The next stage involves summarizing this local spatial variation by a function that best fits the means at the midpoints, which in this example is a 3rd-degree polynomial. The values of neighbouring points are predicted by means of weights derived from the semivariogram. Kriging works by fitting an empirical semivariogram to a typical or model variogram. The lower part of Figure 10.9 shows such a model and identifies three main sections of the fitted variogram. The nugget quantifies the uncertainty of the Z values, the range is the section over which there is strong correlation between distance and the Z values and thus represents the distance over which reliable prediction can be made and the semivariance is constant beyond the sill. Kriging is a geostatistical procedure with error estimates being produced that can provide a useful guide as to where more detailed empirical data might be required. This contrasts with the geometrical basis of moving average and inverse distance weighting techniques.

10.4 Concluding remarks

Concern over the presence of significant spatial autocorrelation and contravention of standard assumptions formerly led to a feeling that regression analysis was inappropriate with spatially distributed data. However, over the last few decades three different possibilities have found favour: ignore the spatial autocorrelation especially if it is demonstrably insignificant and carry on with the analysis as normal; acknowledge that the slope coefficient parameter (β) in regression may not apply globally and employ a strategy that allows local best-fit regression lines to be stitched together to produce an overall surface; and incorporate other components in the model that measure the spatial distribution of autocorrelation. This chapter has introduced a selection of spatial analytic and geostatistical techniques that come under the broad heading of **Exploratory Spatial Data Analysis** (ESDA). A focus of attention is the distribution of spatial autocorrelation and recognition that spatial patterns of phenomena and their measured variables often contravene the assumptions of classical statistics. Although trend surface and residuals analysis utilize some of the computation procedures of classical statistics, such as fitting a regression equation, the failure of many standard assumptions to be satisfied, such as independence and conforming to the Normal probability distribution means that it is not necessary to calculate confidence limits for the fitted surface or to apply inferential statistics.

The second of these possibilities, producing and stitching together local best-fit regression equations, has been termed **Geographically Weighted Regression** (GWR) (Fotheringham, Charlton and Brunsdon, 1998; Fotheringham, Brunsdon and Charlton, 2002) and merits further brief discussion, although full details are beyond the scope of this text. GWR allows the estimated slope coefficients to vary spatially across the study area. Unlike the global and local techniques for smoothing data or fitting a surface examined above, GWR seeks to explain a dependent variable in relation to one or more often several independent variables. It therefore corresponds to the classical statistical technique of multiple or multivariate regression. The principal difference is that classical multiple regression produces one equation, with slope coefficients or parameters and an error term that applies across the complete dataset. GWR has the potential to generate separate equations with all these components for each point or area in the analysis. The analysis proceeds by including the neighbours of each point within a neighbourhood zone usually defined by means of a distance decay function. A weights matrix is defined for each point and the least squares regression is computed. GWR presents a way of quantifying and disentangling complex spatial patterns across a study area and enhances thinking about spatial processes in comparison with applying simple linear regression.

References

Bryman, A. and Cramer, D. (1999) *Quantitative Data Analysis for Social Scientists*, Routledge, London.

Campbell, J. (2007) *Map Use and Analysis*, Wm. C. Brown, Oxford.

Crackell, A.P. and Hayes, L.W.B. (1993) *Introduction to Remote Sensing*, Taylor and Francis, London.

Curran, P.J. (1981) Remote Sensing: the Role of Small Format Light Aircraft Photography. Geog. Papers, No. 75, Dept. of Geography, Univ. of Reading.

Curran, P.J. (1985) *Principles of Remote Sensing*, Longman, New York.

Department of the Environment, Food and Rural Affairs (2003) *2002 June Agricultural Census*. DEFRA: York.

Dewdney, J. (1981) *The British Census*, Invicta Press, Ashford, Kent.

Dorling, D. and Rees, P.H. (2003) A nation still dividing: the British Census and social polarisation 1971-2001. *Environment and Planning A*, **35**, 1287–1313.

Fotheringham, A.S., Charlton, M.E. and Brunsdon, C. (1998) Geographically weighted regression: a natural evolution of the expansion method for spatial data analysis. *Environment and Planning A*, **30**, 1905–1927.

Fotheringham, A.S., Brunsdon, C. and Charlton, M.E. (2002) *Geographically Weighted Regression: the Analysis of Spatially Varying Relationships*, John Wiley & Sons, Inc., New York.

Hardisty, J., Taylor, D.M. and Metcalfe, S.E. (1995) *Computerised Environmental Modelling*, John Wiley & Sons, Ltd, Chichester.

Johnston, R.J. (1979) *Geography and Geographers: Anglo-American Geography since 1945*, Edward Arnold, London.

Kirkby, M.J., Nadem, P.S., Burt, T.P. and Butcher, D.P. (1992) *Computer Simulation in Physical Geography*, John Wiley & Sons, Ltd, Chichester.

Lane, S.N. (2003) Numerical modelling in physical geography: understanding explanation and prediction, in *Key Methods in Geography* (eds N.J. Clifford and G. Valentine), Sage, London, pp. 263–290.

Practical Statistics for Geographers and Earth Scientists   Nigel Walford
© 2011 John Wiley & Sons, Ltd

Layder, D. (1994) Understanding Social Theory. Sage: London.

Lillesand, T.M., Kiefer, R.W. and Chipman, J.W. (2004) *Remote Sensing and Image Interpretation*, John Wiley & Sons, Inc., New York.

Mather, P.M. (1987) *Computer Processing of Remotely Sensed Images*, John Wiley & Sons, Ltd, Chichester.

Mather, P.M. (2004) *Computer Processing of Remotely-Sensed Images – An Introduction (with CD)*, 3rd edn, John Wiley & Sons, Ltd, Chichester.

Neft, D. (1966) *Statistical Analysis for Areal Distributions*, Regional Science Institute Monograph 2, Philadelphia.

Ordnance Survey (2007) GIS Files, http://www.ordnancesurvey.co.uk/oswebsite/gisfiles/ (accessed 02/03/08).

Reichman, W.J. (1961) *Use and Abuse of Statistics*, Penguin, Harmondsworth.

Reid, I. (2003) Making observations and measurements in the field: an overview, in *Key Methods in Geography* (eds N.J. Clifford and G. Valentine), Sage, London, pp. 209–222.

de Smith, M.J., Goodchild, M.F. and Longley, P.A. (2007) *Geospatial Analysis: A Comprehensive Guide to Principles, Techniques and Software Tools*, Matador, Leicester.

Tobler, W. (1970) A computer movie simulating urban growth in the Detroit region. *Economic Geography*, **46** (2), 234–240.

United Nations (1993) Our Common Future: Brundtland Report, http://www.worldinbalance. net/intagreements/1987-brundtland.php (accessed 12/11/07).

Valentine, J.C. and Cooper, H. (2003) What Works Clearinghouse study design and implementation device (Version 1.0). Washington, DC: US Department of Education.

Walford, N.S. (2v002) *Geographical Data: Characteristics and Sources*, John Wiley & Sons, Ltd, Chichester.

Williams, R.B.G. (1984) *Introduction to Statistics for Geographers and Earth Scientists*, Macmillan, London.

Further Reading

Agresti, A. (1996) *An Introduction to Categorical Data Analysis*, John Wiley & Sons, Inc., New York.

Anselin, L. (1988) *Spatial Econometrics: Methods and Models*, Kluwer Academic Publishers, Dordrecht, NL.

Anselin, L. (1995) Local indicators of spatial association – LISA. *Geographical Analysis*, **27**, 93–115.

Asher, H. (1998) *Polling and the Public. What Every Citizen Should Know*, 4th edn, CQ Press, Washington, D.C..

Bachi, R. (1963) Standard distance measures and related methods for spatial analysis. *Papers of the Regional Science Association*, **10**, 83–132.

Bailey, T.C. and Gatrell, A.C. (1995) *Interactive Spatial Data Analysis*, Longman, Harlow (published in the USA by Wiley).

Berry, B.J.L. and Marble, D.F. (1968) *Spatial Analysis – A Reader in Statistical Geography*, Prentice Hall, New Jersey.

Besag, J. and Newell, J. (1991) The detection of clusters in rare diseases. *Journal of the Royal Statistical Society Series A*, **154**, 143–155.

Boots, B.N. and Getis, A. (1988) *Point Pattern Analysis*, Sage, Newbury Park, CA.

Brunsdon, C., Fotheringham, A.S. and Charlton, M.E. (1999) Some notes on parametric significance tests for geographically weighted regression. *Journal of Regional Science*, **39**, 497–524.

Bryman, A. (1988) *Quantity and Quality in Social Research*, Unwin Hyman, London.

Clark, P.J. and Evans, F.C. (1954) Distance to nearest-neighbour as a measure of spatial relationships in populations. *Ecology*, **35**, 445–453.

Clark, R. and Evans, F.C. (1954) Distance to nearest neighbour as a measure of spatial relationships in populations. *Ecology*, **35**, 445–453.

Cliff, A.D. and Ord, J.K. (1973) *Spatial Autocorrelation*, Pion, London.

Cliff, A.D. and Ord, J.K. (1975) The comparison of means when samples consist of spatial autocorrelated observations. *Environment and Planning A*, **7**, 725–734.

Cliff, A.D. and Ord, J.K. (1981) *Spatial Processes: Models and Applications*, Pion, London.

Clifford, P. and Richardson, 5. (1985) Testing the association between two spatial processes. *Statistics and Decisions*, Supplement No. 2, 155–160.

Clifford, N. and G. Valentine (eds) (2010) *Key Methods in Geography*, 2nd edn, Sage, London.

Cochran, W.G. (1997) *Sampling Techniques*, 3rd edn, John Wiley & Sons, Inc., New York.

Cressie, N.A.C. (1991) *Statistics for Spatial Data*, John Wiley & Sons, Inc., New York.

Cressie, N.A.C. and Chan, N.H. (1989) Spatial modelling of regional variables. *Journal of American Statistical Association*, **84**, 393–401.

Davies Withers, S. (2002) Quantitative methods: Bayesian inference, Bayesian thinking. *Progress in Human Geography*, **26**, 553–566.

Diamond, I. and Jeffries, J. (1999) *Introduction to Quantitative Methods*, Sage, London.

Diggle, P.J. (1983) *Statistical Analysis of Point Patterns*, Academic Press, London.

Draper, N.R. and Smith, H. (2003) *Applied Regression Analysis*, 2nd edn, John Wiley & Sons, Ltd, Chichester.

Ebdon, D. (1984) *Statistics in Geography*, 2nd edn, Blackwell, Oxford.

England, K. (2002) Interviewing elites: cautionary tales about researching women managers in Canada's banking industry, in *Feminist Geography in Practice: Research and Methods* (ed. P. Moss), Blackwell, Oxford, pp. 200–214.

Everitt, B.S. and Dunn, G. (2001) *Applied Multivariate Data Analysis*, Arnold, London.

Fischer, M. and Getis, A. (2009) *Handbook of Applied Spatial Analysis: Software Tools, Methods, and Applications*, Springer, New York.

Fotheringham, A.S. and Rogerson, P.A. (2008) *Handbook of Spatial Analysis*, Sage, London.

Fotheringham, A.S., Brunsdon, C. and Charlton, M.E. (2000) *Quantitative Geography: Perspectives on Spatial Data Analysis*, Sage, London.

Fowler, F. (1995) *Improving Survey Questions: Design and Evaluation*, Sage, Thousand Oaks, CA.

Gao, X., Asami, Y. and Ching, C.-J.F. (2006) An empirical evaluation of spatial regression models. *Computers and Geosciences*, **32**, 1040–1051.

Gatrell, A.C., Bailey, T.C., Diggle, P. and Rowlingson, B.S. (1996) Spatial point pattern analysis and its application in geographical epidemiology. *Transactions of the Institute of British Geographers*, **2**, 256–274.

Getis, A. (1984) Interaction modelling using second-order analysis. *Environment and Planning A*, **16**, 173–183.

Getis, A. and Franklin, J. (1987) Second-order neighbourhood analysis of mapped point patterns. *Ecology*, **68** (3), 473–477.

Getis, A. and Ord, J.K. (1992) The analysis of spatial association by use of distance statistics. *Geographical Analysis*, **24**, 189–206.

Getis, A. and Ord, J.K. (1996) Local spatial statistics: an overview, in *Spatial Analysis: Modelling in A GIS Environment* (eds P. Longley and M. Batty), Geoinformation International, Cambridge (distributed by John Wiley & Sons, Inc.: New York), pp. 261–278.

Goodchild, M.F. and Longley, P.A. (1999) The future of GIS and spatial analysis, *Geographical Information Systems: Principles, Techniques, Management and Applications*, 2nd edn (eds P.A. Longley, M.F. Goodchild, D.J. Maguire and D.W. Rhind), John Wiley & Sons, Inc., New York, pp. 235–248.

Griffith, D.A. (1978) A spatially adjusted ANOVA model. *Geographical Analysis*, **10**, 296–301.

Griffith, D.A. (1987) *Spatial Autocorrelation: A Primer*, Association of American Geographers, Washington.

Griffith, D.A. (1996) Computational simplifications for space-time forecasting within GIS: the neighbourhood spatial forecasting model, in *Spatial Analysis: Modelling in A GIS Environment* (eds P.A. Longley and M. Batty), Geoinformation International, Cambridge (distributed by John Wiley & Sons, Inc.: New York), pp. 247–260.

Haining, R. (1990) *Spatial Data Analysis in the Social and Environmental Sciences*, Cambridge University Press, Cambridge.

Haining, R. (2003) *Spatial Data Analysis: Theory and Practice*, Cambridge University Press, Cambridge.

Hammond, R. and McCullagh, P.S. (1978) *Quantitative Techniques in Geography*, Clarendon Press, Oxford.

Haynes, R. (1982) *Environmental Science Methods*, Chapman and Hall, London.

Heiman, G.W. (1992) *Basic Statistics for the Behavioural Sciences*, Houghton Mifflin, Boston, MA.

Hetz, R. and Luber, J. (1995) *Studying Elites: Using Qualitative Methods*, Sage, London.

Keylock, C.J. and Dorling, D. (2004) What kind of quantitative methods for what kind of geography? *Area*, **36**, 358–366.

Kitchin, R. and Tate, N. (2000) *Conducting Research into Human Geography*, Prentice Hall, Harlow.

Kleinbaum, D., Kupper, L.L. and Muller, K.E. (1988) *Applied Regression Analysis and Other Multivariate Methods*, 2nd edn, PWS-Kent Publishing Company, Boston, MA.

Limb, M. and Dwyer, C. (eds) (2001) *Qualitative Methodologies for Geographers*, Arnold, London.

Longley, P., Brooks, S.M., McDonnell, R. and Macmillan, B. (1998) *Geocomputation: A Primer*, John Wiley & Sons, Ltd, Chichester.

Lynn, P. and Lievesely, D. (1991) *Drawing General Population Samples in Great Britain*, Social and Community Planning Research, London.

Marsh, C. (1982) *The Survey Method*, George Allen and Unwin, London.

McGrew, J.C.J. and Monroe, C.B. (2000) *Introduction to Statistical Problem Solving in Geography*, McGraw-Hill, Boston, MA.

Mitchell, A. (2008) *The ESRI Guide to GIS Analysis, Volume 2: Spatial Measurements and Statistics*, ESRI Press, Redlands, CA.

Moran, P.A.P. (1948) The interpretation of statistical maps. *Journal of the Royal Statistical Society Series B*, **10**, 245–251.

Moran, P.A.P. (1950) Notes on continuous stochastic phenomena. *Biometrika*, **37**, 17–23.

Moser, C. and Kalton, G. (1971) *Survey Methods in Social Investigation*, 2nd edn, Heinemann, London.

Oppenheim, A.N. (1992) *Questionnaire Design, Interviewing and Attitude Measurement*, Continuum, London.

Ord, J.K. and Getis, A. (1995) Local spatial autocorrelation statistics: distribution issues and an application. *Geographical Analysis*, **27** (4), 286–306.

Oskamp, S. and Schultz, P.W. (2004) *Attitudes and Opinions*, 3rd edn, Erlbaum, Mahwah, NJ.

Rogerson, P.A. (2006) *Statistical Methods for Geographers: A Student's Guide*, Sage, Los Angeles, CA.

Rose, D. and Sullivan, O. (1996) *Introductory Data Analysis for Social Scientists*, 2nd edn, Open University Press, Buckingham.

Sachs, L. (1984) *Applied Statistics: A Handbook of Techniques*, Springer-Verlag, New York.

Sawada, M. (1999) ROOKCASE: An Excel 97/2000 Visual Basic (VB) Add-in for exploring global and local spatial autocorrelation. *Bulletin of the Ecological Society of America*, **80**, 231–234.

Scheffé, H. (1959) *The Analysis of Variance*, John Wiley & Sons, Inc., New York.

Simpson, L. and Dorling, D. (eds) (1998) *Statistics in Society*, Arnold, London.

Skelton, T. (2001) Cross-cultural research: issues of power, positionality and 'race', in *Qualitative Methodologies for Geographers* (eds M. Limb and C. Dwyer), Arnold, London, pp. 87–100.

Slocum, T. (1990) The use of quantitative methods in major geographical journals, 1956–1986. *Professional Geographer*, **42**, 84–94.

Smith, F. (2003) Working in different cultures, in *Key Methods in Geography* (eds N. Clifford and G. Valentine), Sage, London, pp. 179–193.

Sudman, S. and Bradburn, N. (1982) *Asking Questions: A Practical Guide to Questionnaire Design*, Jossey-Bass, San Francisco, CA.

Tukey, J.W. (1972) Some graphic and semigraphic displays, in *Statistical Papers in Honour of George W. Snedecor* (ed. T.A. Bancroft), Iowa State University Press, Ames, IA, pp. 293–316.

Tukey, J.W. (1977) *Explanatory Data Analysis*, Addison-Wesley, Reading, MA.

Walford, N.S. (1995) *Geographical Data Analysis*, John Wiley & Sons, Ltd, Chichester.

Williams, R.B.G. (1985) *Intermediate Statistics for Geographers and Earth Scientists*, Macmillan, London.

Wrigley, N. (1985) *Categorical Data Analysis for Geographers and Environmental Scientists*, Longman, Harlow.

Index

Note: Page numbers in *italics* refer to Figures; those in **bold** to Tables

---

Practical Statistics for Geographers and Earth Scientists   Nigel Walford
© 2011 John Wiley & Sons, Ltd

Statistical Analysis Planner and Checklist

The selection of appropriate techniques to use when carrying out research is not an easy task and there are many pitfalls for novice and even reasonably experienced researchers. Deciding what analysis to carry out involves delving into a large and possibly confusing collection of statistical techniques. Nowadays, there is a range of proprietary software, such as Minitab, SAS, PASW (formerly SPSS) and MS Excel, as well as freely available shareware and websites on the internet that enable researchers to apply different techniques. Most of these resources include useful Help information and have made the process of producing statistical results and getting an answer relatively easy. However, these resources rarely provide an overview of how to plan your statistical analysis so that you can chose the techniques that best fit into the overall purpose of your research.

This short interlude between the previous five introductory chapters and the following five explanatory chapters is intended to help you find your way through this complex collection of statistical techniques and to identify those that might be used in your research. Over the next two pairs of pages the entry-level techniques covered in the following chapters have been divided into two broad groups: those dealing with hypothesis testing and those exploring relationships. There are diagrams on these pages that summarize the connections between the techniques and how decisions about such matters as the number of samples and whether you want to analyse spatial patterns lead you towards applying different statistical techniques. The idea is that by considering your answers to a short series of questions, which are included afterwards, you will be able to follow a route through these diagrams to discover which techniques you should consider using. Referring to the index you can then locate where to find explanations and worked examples from the geographical, Earth and environmental sciences of these techniques in the following chapters.

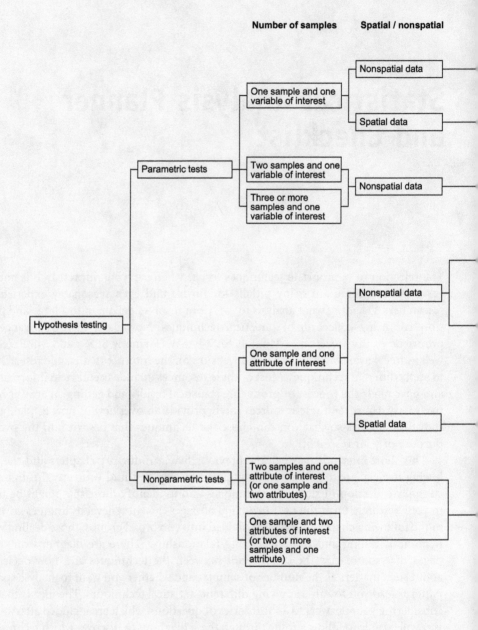

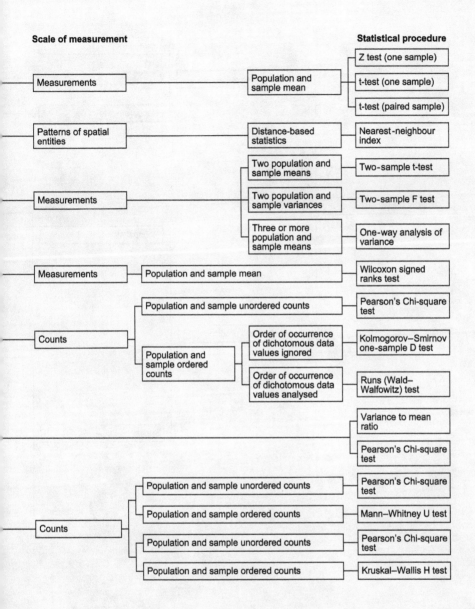

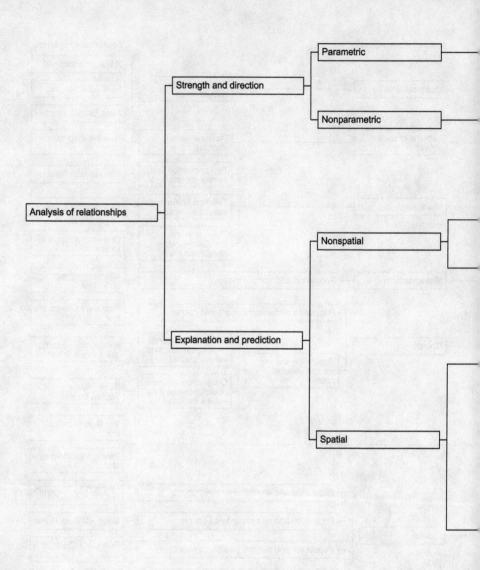

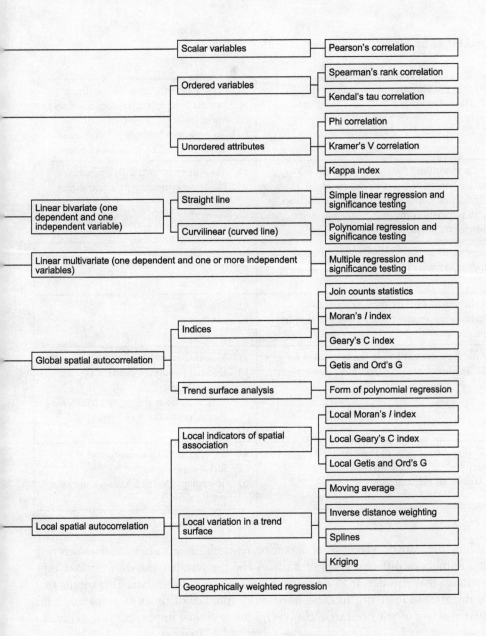

	Scalar variables		Pearson's correlation
	Ordered variables		Spearman's rank correlation
			Kendal's tau correlation
	Unordered attributes		Phi correlation
			Kramer's V correlation
			Kappa index
Linear bivariate (one dependent and one independent variable)	Straight line		Simple linear regression and significance testing
	Curvilinear (curved line)		Polynomial regression and significance testing
Linear multivariate (one dependent and one or more independent variables)			Multiple regression and significance testing
Global spatial autocorrelation	Indices		Join counts statistics
			Moran's *I* index
			Geary's C index
			Getis and Ord's G
	Trend surface analysis		Form of polynomial regression
Local spatial autocorrelation	Local indicators of spatial association		Local Moran's *I* index
			Local Geary's C index
			Local Getis and Ord's G
	Local variation in a trend surface		Moving average
			Inverse distance weighting
			Splines
			Kriging
	Geographically weighted regression		

Checklist of questions for planning statistical analysis

What is the objective of the statistical analysis?	a) To test hypotheses and/or estimate population parameters for attributes/variables b) To explore relationships between attributes and/or variables of interest
Is the intention to estimate population parameters with a certain degree of confidence and/or compare sample statistics with population parameters for normally distributed variables?	a) Yes – use parametric techniques b) No – use nonparametric techniques
How many samples provide the attributes and/or variables of interest?	a) 1 b) 2 c) 3 or more
Is the aim of the statistical analysis to examine the spatial patterns and location of phenomena?	a) Yes b) No
What measurement scale has been used for the attributes and/or variables of interest?	a) Nominal (counts) b) Ordinal (counts or numerical measurements) c) Scalar (Ratio and Interval) (numerical measurements) d) Mixture
What is the purpose of exploring the relationship between attributes and/or variables of interest?	a) To determine its strength and direction b) To explain the relationship and make predictions

Planning the statistical analysis of data for a research project often involves carrying out a number of different types of analyses and the results obtained from one type may mean that you decide to do some more analysis of your data. This means that you might need to return to these questions a number of times in preparation for different stages of your research. However, what is almost impossible is to return to the same set of observations under exactly the same circumstances and collect some more data.